MW00759975

ADVANCED ENERGY SYSTEMS

Applied Energy Technology Series

James G. Speight, Ph.D., *Editor*

ADVANCED ENERGY SYSTEMS

Edited by

Nikolai V. Khartchenko
Institute of Energy Engineering
Technical University Berlin

Taylor & Francis
Publishers since 1798

USA	Publishing Office:	Taylor & Francis 1101 Vermont Avenue, NW, Suite 200 Washington, DC 20005-3521 Tel: (202) 289-2174 Fax: (202) 289-3665
	Distribution Center:	Taylor & Francis 1900 Frost Road, Suite 101 Bristol, PA 19007-1598 Tel: (215) 785-5800 Fax: (215) 785-5515
UK		Taylor & Francis Ltd. 1 Gunpowder Square London EC4A 3DE Tel: 0171 583 0490 Fax: 0171 583 0581

ADVANCED ENERGY SYSTEMS

This book was set in Times Roman. The editors were Heather Worley and Carol Edwards. The Acquisitions Editor was Lisa Ehmer. Cover design by Michelle Fleitz.

A CIP catalog record for this book is available from the British Library.
∞ The paper in this publication meets the requirements of the ANSI Standard Z39.48-1984 (Permanence of Paper)

Library of Congress Cataloging-in-Publication Data

Khartchenko, N. V. (Nikolai Vasil'evich)
    Advanced energy systems / Nikolai V. Khartchenko.
      p.   cm. — (Advanced energy technology series)
    Includes bibliographical references and index.

    1. Power (Mechanics)   1. Title. II. Series.
TJ163.9.K48   1997
621.042—dc21   97-23623
                                                     CIP

To my children, Vadim and Lara

CONTENTS

LIST OF SYMBOLS

A	area, m^2
A	ash mass fraction in a coal, kg ash/kg coal
AF	air-fuel ratio, kg air/ kg fuel
B	magnetic field strength, Wb/m^2 or T
c	flue gas pollutant concentration, mg/m^3
c	specific heat, J/(kg K)
c	velocity of light $(= 2.998 \times 10^8$ m/s)
c_p	specific heat at constant pressure, J/(kg K)
c_v	specific heat at constant volume, J/(kg K)
C	mass fraction of carbon in the fuel, kg carbon/kg fuel
CF	capacity factor, dimensionless
CH_4	percentage by volume of methane in the fuel gas
C_nH_m	percentage by volume of higher hydrocarbons in the fuel gas
CO	percentage by volume of carbon monoxide in the fuel gas
CO_2	percentage by volume of carbon dioxide in the fuel gas
d	diameter, m
DGL	dry gas boiler loss, J/kg fuel
e	elementary charge of an electron $(= 1.602 \times 10^{-19}$ J/V)
E	energy, J
E	electrical potential, emf, V
EUF	energy utilization factor
F	Faraday's constant $[= 9.649 \times 10^7$ coulombs/(kg mol of electrons)]
F	force, N
F_L	Lorentz force, N
FESR	fuel energy savings ratio
g	acceleration of gravity, m/s^2; standard gravity is 9.81 m/s^2
G	Gibbs free energy, J
g_{CO_2}	specific CO_2 emissions of the power plant, kg/MJ
g'_{CO_2}	specific CO_2 emissions of the power plant, kg/kWh
h	heat transfer coefficient, $W/(m^2$ K)
h	Planck's constant, 6.626×10^{-34} J s $= 4.136 \times 10^{-21}$ MeV s
h	specific enthalpy, J/kg
H	enthalpy, J
H	hydrogen mass fraction in the fuel, kg H_2/kg fuel
H_2	percentage by volume of hydrogen in the fuel gas
HHV	higher heating value of the fuel, J/kg or J/m^3
HR	heat rate, J/kWh
i	electric current, A
I_{sc}	short circuit current, A
J	current density, A/m^2
k	Boltzmann's constant $(=1.551 \times 10^{-23}$ J/K $= 8.618 \times 10^{-11}$ MeV/K)
k	isentropic exponent or ratio of specific heats, c_p/c_v
k	thermal conductivity of a material, W/(m K)

K	loading factor for MHD generator
KE	kinetic energy, J
L	length, m
LHV	lower heating value of the fuel, J/kg or J/m^3
m	mass, kg
m	mass flow rate, kg/s
m_a	mass flow rate of air, kg/s
m_c	mass flow rate of condensate, kg/s
m_f	mass flow rate of fuel, kg/s
m_s	mass flow rate of steam, kg/s
m_{ash}	amount of fly ash in kg per MJ of input fuel energy
m_{CaO}	lime requirements for the FGD system, kg/h
m_{NH_3}	hourly ammonia requirements for NO_x removal, kg/h
m_{slag}	amount of slag in kg per MJ of input fuel energy
m_{SO_2}	hourly amount of SO_2 to be removed from the flue gas in the FGD plant, kg/h
m_{sr}	amount of solid residuals in kg per MJ of input fuel energy
m'_{sr}	amount of solid residuals in kg per kWh of generated electrical energy
M	molar mass, kg/kmol
M	moisture mass fraction in the fuel, kg water/kg fuel
MCAL	Moisture -in-combustion-air loss, J/kg fuel
ML	Moisture boiler loss, J/kg fuel
n	angular velocity of engine, r/min
n	moles of a constituent (e.g., carbon, hydrogen) per mole of fuel
N	nitrogen mass fraction, kg N_2/kg fuel
N_2	percentage by volume of nitrogen in the fuel gas
O	oxygen mass fraction, kg O_2/kg fuel
p	pressure, kPa
p_{CO_2}	partial pressure of CO_2 in the flue gas, Pa
p_{H_2O}	partial pressure of the water vapor in the flue gas, Pa
p_0	reference pressure of 1 atm [= 101.3 kPa (14.7 psi)]
P	power, W
P_{in}	input power, W
P_{max}	maximum (peak) power, W
P_{net}	net power output, W
P_{out}	output power, W
PE	potential energy, J
PP	pinch point, K
q	heat added or rejected per kg of working fluid, J/kg
Q	heat, J
Q	heat transfer rate (flux), W
Q_{in}	heat input, J
Q_u	useful heat, J
r	radius, m
r_c	cutoff ratio for diesel cycle, dimensionless
r_v	compression ratio, dimensionless
R	gas constant, J/(kg K)
R	radius, m
R	refuse mass fraction, kg refuse/kg fuel
R_g	total generator resistance, Ω
R_i	internal resistance of fuel cell, Ω
R_L	resistance of external load resistor, Ω
R_t	total resistance of circuit, Ω
RL	radiation and unaccounted boiler loss, J/kg fuel
s	specific entropy, J/(kg K)
S	sulfur mass fraction in the fuel, kg S/kg fuel

t	temperature, °C
t	time, s
t_a	ambient temperature, °C
t_{exh}	temperature of exhaust gas, °C
t_{stack}	stack temperature, °C
T	absolute temperature, K
u	turbine blade velocity, m/s
U	overall heat transfer coefficient, $W/(m^2\ K)$
UCL	unburned-carbon loss, J
v	specific volume, m^3/kg
V	electrical potential, V
V	volume, m^3
V_{CO}	carbon dioxide volume in the flue gas, m^3/kg fuel
V_{CO_2}	volume of carbon dioxide in the flue gas, m^3/kg or m^3 of fuel
V_g	voltage drop across generator, V
V_g	volume of wet flue gas, m^3/kg or m^3 of fuel
V_L	voltage drop across external load resistor, V
w	humidity ratio of air, kg water vapor/kg dry air
w	specific work, J/kg
w_{net}	net specific work of cycle, J/kg
W	work, J
x	linear distance, m

SUBSCRIPTS

a	actual, air, ambient
ash	ash
b	boiler
B	bottoming cycle
c	cold (heat sink), condensate, convection
C	elemental carbon
CaO	lime
CC	combined cycle
CG	cogeneration
cm	cooling medium
CO	carbon monoxide
CO_2	carbon dioxide
cond	condenser
conv	conventional power plant
cw	cooling water
d	dry
dp	dew point
e	effective, equivalent, exit
el	electrical
exh	exhaust
f	fluid, fuel
fa	fly ash
g	gas (wet), generator
h	hot, heat source
H	elemental hydrogen
H_2O	water vapor
i	inlet, internal
ic	intercooler
in	input

l	loss
L	load
m	mean
max	maximum
mf	minimum fluidization
min	minimum
net	net
N	elemental nitrogen
N_2	nitrogen
NH_3	ammonia
o	overall, outlet, reference condition
O	elemental oxygen
O_2	oxygen
out	output
p	constant pressure
r	radiation, refuse
rev	reversible
s	saturation, steam, storage
S	elemental sulfur
sr	solid residuals
SO_2	sulfur dioxide
t	terminal, theoretical, total
T	topping cycle
th	thermal
t,w	theoretical, wet
u	useful
v	constant volume
w	wall, wet

GREEK SYMBOLS

α	angle of attack, degrees
β	compressor pressure ratio, dimensionless
δ	thickness, m
Δ	difference, change
ΔG	change in Gibbs free energy for a chemical reaction, J/(kg mol)
ΔH	change in the enthalpy of formation for a chemical reaction, J/(kg mol)
Δp	pressure drop, Pa
ΔT	temperature difference, K or °C
Δs	change in specific entropy, J/K
Δz	change in elevation, m
ε	effectiveness, dimensionless
ε	emissivity, dimensionless
η	dynamic viscosity, Pa s
η	efficiency, dimensionless
η_b	boiler efficiency, dimensionless
η_C	thermal efficiency of the Carnot cycle, dimensionless
η_{coll}	particulate matter collection efficiency, dimensionless
η_{comb}	combustor efficiency, dimensionless
η_{el}	electrical efficiency, dimensionless
η_g	generator efficiency, dimensionless
η_{gas}	gasification efficiency, dimensionless
η_{ic}	compressor isentropic efficiency, dimensionless
η_{it}	turbine isentropic efficiency, dimensionless
η_m	mechanical efficiency, dimensionless

η_{p}	pump efficiency, dimensionless
η_{th}	thermal efficiency of power plant, dimensionless
λ	heat-to-power ratio in a cogeneration plant, dimensionless
λ	wavelength, μm
ν	frequency, 1/s
ν	kinematic viscosity, m^2/s
ρ	density, kg/m^3
ρ	electrical resistivity of a material, W m
ρ	reflectivity, dimensionless
σ	Stefan-Boltzmann constant $[= 5.67 \times 10^{-8} \text{ W/(m}^2 \text{ K}^4)]$
ϕ	relative humidity of the air, %
ω	angular velocity, rad/s

PREFACE

The scope of this book covers mature state-of-the-art advanced energy systems and those emerging energy technologies that have already proven their feasibility in large-scale demonstration projects and are therefore on the brink of full-scale commercialization.

The forthcoming millennium will set new demands for more energy production at lower cost and reduced environmental impact. Ecological problems have become more severe primarily because of steadily increasing CO_2 emissions from fossil fuel burning plants, including those used for power production, transport, and industry. The ultimate global warming effect can cause dangerous climatic changes on Earth. Steadily increasing emissions of other atmospheric pollutants such as sulfur and nitrogen oxides are also very damaging to the environment. Therefore reduction of all emissions from the energy sector is of the utmost importance.

These concerns create new challenges in the area of energy technology. The existing power production technology cannot fulfill these requirements in the most sensitive areas of efficiency and pollutant emissions. At present, energy technology is marked by tremendous efforts aimed at achieving higher efficiencies in energy conversion with reduced environmental impact and at lower costs. Therefore advanced energy systems that have been developed in the last decades become mature contenders to the conventional energy technology based predominantly on coal-fired central steam power stations. Although conventional energy technologies have been consistently improving in terms of efficiency, economics, and reduction of environmental impact in the last decade, modern power stations have achieved the ultimate degree of intrinsically limited efficiency. Further improvements are possible only by revolutionary changes of the energy technology. The 21st century will drastically change energy production and will become the era of more efficient combined power plants. The development trends are directed at the changeover from basically steam power technology to combined-cycle power based on such plants as gas turbines, fuel cells, MHD generators, as well as the advanced energy systems based on the utilization of renewable energy sources.

The book deals with all those advanced energy systems that have reached a sufficiently mature state of development to conform to the requirement for commercial application before the end of this century. These are predominantly combined-cycle gas turbine and steam turbine power plants, various improved gas turbine configurations, second- and third-generation gas turbines with intercooling, reheat, water/steam injection, air-conditioning, etc., coal-based low-emission power systems such as integrated gasification combined cycle (IGCC) and fluidized bed combustion (including second-generation pressurized fluidized bed combustion, i.e., PFBC), means of reduction of environmental impact of energy systems, low-emission combustors and furnaces, advanced cogeneration plants, advanced solar and wind systems, as well as advanced energy storage systems.

The book comprises 10 Chapters and appendices.

Chapter 1 gives an insight into the fundamentals of energy manifested in different forms. It also discusses the issues of energy conversion, especially the conversion of heat to work, the issues of energy conservation, reserves and resources of fossil fuel energy, renewable energy, primarily solar energy, as well as the issue of world nuclear energy conversion production.

Chapter 2 features the fundamentals of fuel combustion and coal gasification. Its primary purpose is to present fuel combustion calculation methods in a concise form. The applications of these methods are illustrated by a number of solved examples. Its secondary purpose is to acquaint the reader with the basics of coal gasification as an introduction to integrated gasification combined cycles, which are considered in Chapter 9. This includes the description of modern gasification processes and gasifier types, namely, fixed bed, fluidized bed, and entrained flow gasifiers.

Chapter 3 deals with the problem of reducing the adverse impact of fossil fuel fired power plants on the environment. The mechanisms of pollutant formation during fuel combustion are first considered. Then, emissions control methods are described for all kinds of pollutants (fly ash, sulfur dioxide, and nitrogen oxides). Chapter 3 describes the modern abatement techniques that are generally applied for flue gas cleaning in power plants and features advanced multipollutant control methods such as two-sorbent, urea injection, and char-based processes. Special issues such as NO_x control in advanced gas turbine combustors and fluidized bed combustion technology with its intrinsic capability of reducing SO_2 and NO_x emissions are considered in chapters 5 and 9.

Chapter 4 reviews major issues related to advanced steam power generation technology with emphasis on methods for performance enhancement. Advanced coal-fired supercritical steam power plants apply enhanced steam parameters, low condenser pressure, single or double reheat, and multiple regenerative feedwater heating. Thermodynamic considerations are supported by numerous solved problems, each analyzing the effect of these parameters on power plant performance. Net efficiencies of state-of-the-art steam power plants of about 45% for hard coal fired power plants and about 43% for lignite-fired power plants have currently been achieved. Attainment of efficiencies up to 47–48% are predicted within the next few years with new advanced materials for the hottest components of the steam generator, steam conduits, and steam turbine.

The most impressive progress has been made in the last decade in the area of power generation with gas turbines. Chapter 5 discusses the major issues of modern gas turbine power generation technology, including thermodynamic analysis of advanced gas turbine cycles. Techniques used to improve gas turbine performance and economic and environmental characteristics of gas turbine power plants are featured in this chapter. These techniques include utilization of high gas turbine temperatures and compressor pressure ratios, intercooling and reheating, and conditioning the compressor inlet. The most advanced industrial gas turbines, particularly aeroderivative gas turbines, achieve efficiencies of about 40%. Also discussed are methods of reducing the NO_x emissions, such as utilization of dry low-NO_x combustors and burners with precise control of fuel and air addition and mixing, or water or steam injection. The most effective cycles such as recuperated water injected (RWI) and humid air turbine (HAT) cycles are also featured here.

Chapter 6 forms, along with Chapter 5, the central part of this book. They discuss the main direction of advance in the field of efficient power production with low environmental impact. Chapter 6 features all major issues related to advanced combined-cycle power plants including thermodynamics, design, environmental impact, and economics. A thermodynamic analysis of a single-pressure combined-cycle power plant is given. Based on the most efficient gas turbines, combined-cycle power plants using a rather sophisticated heat recovery steam generator (HRSG) and an advanced reheat steam cycle can achieve the highest efficiency, around 58–60%. The newest trends in the development of even more efficient combined-cycle power plants are described. Numerous tables contain current specifications and performance data of advanced combined-cycle power plants.

Chapter 7 deals with advanced cogeneration plants based on the utilization of back-pressure or extraction steam turbines, and gas turbines or gas/diesel engines with waste heat boilers or HRSGs. In contrast to plants for power generation only, the efficiency of cogeneration plants is characterized by such criteria as the energy utilization factor (EUF), heat-to-power ratio, and fuel energy savings ratio, in addition to the electrical efficiency, which characterizes the heat-to-power conversion. Maximum EUF values of 85–90% can be attained in advanced cogeneration plants.

Chapter 8 reviews the fundamentals and features the recent progress in the fuel cell and magnetohydrodynamic (MHD) power generation technologies. Major types of fuel cells are described, including phosphoric acid (PAFCs), molten-carbonate (MCFCs), and solid electrolyte (SOFCs) fuel cells, along with their component electrodes and electrolytes. Combined cycles based on high-temperature fuel cells (650°C for MCFCs and about 1000°C for SOFCs) promise to attain overall efficiencies of more than 60%. Chapter 8 also describes the current state of the development of MHD technology that can be used in advanced combined-cycle power plants with efficiencies over

60%. Critical technological problems, which prevent a rapid realization of large-scale combined-cycle plants, are also discussed.

Chapter 9 features the fluidized bed technology applied to both coal combustion and gasification and the IGCC technology. They are the major clean coal technologies, which are now undergoing the commercialization stage of development. Fluidized bed combustion technology has an intrinsic capability of reducing sulfur and nitrogen oxide emissions. Both the currently used circulating fluidized bed and the newly developed pressurized fluidized bed combustion technology are discussed. One IGCC technology integrates a coal gasification unit with a gas turbine and steam turbine combined-cycle power plant into one unit. This technology enhances energy conversion efficiency and reduces environmental impact in comparison to conventional coal-fired power plants. Critical issues such as hot gas cleaning are also considered. Emphasis is on the PFBC technology, which has advantages as compared to the atmospheric fluidized bed technology because it can be directly applied to combined-cycle power plants.

Chapter 10 deals with advanced energy storage technologies such as pumped hydro power (PHPS), compressed-air energy (CAES), and electric battery for the storage of electric energy, and with sensible and latent heat storages in solar energy systems. Also presented are equations for calculation of energy storage capacity and of mass and volumes of energy storage media required to store a certain amount of energy. A special feature is the energy storage technology for solar power plants.

This book may be used as a text for undergraduate and postgraduate students studying in the area of energy technology and environmental conservation. Many examples of the utilization of the theoretical fundamentals are presented in a very reader-friendly manner. Problems given at the end of each chapter may be used as homework assignments.

The book may also be of interest to practical engineers and scientific workers, as it presents a review of the state-of-the-art energy technology, analyzes major advanced power generation technologies, outlines the development trends, and gives an outlook as to the future of power engineering.

Chapter One

INTRODUCTION: FUNDAMENTALS OF ENERGY

ENERGY FORMS

Energy is the capability of matter to do work. All available energy forms may be classified as accumulated (stored) energy or transitional energy. Examples of accumulated energy are chemical energy of fossil fuels, internal energy of a substance, potential energy associated with position of a mass in a force field, such as the gravitational field of Earth or an electrostatic field. Transitional energy is energy transferred between a system and its surroundings. In the case of conversion of thermal energy, heat and work, are the transitional energies.

Energy is manifested in various forms (Culp, 1991; Decher, 1994):

- energy of electromagnetic radiation
- chemical energy (chemical reaction energy)
- nuclear energy (binding energy of nuclei)
- mechanical energy (potential energy, kinetic energy, work)
- internal energy
- thermal energy (heat)
- electrical energy

Any one form of energy can be converted to any other form. The extent of energy conversion can be complete or partial. Mechanical, chemical, and electrical energy can be completely converted to thermal energy (heat). The conversion of heat to mechanical energy, however, is only partial and occurs in a conversion system such as a turbine or internal combustion engine using a working fluid (gas, steam) with a cyclic change of its state. The efficiency of this energy conversion depends on the temperature difference between the working fluid in the system and the surroundings. Typical efficiencies of coal-fired power plants are 38–44%, and those internal combustion engines and gas turbines are 32% to ~40%.

When considering energy production, conversion, and utilization, one uses such terms as conventional and alternative energy sources, primary, secondary, end use energy, and useful energy. Conventional energy sources are fossil fuels (coal, lignite, pit, fuel oil, natural gas, and wood) as well as artificially produced fuel types such as coal gas, liquefied gas, coke, char, as well as combustible wastes. Renewable (also called alternative, nonconventional) energy sources are solar, wind, hydro, geothermal, wave and tidal, and biomass energy. Primary energy is the energy of an energy source without conversion, e.g., solar energy and chemical energy of fossil fuels. Secondary energy forms are the energy forms produced by conversion of primary energy. These are electrical energy, work, and thermal energy for heating, cooking, and cooling. To calculate the amount of secondary energy produced from a given amount of primary energy, a conversion efficiency of a converter facility such as a steam or gas turbine in a power station, an internal combustion engine is used. The end use energy is the energy that is available for the energy user. Its amount for a particular application is equal to the secondary energy less the energy losses in the energy transport and distribution systems. The useful energy is the energy that is required to be supplied for a particular purpose, such as vehicles, motors, lighting, heating, cooling, cooking, or industrial technological processes.

The basic energy unit is the joule: $1 \text{ J} = 1 \text{ N} \times 1 \text{ m} = 1 \text{ kg m/s}^2 \times 1 \text{ m} = 1 \text{ kg m}^2/\text{s}^2$. For large quantities of energy, the following units are used:

$1 \text{ EJ (exajoule)} = 10^{18} \text{ J}$ $1 \text{ PJ (petajoule)} = 10^{15} \text{ J}$ $1 \text{ TJ (terajoule)} = 10^{12} \text{ J}$

$1 \text{ GJ (gigajoule)} = 10^{9} \text{ J}$ $1 \text{ MJ (megajoule)} = 10^{6} \text{ J}$ $1 \text{ kJ (kilojoule)} = 10^{3} \text{ J}$

Energy, particularly electrical energy, is also measured in kilowatt-hours (kWh). These energy units are related as follows: $1 \text{ kWh} = 3.6 \text{ MJ}$ and $1 \text{ MJ} = 0.278 \text{ kWh}$.

In statistical reports, two energy equivalents may be used for the purpose of comparison of different types of fuel and energy: coal equivalent and mineral oil equivalent. Thereby, 1 tonne of coal equivalent is 1 tce $= 29.308 \text{ MJ}$ and 1 tonne of mineral oil equivalent is 1 toe $= 41,868 \text{ MJ}$ (1 tonne $= 1000 \text{ kg}$). For large (fuel) energy quantities, a million tonnes of oil equivalent (1 Mtoe) is used: $1 \text{ Mtoe} = 41,868 \times 10^{12} \text{ J} = 41,868 \text{ TJ} = 11,630 \times 10^{6} \text{ kWh}$.

Short descriptions of the different forms of energy follow.

Energy of Electromagnetic Radiation

The whole spectrum of electromagnetic radiation includes γ radiation (wavelength range $\lambda \leq 10^{-11}$ m), X-rays ($\lambda \leq 10^{-8}$ m), solar and thermal radiation ($\lambda < 10^{-4}$ m), radio, TV, and radar waves ($\lambda > 10^{-2}$ m).

Solar Energy

Solar energy is electromagnetic energy in the wavelength range from 0.25 to $\sim$4 μm (1 μm $= 10^{-6}$ m). The solar spectrum includes ultraviolet radiation ($\lambda = 0.25 - 0.38$ μm), visible light ($\lambda \leq 0.78$ μm), and near-infrared radiation ($\lambda \leq 4$ μm).

The quantity of energy carried by a photon of light (solar radiation) is given by

$$E = h\nu = hc/\lambda \quad \text{J} \tag{1.1}$$

where h is Planck's constant (6.626×10^{-34} Js), ν is the wave frequency in 1/s, c is the light velocity (2.998×10^{8} m/s), and λ is the wavelength in m.

Energy of a photon of a wave with a shorter wavelength, e.g., in the range of ultraviolet radiation, is higher than that at a longer wavelength, say, in the range of near-infrared radiation.

The total emissive power of the sun is 4×10^{23} kW (Khartchenko, 1995). Only a very small fraction of the total solar energy flux is emitted to the Earth at a solid angle of 0.53°. However, the solar energy flux that reaches the Earth's surface is huge (3.7×10^{24} J $= 1.03 \times 10^{18}$ kWh) and surpasses annual world primary energy consumption by a factor of about 13,000.

The sun is the most abundant undepletable source of nonpolluting energy. However, the utilization of solar energy is complicated by its relatively low intensity and the stochastic character of incident solar radiation, which is strongly influenced by season, day-night cycle, and cloudiness of the sky. These peculiarities of incident solar energy are accounted for by sizing solar systems. Thus solar collectors must be large enough to collect a quantity of solar energy not only for current use but also for use during periods of low insolation and at night. This additional amount of solar energy is stored. To guarantee the energy supply, an additional energy (backup) system is used in practically all solar plants. One square meter of collector surface area can produce about 500 kWh of useful heat with a temperature of the heat transfer fluid of 45–60°C at a locality with an annual solar flux of about 1200 kWh/m^2. A flat plate collector of 4–6 m^2 and a hot water tank with a capacity of 200–300 liters are required for domestic water heating for a family of four. The capital cost of such a plant in Germany is about $7000. The amortization period lies between 12 and 15 years (Khartchenko, 1995).

Chemical Energy

Chemical energy is energy stored in a chemical compound. For thermal energy conversion, energy (heat) of formation and energy (heat) of chemical reactions are important. Energy (heat) is released in exothermic reactions, e.g., those of fuel combustion, and is absorbed in endothermic reactions, e.g., those of coal gasification.

The heat (enthalpy) ΔH_f of formation of a compound is the heat required to form 1 mol of the product from the reactants in a thermodynamic standard state. Usually, ΔH_f is given at standard temperature and pressure (25°C and 1 atm). For example, the enthalpy of formation ΔH_f of water vapor from hydrogen and oxygen is $-57{,}798$ kcal/mol (or $\times\ 4.187$ kJ/kcal $= -242.0$ kJ/mol). For simple substances in natural form such as H_2, O_2, and C (solid), ΔH_f is zero. The negative sign of ΔH_f means that the heat is released and the reaction is exothermic.

The heat (enthalpy) ΔH_r of reaction is the difference between the sum of heats of formation of products ΔH_{fp} and the sum of heats of formation of reactants (Culp, 1991; Decher, 1994):

$$\Delta H_r = \sum \Delta H_{fp} - \sum \Delta H_{fr} \tag{1.2}$$

As an example, consider the reaction of hydrogen combustion:

$$H_2 + 1/2 O_2 = H_2O(\text{vapor}) \tag{1.3}$$

Then,

$$\Delta H_r = \Delta H_f(H_2O) - [\Delta H_f(H_2) + 1/2\Delta H_f(O_2)] = -242.0 \text{ kJ/mol} - 0 = -242.0 \text{ kJ/mol}$$

and with the molar mass of hydrogen of 2.016 g/mol,

$$\Delta H_r = -242.0 \text{ kJ}/2.016 \text{ g} = -120.04 \text{ kJ/g} = -120.04 \text{ MJ/kg}$$

The heating value of a fuel is equal to $-\Delta H_r$; thus for hydrogen it is 120.04 MJ/kg.

Nuclear Energy

Nuclear energy is released by radioactive decay, fission, or fusion of nuclei. A nucleus of an atom with a mass number A is composed of z protons and $n = A - z$ neutrons. The mass of a nucleus is less than the total mass of its nucleons. This difference is called the mass defect:

$$\Delta m = z m_p + n m_n - m_{\text{nucleus}} \tag{1.4}$$

The total binding energy is equivalent to the mass defect Δm:

$$E = \Delta m c^2 \tag{1.5}$$

where c is the light velocity (2.998×10^8 m/s). Usually E is given in MeV and Δm is in atomic mass units (amu):

$$1 \text{ MeV} = 1.602 \times 10^{-13} \text{ J} \qquad 1 \text{ amu} = 1.66 \times 10^{-24} \text{ g} \qquad 1 \text{ amu} = 931.5 \text{ MeV}$$

Thus

$$E = 931.5\Delta m \quad \text{MeV} \tag{1.6}$$

The excess binding energy is released either in fission or fusion processes. The fission process taking place in a nuclear reactor occurs when a heavy nucleus absorbs a neutron and splits into two or more light nuclei. The amount of energy released by fission of 1 g of uranium 235 is 490×10^{21} MeV $= 78.5$ GJ. This energy is equivalent to the amount of heat that is released by combustion of 2700 kg of coal with a heating value of 29.3 MJ/kg.

In a fusion process, two light nuclei combine to form a heavier nucleus with release of excess binding energy. In the center of the sun, two hydrogen nuclei produce one helium nucleus with release of 23.85 MeV energy.

Mechanical Energy

The forms of mechanical energy are kinetic energy, potential energy, and work. Kinetic energy is associated with motion of translation, rotation, or oscillation. For a body of a mass m moving with a velocity w, the kinetic energy is

$$E_k = 1/2mw^2 \quad \text{J} \tag{1.7}$$

where m is mass in kg and w is velocity in m/s. The change of velocity from w_1 to w_2 causes a corresponding change in its kinetic energy:

$$\Delta E_k = E_2 - E_1 = 1/2m\left(w_2^2 - w_1^2\right) \quad \text{J} \tag{1.8}$$

Potential Energy

Potential energy is associated with a body in a force field. For a body of a mass in a gravitation field, the potential energy is

$$E_p = gmz = g\rho Vz \quad \text{J} \tag{1.9}$$

where g is the acceleration due to gravity in m/s^2, m is the mass of a body in kg, ρ is the density of the substance in kg/m^3, V is the volume of the body in m^3, and z is the height of the center of gravity of the body above a reference plane in m.

The change of the body position in the gravitation field causes a corresponding change in its potential energy. Thus

$$\Delta E_p = gm(z_1 - z_2) \quad \text{J} \tag{1.10}$$

Internal Energy

Internal energy is a thermodynamic parameter of the state of a substance. It consists of the kinetic energy of motion of molecules, atoms, and subatomic particles and the potential energy due to their relative position. For an ideal gas the internal energy U depends only on the temperature. For real working fluids such as steam or refrigerant vapor, it depends on the temperature and pressure.

The change in the internal energy of an ideal gas is given by

$$\Delta U = U_2 - U_1 = mc_v(T_2 - T_1) \quad \text{J} \tag{1.11}$$

where m is the mass of gas in kg, c_v is the specific heat of the gas at constant volume in J/(kg K), and T_1 and T_2 are the initial and final temperatures of the gas in K.

Enthalpy

Like internal energy, enthalpy is a thermodynamic parameter of the state of a substance. It is the sum of the internal energy and the pressure energy. Thus the specific enthalpy is

$$h = u + pv \quad \text{J/kg} \tag{1.12}$$

The change in the total enthalpy of an ideal gas is given by

$$\Delta H = H_2 - H_1 = mc_p(T_2 - T_1) \quad \text{J} \tag{1.13}$$

where c_p is the specific heat of the gas at constant pressure in J/(kg K).

Electrical Energy

Electrical energy is one form of end use energy. It is associated with the flow of electrons (current) through a conductor. The electrical energy is generated in power plants by conversion of primary energy of fossil or nuclear fuels, solar energy, hydroenergy, geothermal energy, or wind energy in power plants of various types.

WORK AND POWER

Displacement Work

Work is mechanical energy in transition. Let us consider displacement work, flow work, and shaft work. Displacement work is work that occurs at the boundary of a thermodynamic system when the volume of a working fluid changes. Thus it is associated with the change in volume of a working fluid either in an expansion or compression process. It is positive in an expansion process and negative in a compression process. A typical example is a closed system shown in Figure 1.1 that comprises a cylinder-piston arrangement containing a working fluid.

Defining work as the product of a force acting through a distance yields for the infinitesimally small displacement work (Figure 1.1) the following equation:

$$dw = Fdx = pAdx = pdv \quad \text{J/kg} \tag{1.14}$$

where F is force in N, dx is distance in the force direction in m, p is pressure in Pa, A is cross-sectional area of the system in m^2, and v is the specific volume of the working fluid in m^3/kg. Integrating yields the specific displacement work:

$$w = \int_{v_1}^{v_2} pdv \quad \text{J/kg} \tag{1.15}$$

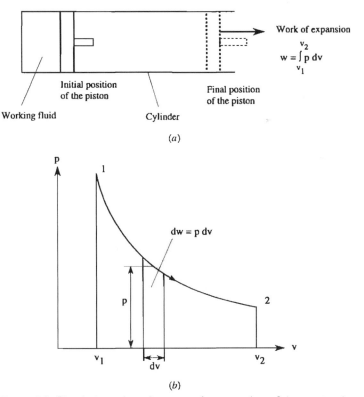

Figure 1.1. Closed thermodynamic system for conversion of heat to work: (a) cylinder-piston arrangement with a working fluid and (b) displacement work of an expansion process on a p-v diagram.

For a mass m the total displacement work is

$$W = mw = m \int_{v_1}^{v_2} p \, dv \quad \text{J} \tag{1.16}$$

Thus it is seen that displacement work depends on the relationship between the two parameters of the working fluid—the pressure p and the specific volume v of the working fluid in a thermodynamic process (isochoric, isobaric, isothermal, adiabatic, or polytropic process).

For a given thermodynamic process of expansion or compression of the working fluid, the integration of Eq. (1.15) or (1.16) yields Eqs. (1.17)–(1.20).

For an isobaric process, i.e., at $p = \text{const}$,

$$w = p(v_2 - v_1) \quad \text{J/kg} \tag{1.17}$$

For a reversible adiabatic expansion or compression process, i.e., at $s = \text{const}$ ($pv^k = \text{const}$),

$$w = (p_2 v_2 - p_1 v_1)/(k - 1) \quad \text{J/kg} \tag{1.18}$$

For an isothermal expansion or compression process ($pv = \text{const}$),

$$w = p_1 v_1 \ln (v_2/v_1) = p_1 v_1 \ln (p_1/p_2) \quad \text{J/kg} \tag{1.19}$$

For a polytropic expansion or compression process ($pv^n = \text{const}$),

$$w = (p_2 v_2 - p_1 v_1)/(n - 1) \quad \text{J/kg} \tag{1.20}$$

To get the total work displacement, the values of the specific work w calculated from Eqs. (1.17)–(1.20) should be multiplied by the mass m of the working fluid in the system. For an isochoric process, i.e., at $v = \text{const}$, work is zero.

Flow Work

Flow work is work required to move the working fluid into and out of an open system across its boundaries (Figure 1.2). It is given by

$$dw_{\text{flow}} = d(pv) \quad \text{J/kg} \tag{1.21}$$

$$W_{\text{flow}} = p_2 V_2 - p_1 V_1 \quad \text{J} \tag{1.22}$$

Shaft Work

Shaft work is work measured on the shaft of a heat engine, e.g., turbine, or of a machine such as a compressor, pump, or fan that is used to increase the pressure of the fluid. Shaft work is positive for turbines and negative for compressors, pumps, or fans.

Shaft work is the difference between displacement work and flow work. Thus the shaft work for infinitesimally small change in the state of the working fluid is given by

$$dw_{\text{shaft}} = dw - dw_{\text{flow}} = dw - d(pv) = pdv - (pdv + vdp) = -vdp \quad \text{J/kg} \tag{1.23}$$

For a finite change in the state of the working fluid,

$$w_{\text{shaft}} = - \int_{p_1}^{p_2} vdp \quad \text{J/kg} \tag{1.24}$$

$$W_{\text{shaft}} = m w_{\text{shaft}} = - \int_{p_1}^{p_2} V dp \quad \text{J} \tag{1.25}$$

Otherwise,

$$w_{\text{shaft}} = w - (p_2 v_2 - p_1 v_1) \quad \text{J/kg} \tag{1.26}$$

$$W_{\text{shaft}} = m w_{\text{shaft}} = W - (p_2 V_2 - p_1 V_1) \quad \text{J} \tag{1.27}$$

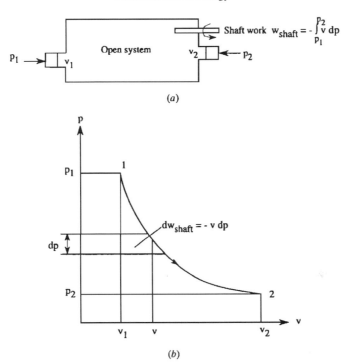

Figure 1.2. Open thermodynamic system for conversion of heat to work:
(a) schematic and (b) shaft work of an expansion process on a p-v diagram.

where w_{shaft}, w, and w_{flow} are specific shaft work, displacement work, and flow work in J/kg, v is specific volume of the working fluid in m³/kg, and dp is change in pressure in Pa. For example, for an adiabatic expansion or compression of the working fluid, the shaft work is

$$W_{shaft} = (p_2 V_2 - p_1 V_1)k/(k-1) = mR(T_2 - T_1)k/(k-1) = mc_p(T_2 - T_1) \quad \text{J} \qquad (1.28)$$

Power

Power is the rate of energy transfer or the work done per second:

$$P = dW/dt \quad \text{W} \qquad (1.29)$$

The basic power unit is the watt (1 W = 1 J/s). Other power units are 1 kW = 10^3 W, 1 MW = 10^6 W, and 1 GW = 10^9 W.

HEAT

Heat is thermal energy in transition. It can be either added to a system or rejected. In general, for an infinitesimally small change in entropy of the working fluid, heat added to a system or rejected per one or m kg of the working fluid is given by

$$dq = Tds \quad \text{J/kg} \qquad (1.30)$$

$$dQ = mdq = TdS \quad \text{J} \qquad (1.31)$$

where T is temperature in K, ds is change in specific entropy of the working fluid in J/(kg K), and $dS = mds$ is entropy change in J/K.

Similar to work, the heat quantity depends on the thermodynamic process, i.e., on the relationship between the two parameters of the working fluid—the temperature and the entropy:

$$q = \int_{s_1}^{s_2} T ds \quad \text{J/kg} \tag{1.32}$$

$$Q = mq = \int_{s_1}^{s_2} T dS \quad \text{J} \tag{1.33}$$

The heat added to the working fluid is positive; the rejected heat is negative. Thus for an adiabatic process ($pv^k = \text{const}$) entropy s is constant and heat is zero. For an isothermal process ($T = \text{const}$) the heat quantity is given by

$$q = T(s_2 - s_1) \quad \text{J/kg} \tag{1.34}$$

$$Q = mq = T(S_2 - S_1) \quad \text{J} \tag{1.35}$$

where T is temperature in K, s_1 and s_2 are specific entropy of the working fluid at initial and end state in J/(kg K), and S_1 and S_2 are entropy of the working fluid at initial and end state in J/K. If the temperature change in a thermodynamic process such as an isochoric or isobaric process is known (and is not equal to zero), the heat quantity is given by Eqs. (1.36) and (1.37). For an isochoric process with an ideal gas as the working fluid, i.e., at $v = \text{const}$,

$$Q = mc_v(T_2 - T_1) \quad \text{J} \tag{1.36}$$

where m is mass of the working fluid in kg, c_v is constant volume specific heat of the working fluid in J/(kg K), and T_1 and T_2 are initial and final temperature of the working fluid, respectively, in K. For an isobaric process with an ideal gas as the working fluid, i.e., at $p = \text{const}$,

$$Q = mc_p(T_2 - T_1) \quad \text{J} \tag{1.37}$$

where c_p is constant pressure specific heat of the working fluid in J/(kg K). Thereby,

$$c_p = c_v + R \quad \text{J/(kg K)} \tag{1.38}$$

$$c_p/c_v = k \tag{1.39}$$

where R is the gas constant of the working fluid (ideal gas) in J/(kg K), and k is the isentropic exponent (1.4 for air and other gases with two atoms in the molecule).

ENERGY CONSERVATION

The principle of energy conservation states that energy cannot be created or destroyed; it can only be converted to other forms of energy in equivalent quantities. On this principle, the operation of all energy conservation devices is based. The energy equation is the mathematical description of the principle of energy conservation for a thermodynamic system that is used to convert heat to mechanical energy. It is the First Law of Thermodynamics for a control volume of an open steady-flow system. The First Law of Thermodynamics states that the heat Q transferred to the control volume is equal to the shaft work W_s and the sum of changes in the internal energy ΔU, in the flow work $\Delta(pV)$, in the kinetic energy ΔE_k, and in the potential energy ΔE_p. Thus the energy equation for the control volume shown in Figure 1.3 is given by

$$Q = W_s + \Delta U + \Delta(pV) + \Delta E_k + \Delta E_p \quad \text{J} \tag{1.40}$$

The sum of changes in the internal energy ΔU, in the flow work $\Delta(pV)$, can be replaced by the change in enthalpy,

$$\Delta H = \Delta U + \Delta(pV) = U_2 - U_1 + p_2 V_2 - p_1 V_1 \quad \text{J} \tag{1.41}$$

The change in kinetic energy of the working fluid is

$$\Delta E_k = 1/2 \, m \left(w_2^2 - w_1^2 \right) \quad \text{J} \tag{1.42}$$

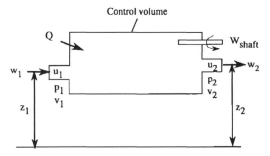

Figure 1.3. Energy flows in the control volume of an open thermodynamic system.

where w_1 and w_2 are the working fluid velocity at the system inlet and outlet, respectively, in m/s. The change in its potential energy is

$$\Delta E_p = gm(z_2 - z_1) = gm\Delta z \quad J \tag{1.43}$$

where z_1 and z_2 are the elevation of the system inlet and outlet measured above an arbitrary reference plane, respectively, in m.

In most cases in heat engines (gas and steam turbines), Δz, and therefore ΔE_p, may be neglected. Thus Eq. (1.40) may be rearranged as follows:

$$Q = W_{shaft} + \Delta H + \Delta E_k \quad J \tag{1.44}$$

For a very important practical case of an adiabatic expansion process in a gas or steam turbine $(Q = \Delta E_k = 0)$, the turbine specific shaft work and the theoretical power output are correspondingly

$$w_{shaft} = -\Delta h = h_1 - h_2 \quad J/kg \tag{1.45}$$

$$P_t = m w_{shaft} = m(h_1 - h_2) \quad W \tag{1.46}$$

where m is the mass flow rate of the working fluid in kg/s and h_1 and h_2 are the specific enthalpy of the working fluid at the system inlet and outlet, respectively, in J/kg.

ENERGY RESERVES AND RESOURCES

There are conventional and nonconventional energy sources. Conventional energy sources are associated with fossil fuels and artificially produced fuels. Nonconventional energy sources include nuclear energy and renewable energy sources. The renewable (also called alternative) energy sources are solar, wind, hydro, geothermal, wave and tidal, and biomass energy. The data on world primary energy consumption are presented in Table 1.1 (status 1993) (*BP Statistical Review of World Energy*, 1994).

Table 1.1. World primary energy consumption

Region	Mtoe	EJ	PWh
World	7804.3	337.14	93.65
Africa	220.9	9.54	2.65
Asia and Australasia	1924.7	83.16	23.10
Europe (non-OECD countries)	1387.7	59.94	16.65
Europe (OECD countries)	1411.9	60.98	16.94
Latin America	387.4	16.74	4.65
Middle East	261.0	11.27	3.13
North America	2210.7	95.50	26.53
United States	1996.0	86.22	23.95

Mtoe = million tonnes of oil equivalent, 1 EJ = 10^{18} J, 1 PWh = 10^{15} Wh.

Table 1.2. World energy reserves, consumption, and resources of
fossil fuels

Fuel	Energy reserves, EJ	Consumption, EJ	Resources, years
Coal	30,226	135	224
Oil	5,790	129	45
Natural gas	4,970	76	65

1 EJ = 10^{18} J.

Table 1.3. Proven reserves to production ratio for fossil fuels in years

Fuel	Coal	Oil	Natural gas
World	224	45	65
Non-OECD Europe	325	18	68
OECD Europe	230	10	15
LDCs	160	63	153

LDCs, Latin America, Middle East, Africa, and non-OECD Asia.

Table 1.4. Annual production of nuclear
energy in the world

Region	Nuclear energy, 10^9 kWh
World	6700
Non-OECD Europe	830
OECD Europe	2440
LDCs	380
North America	2260

LDCs, Latin America, Middle East, Africa, and non-
OECD Asia.

Fossil Fuel Reserves

World reserves of fossil fuels are limited. Table 1.2 contains data on world reserves and consumption of fossil fuel (coal, oil, and natural gas) energy as well as their resources (*BP Statistical
Review of World Energy*, 1994). Table 1.3 contains data on the proven reserves to production ratio
for fossil fuels in years (*BP Statistical Review of World Energy*, 1994).

Nuclear Energy Production

Data on annual production and consumption of nuclear energy in the world are given in Table 1.4
(status 1993) (*BP Statistical Review of World Energy*, 1994).

Renewable Energy Resources

Solar energy is the most important source of energy for humans (Boyle, 1966; Burnham, 1993).
The annual amount of solar energy incident on the Earth is equivalent to about 13,000 times the
world's annual consumption of energy of fossil and nuclear fuels and hydroenergy. It is about 160
times the proven reserves of energy of fossil fuels.

Data on extraterrestrial solar radiation above the Earth's atmosphere and solar radiation incident
on the Earth are given in Table 1.5 (Boyle, 1966).

Hydro energy is the most utilized energy of all the renewable energy sources at the present
time. Data on annual production and consumption of hydro electricity in the world are presented
in Table 1.6 (status 1993) (*BP Statistical Review of World Energy*, 1994).

Table 1.5. Extraterrestrial and terrestrial solar radiation and wind energy

| | Annual energy flux | | Power, |
	10^{24} J	10^{18} kWh	10^{14} kW
Extraterrestrial solar radiation	5.6	1.56	1.78
Solar radiation at the Earth's surface	3.7	1.03	1.20

Table 1.6. Annual production and consumption of hydro electricity in the world

Region	Hydro electricity, 10^9 kWh
World	2400
Non-OECD Europe	300
OECD Europe	470
LDCs	840
North America	615

LDCs, Latin America, Middle East, Africa, and non-OECD Asia.

Table 1.7. Estimated renewable energy resources

Energy source	Resource potential, TW	Recoverable energy flux, TW
Solar	90,000	1,000
Wind	1,200	10
Hydro power	3,000	
Geothermal	30	n.a.
Tidal	30	0.1
Wave	3	0.5
Stored energy, PWh		
Biomass (plants)	3,940	n.a.
Geothermal	8.8×10^{11}	>400

Source: Boyle (1996).

There are various estimates of the world's renewable energy resources. Table 1.7 contains some data on estimated renewable energy resources and recoverable energy fluxes presented in terawatts (1 TW = 10^{12} W) as well as on the energy stored in biomass and as geothermal energy both given in petawatt-hours (1 PWh = 10^{15} Wh).

PROBLEMS

1.1. One kilogram of air is isentropically compressed from an initial state of 12°C and 0.97 bar to a final pressure of 14 bars. Calculate (a) the air final temperature, (b) the displacement work, (c) the shaft work, (d) the change in the internal energy of the air, and (e) the change in the enthalpy of the air. Assume an air gas constant of 287 J/(kg K), an isentropic exponent k of 1.4, and the specific heat of the air at constant volume of 718 J/(kg K) and constant pressure of 1005 J/(kg K).

1.2. An amount of air, 200 kg/s, is heated at constant pressure of 1.47 MPa from 330°C to 1150°C. Calculate (a) the displacement work rate, (b) the heat input rate, (c) the rate of change in the internal energy of the air, and (d) the rate of change in the enthalpy of the air. Assume an air gas constant of 287 J/(kg K), an isentropic exponent k of 1.4, and specific heat of the air at constant volume 718 J/(kg K) and constant pressure 1005 J/(kg K).

1.3. Determine the effect of changes in the average heat input and heat rejection temperatures on the thermal efficiency of an arbitrary heat engine cycle. Assume (a) that the average heat input temperature increases from 450°C to 620°C at a constant heat rejection temperature of 300°C, and (b) that the average heat rejection temperature changes from 300°C to 350°C at a constant heat input temperature of 570°C.

REFERENCES

Boyle, G., ed. 1966. *Renewable energy: Power for a sustainable future*. Oxford University Press.
BP Statistical Review of World Energy. June 1994.
Burnham, L., ed. 1993. *Renewable energy: Sources for fuels and electricity*. Washington, D.C.: Island Press.
Culp, A. 1991. *Principles of energy conversion*. New York: McGraw-Hill.
Decher, R. 1994. *Energy conversion*. Oxford University Press.
Khartchenko, N. V. 1995. *Thermische Solaranlagen* (in German). Berlin: Springer.

Chapter Two

FUEL COMBUSTION AND GASIFICATION

This chapter covers two important issues—the fundamentals of fuel combustion and of coal gasification. The information related to the composition of solid, liquid, and gaseous fuels and their heating values is presented first. Then fuel combustion calculations, both for stoichiometric and actual fuel combustion, are described. These include the calculation of combustion air requirements and volumes of products of combustion as well as the calculation of their composition in terms of mole fractions of flue gas constituents. In addition, formulae are given for the calculation of the adiabatic combustion temperature and the excess air ratio.

The second part of this chapter reviews the fundamentals of coal gasification, including the stoichiometry, as well as carbon conversion, gasification, and thermal efficiencies. The state-of-the-art coal gasification processes and gasifier types are described, including fixed bed Lurgi and British Gas/Lurgi (BGL) gasifiers, fluidized bed Winkler and high-temperature Winkler gasifiers, and Koppers-Totzek entrained flow gasifiers along with other proprietary gasifier types.

FUEL COMPOSITION AND HEATING VALUE

Solid Fuel Constituents

Anthracite, bituminous coals, lignite, wood, peat, and solid wastes are used as solid fuels. Combustion of a solid fuel has several stages: heating and drying, devolatilization and forming of char, and igniting and burning of volatiles and char. When a coal is heated to relatively low temperatures, the volatile matter is released, and char is formed. The volatile matter content lies between 3 and 10% for anthracite and 28 and 50% for high-volatile bituminous coal. It contains tars, oils, phenols, and hydrocarbon gases. At high temperatures they are cracked and oxidized to CO, H_2, CO_2, and H_2O.

All fuels contain combustible and noncombustible constituents. Carbon, hydrogen, and sulfur are combustible constituents of a fuel, whereas nitrogen, ash, and moisture are its noncombustible constituents. Ash, sulfur, and moisture are the unwanted constituents of a coal, as they lower its quality. The moisture content of a solid fuel varies widely, depending on the type of fuel and its handling. Raw lignites have the highest moisture content among all coals, being in the range of 40–60%. The moisture content of bituminous coals is typically 1–3%, and that of coke is 3–8%. The ash content of bituminous coals is 8–10%, of anthracite is 3–6%, and of lignite is 3–8%. The choice of combustion method and combustor type as well as the furnace outlet temperature are strongly affected by the ash composition and properties, especially by the ash softening temperature. Steam generators and boilers can be fired practically with all types of fuels including lignites, bituminous coals, biomass and combustible wastes, heavy fuel oils, natural gas, and other types of fuel gases.

Proximate and Ultimate Analyses of Coals

Coal composition is usually presented by means of proximate and ultimate analyses (Combustion Engineering, 1981). Both these analyses may be given on an as-received or as-fired basis as well as on a dry, ash-free basis. Before the actual combustion of a coal begins, the devolatilization of the coal occurs by heating to a higher temperature. Thus the volatile matter will be released from the coal.

The proximate analysis of a coal gives the contents of fixed carbon (FC), volatile matter (VM), ash (A), and moisture (M) in the coal as mass fractions. The as-fired (af) proximate analysis of a coal is given by

$$(FC + VM + A + M)_{af} = 1 \tag{2.1}$$

In addition, the sulfur content (S) and the higher heating value (HHV) of coal are given as mass fraction and in kJ/kg, respectively. The definition of the heating value of a fuel is given in the next section.

To convert the af proximate analysis on a dry, ash-free (daf) basis, one should divide the above values of FC_{af} and VM_{af} by $(1 - A - M)$. Thus, for the proximate analysis on a daf basis,

$$FC_{daf} = FC_{af}/(1 - A - M) \tag{2.2}$$
$$VM_{daf} = VM_{af}/(1 - A - M) \tag{2.3}$$
$$(FC + VM)_{daf} = 1 \tag{2.4}$$

The sulfur content and the HHV of the coal should be converted as follows:

$$S_{daf} = S_{af}/(1 - A - M) \tag{2.5}$$
$$HHV_{daf} = HHV_{af}/(1 - A - M) \tag{2.6}$$

The coal ultimate analysis is the elemental analysis. The ultimate analysis on a daf basis gives the mass fractions of carbon (C), hydrogen (H), sulfur (S), oxygen (O), and nitrogen (N). Therefore

$$(C + H + S + O + N)_{daf} = 1 \tag{2.7}$$

To convert the daf values into the af values, one should multiply them by $(1 - A - M)$. Thus

$$C_{af} = C_{daf}(1 - A - M) \tag{2.8}$$
$$H_{af} = H_{daf}(1 - A - M) \tag{2.9}$$
$$S_{af} = S_{daf}(1 - A - M) \tag{2.10}$$
$$O_{af} = O_{daf}(1 - A - M) \tag{2.11}$$
$$N_{af} = N_{daf}(1 - A - M) \tag{2.12}$$

The af ultimate analysis gives the mass fractions of not only these elements, but also of ash and moisture. Hence

$$(C + H + S + O + N)_{af} + A + M = 1 \tag{2.13}$$

Example 2.1

For a bituminous coal, the daf proximate analysis is given as follows (in % by mass): $VM_{daf} = 39.5\%$ and $FC_{daf} = 60.5\%$. The sulfur, ash, and moisture contents are $S = 4\%$, $A = 5\%$, and $M = 10\%$. The daf ultimate analysis of the coal is (in mass fractions) $C_{daf} = 0.83$, $H_{daf} = 0.055$, $S_{daf} = 0.04$, $O_{daf} = 0.06$, and $N_{daf} = 0.015$.

Determine the af proximate and ultimate analyses of the given coal.

Solution

1. To convert the daf proximate analysis to an af basis, the conversion factor is $1 - A - M = 1 - 0.05 - 0.1 = 0.85$.

Thus the af proximate analysis is (in % by mass)

$$VM_{af} = 0.85VM_{daf} = 0.85 \times 39.5 = 33.6\%$$

$$FC_{af} = 0.85FC_{daf} = 0.85 \times 60.5 = 51.4\%$$

$$A = 5\%$$

$$M = 10\%$$

Total $= 100\%$

2. Then, the af ultimate analysis (in mass fractions) is

$$C_{af} = 0.85C_{daf} = 0.85 \times 0.83 = 0.705$$

$$H_{af} = 0.85H_{daf} = 0.85 \times 0.055 = 0.047$$

$$S_{af} = 0.85S_{daf} = 0.85 \times 0.04 = 0.034$$

$$O_{af} = 0.85O_{daf} = 0.85 \times 0.06 = 0.051$$

$$N_{af} = 0.85N_{daf} = 0.85 \times 0.015 = 0.013$$

$$A = 0.05$$

$$M = 0.10$$

Total $= 1$

Heat of Combustion

The First Law of Thermodynamics applied to a combustion process in a control volume is given by (Combustion Engineering, 1981; Bartok and Sarofim, 1991; Smoot and Smith, 1985; Merrick, 1984)

$$H_r = H_{fp} - H_{fr} \tag{2.14}$$

where H_r is heat (enthalpy) of reaction (negative for combustion process); H_{fp} is heat (enthalpy) of formation of products, and is equal to the number of moles times enthalpy per mole; and H_{fr} is heat (enthalpy) of formation of reactants. These values are referred to a reference temperature of 25°C (298 K) and 1 bar.

The heating value (HV) of a fuel is the amount of heat released by a mole of fuel when it is completely burned and the products of combustion are cooled to the original fuel temperature. Thus

$$HV = -H_r \quad kJ/mol \tag{2.15}$$

However, for practical purposes, HV is normally given on a unit mass basis in kJ/kg. For gaseous fuels, it is given on a unit volume basis in kJ/m³ at standard conditions of 0°C and 101.3 kPa. Two heating values of a fuel—higher and lower heating values—are used. If the water in the products of combustion remains in vapor form, the heating value is called the lower heating value (LHV). If the water vapor is condensed within the combustion chamber, the heating value thus obtained is the higher heating value (HHV).

For anthracite and bituminous coals, the HHV may be estimated by Dulong's equation (Combustion Engineering, 1981; Bartok and Sarofim, 1991; Smoot and Smith, 1985; Merrick, 1984):

$$HHV = 33.95C + 144.2(H - O/8) + 9.4S \quad MJ/kg \tag{2.16}$$

where C, H, S, and O are the mass fractions of carbon, hydrogen, oxygen, and sulfur in the ultimate analysis of af fuel.

The LHV is obtained by subtracting from the HHV, the enthalpy of vaporization of the water content of flue gas. As a rule, the products of combustion leave the furnace at a temperature at

Table 2.1. Average composition, higher heating value (HHV), and lower heating value (LHV) of solid fuels

Fuel type	Ash, %	Water, %	Ultimate analysis (daf), % by mass					HHV, MJ/kg	LHV, MJ/kg
			C	H	S	O	N		
Hard coal	3–12	0–10	80–90	4–9	0.7–1.4	4–12	0.6–2	29–35	27–34
Lignite	2–8	50–60	65–75	5–8	0.5–4	15–26	0.5–2	10–13	8–10.5
Anthracite	2–6	0–5	90–94	3–4	0.7–1	0.5–4	1–1.5	33.5–35	32.5–34

which no condensation of the water vapor can occur. Therefore the LHV is often used. If the flue is cooled down below the dew point, at which the condensation of water vapor occurs within the furnace, the HHV must be used. The LHV of any fuel is

$$LHV = HHV - h_{fg}m_{H_2O} = HHV - 2.5(9H + M) \quad MJ/kg \tag{2.17}$$

where h_{fg} is the specific enthalpy of vaporization for water (2.5 MJ/kg at 0°C and 1 bar), m_{H_2O} is the water vapor mass per kilogram of fuel, H is the hydrogen content as given in the fuel ultimate analysis, and M is the moisture content in the fuel. The values of m_{H_2O}, H, and M are all in units of kg/kg.

Table 2.1 gives the composition, HHV, and LHV of typical solid fuel types.

Example 2.2

Calculate the HHV and LHV of a bituminous coal that has the following af ultimate analysis (in mass fractions): $C = 0.705$, $H = 0.047$, $S = 0.034$, $O = 0.051$, $N = 0.013$, $A = 0.05$, and $M = 0.10$.

Solution

$$HHV = 33.95C + 144.2(H - O/8) + 9.4S = 30.11 \text{ MJ/kg}$$

$$LHV = HHV - 2.5(9H + M) = 28.80 \text{ MJ/kg}.$$

Liquid Fuels

Liquid fuels consist of hydrocarbons and small amounts of other compounds. Similar to solid fuels, the composition of liquid fuels is usually reported in the form of an ultimate analysis. Because of the detrimental effect of sulfur dioxide on the environment, sulfur is an undesirable constituent of liquid fuels.

In power plants the mineral oil products such as heavy fuel oil (HFO) and light fuel oil are used. Fuel oils are mixtures of hydrocarbons with chemical compounds of sulfur, nitrogen, and oxygen. At ambient temperature, the HFO has a high viscosity, and therefore it must be preheated (up to 80–140°C) before it is burned. Liquid fuels are atomized in burners by means of air, steam, or mechanical force. Light distillates burn with practically no solid contaminants, and therefore they are used to ignite and fire gas turbines. The typical composition and heating value of liquid fuels are given in Table 2.2.

The LHV for mineral oil products may be calculated as follows:

$$LHV = 33.15C + 94.1H + 10.46(S - O) \quad MJ/kg \tag{2.18}$$

Gaseous Fuels

Gaseous fuels include both the natural gas and the synthetic gaseous fuels such as the product of fuel gas, coke gas, and biogas. The composition of gasesous fuels is usually given in mole fractions or in percent by volume at standard conditions (0°C and 101.3 kPa).

Table 2.2. Density (at 20°C), composition, higher heating value (HHV), and lower heating value (LHV) of liquid fuels

Fuel	Density, kg/m³	Composition, % by mass				HHV, MJ/kg	LHV, MJ/kg
		C	H	O+N	S		
Light fuel oil (No. 2)	0.82–0.86	86–87	13–14	0.5	0.3	45.5	43
Heavy fuel oil	0.90–0.92	84–88	11–12	1–3	2	42.5	40
Gasoline	0.72–0.8	85	15	—	—	47	42.5
Diesel oil	0.84	86	13	0.4	0.6	45	41.5

Table 2.3. Composition, higher heating value (HHV), and lower heating value (LHV) of gaseous fuels

Gaseous fuel	Composition, % by volume						HHV, MJ/m³	HHV, MJ/m³
	CH_4	C_nH_m	CO	H_2	CO_2	N_2		
Natural gas	83–94	0–15	0–3	0–2	0.2–1	0.5–8	35–42	32–38
Fuel gas	0–19	—	12–51	18–28	4–5	5–54	5.3–18.2	5–16.3

Thus

$$H_2 + CO + CH_4 + C_nH_m + CO_2 + N_2 = 1 \qquad (2.19)$$

where H_2, CO, CH_4, C_nH_m, CO_2, and N_2 are the mole fractions of hydrogen, carbon monoxide, methane, higher hydrocarbons, carbon dioxide, and nitrogen in the fuel gas, respectively. Here C_nH_m is a common chemical formula for all hydrocarbons, except for methane CH_4, i.e., for ethane C_2H_6, propane C_3H_8, butane C_4H_{10}, ethylene C_2H_4, and acetylene C_2H_2. The mole fractions correspond to m³ of constituents per m³ of fuel gas (at standard conditions, i.e., 0°C and 101.3 kPa).

The typical composition and heating value of natural gas and fuel gas are given in Table 2.3.

The HHV or LHV of a fuel gas may be calculated from the values for individual gas constituents:

$$HV = \sum r_i\, HV_i \qquad (2.20)$$

where r_i is the mole fraction of the ith constituent of dry fuel gas and HV_i is the higher or lower heating value of the ith constituent. Heating values of the fuel gas constituents are given in Table 2.4.

Table 2.4. Lower heating values (LHV) and higher heating values (HHV) of constituents of fuel gases

Gas	Chemical formula	LHV, MJ/m³	HHV, MJ/m³
Hydrogen	H_2	10.81	12.78
Carbon monoxide	CO	12.64	12.64
Methane	CH_4	35.93	39.87
Acetylene	C_2H_2	56.9	58.9
Ethylene	C_2H_4	59.55	63.5
Ethane	C_2H_6	64.5	70.45
Propane	C_3H_8	93	101
Butane	C_4H_{10}	123.8	134
Hydrogen sulfide	H_2S	28.14	30.3

Table 2.5. Molar mass and density (at 0°C and 101.3 kPa) of air, fuel, and flue gas constituents

Substance	Chemical formula	Molar mass, kg/(kg mol)	Density, kg/m^3
Air	—	28.96	1.293
Carbon	C	12.01	—
Hydrogen	H_2	2.016	0.090
Nitrogen	N_2	28.16	1.257
Oxygen	O_2	32.00	1.429
Sulfur	S	32.06	—
Carbon dioxide	CO_2	44.01	1.977
Sulfur dioxide	SO_2	64.06	2.931
Water vapor	H_2O	18.016	0.804

COMBUSTION STOICHIOMETRY

Stoichiometric Equations for Solid and Liquid Fuels

The combustion of a fuel is the chemical conversion of the fuel into products of combustion with the release of heat of combustion. The combustible constituents of a solid or liquid fuel, i.e., carbon, hydrogen, and sulfur, interact with oxygen of combustion air by means of exothermic chemical reactions. Figure 2.1 shows a schematic of the combustion process. The products of combustion consist of flue gases and solid residuals from combustion of solid fuels. Flue gases contain harmful pollutants such as sulfur and nitrogen oxides and greenhouse gases such as carbon dioxide and water vapor.

The stoichiometric equations present a simplified description of the reactions of complete combustion of fuel-combustable constituents with oxygen. They also give the material balance of reactions on a mole or mass basis. The molar masses and densities of air, fuel, and flue gas constituents required for combustion calculations are given in Table 2.5.

For the sake of simplicity, the subscript af in the as-fired ultimate analysis of a fuel is dropped in the following combustion calculations (Combustion Engineering, 1981; Bartok and Sarofim, 1991; Smoot and Smith, 1985; Merrick, 1984).

The oxidation reaction of carbon with oxygen to carbon dioxide is

$$C + O_2 = CO_2 + 393.5 \text{ MJ/(kg mol)}$$

$$1 \text{ kg mol C} + 1 \text{ kg mol } O_2 = 1 \text{ kg mol } CO_2 \tag{2.21}$$

$$12 \text{ kg C} + 32 \text{ kg } O_2 = 44 \text{ kg } CO_2$$

Consequently, the mass of oxygen required to completely burn a unit mass of carbon is $32/12 = 2.67$ kg O_2 per kg C. The mass of CO_2 per a unit mass of carbon is $44/12 = 3.67$ kg CO_2 per kg C. Similarly, for the combustion of hydrogen,

$$H_2 + 1/2O_2 = H_2O \text{ (water vapor)} + 241.8 \text{ MJ/(kg mol)}$$

$$1 \text{ kg mol } H_2 + 1/2 \text{ kg mol } O_2 = 1 \text{ kg mol } H_2O \tag{2.22}$$

$$2.016 \text{ kg } H_2 + 16 \text{ kg } O_2 = 18.016 \text{ kg } H_2O$$

Figure 2.1. Schematic of combustion process.

It follows that 7.94 kg of oxygen are required to burn 1 kg of hydrogen and that 8.94 kg of water vapor are formed per kg of hydrogen. Likewise,

$$S + O_2 = SO_2 + 296.9 \text{ MJ/(kg mol)}$$

$$1 \text{ kg mol } S + 1 \text{ kg mol } O_2 = 1 \text{ kg mol } SO_2 \tag{2.23}$$

$$32 \text{ kg } S + 32 \text{ kg } O_2 = 64 \text{ kg } SO_2$$

Thus, to burn 1 kg of sulfur, 1 kg of O_2 is required and 2 kg of SO_2 per kg S form by the combustion.

Stoichiometric Equations for Gaseous Fuels

Gaseous fuels contain combustible constituents such as H_2, CO, CH_4, and C_nH_m, including C_2H_6, C_3H_8, and C_4H_{10}. The stoichiometric equations for the combustion of gaseous fuels is written on a molar basis. It is convenient to consider 100 mol of fuel. Then the number of moles of reactants and products is equal to the percent by volume of constituents of fuel and products of combustion, respectively.

For the combustion of hydrogen, carbon monoxide, methane, and higher hydrocarbons, the following stoichiometric equations on the molar basis may be written:

$$H_2 + 1/2O_2 = H_2O + 241.8 \text{ MJ/(kg mol)}$$
$$1 \text{ mol} + 1/2 \text{ mol} = 1 \text{ mol} \tag{2.24}$$

$$CO + 1/2O_2 = CO_2 + 283.0 \text{ MJ/(kg mol)}$$
$$1 \text{ mol} + 1/2 \text{ mol} = 1 \text{ mol} \tag{2.25}$$

$$CH_4 + 2O_2 = CO_2 + 2H_2O$$
$$1 \text{ mol} + 2 \text{ mol} = 1 \text{ mol} + 2 \text{ mol} \tag{2.26}$$

$$C_nH_m + (n + m/4)O_2 = nCO_2 + m/2H_2O$$
$$1 \text{ mol} + (n + m/4) \text{ mol} = n \text{ mol} + m/2 \text{ mol} \tag{2.27}$$

COMBUSTION CALCULATIONS

Theoretical and Actual Air Requirements

Theoretical Air Requirements

Table 2.6 contains the mass and mole fractions of oxygen and nitrogen in the dry air. All other constituents of the air (CO_2 and inert gases) are neglected.

The theoretical or stoichiometric air is the minimum air amount required for the complete combustion of a unit mass of fuel. It may be calculated on a mass or molar basis. For solid and liquid fuels the mass basis and for gaseous fuels the molar basis is preferred.

The ratio of the mass of combustion air to the mass of fuel is called air-fuel ratio:

$$AF = m_{air}/m_{fuel} \quad \text{kg air/kg fuel} \tag{2.28}$$

Based on the stoichiometric relations, Eqs. (2.21)–(2.23), the theoretical air-fuel ratio in terms of kg of dry air per kg of a fuel is given by

$$AF_t = 1/0.232[2.67C + 7.94H + S - O] = 11.49C + 34.22H + 4.31(S - O) \tag{2.29}$$

where C, H, S, and O are the mass fractions in the ultimate analysis of solid or liquid fuel.

Table 2.6. Dry air composition

	Oxygen	Nitrogen	Ratio of oxygen to nitrogen
Mass fraction	0.232	0.768	3.31
Mole fraction	0.21	0.79	3.76

The theoretical volume of dry air required to burn 1 kg of a solid or liquid fuel is given by

$$V_{a,t} = AF_t/\rho_{air} \quad m^3/kg \tag{2.30}$$

where ρ_{air} is the density of air at standard conditions (0°C and 101.3 kPa), i.e., 1.293 kg/m^3.

For gaseous fuels the theoretical number of moles of oxygen required for the complete combustion of 100 mol of gaseous fuel is calculated from the stoichiometric equations given above. Multiplying by $100/21 = 4.76$ yields the theoretical air requirements:

$$AF_{t,m} = 4.76[0.5(H_2 + CO) + 2CH_4 + (n + m/4)C_nH_m - O_2] \quad \text{mol air/mol fuel} \tag{2.31}$$

where H_2, CO, CH_4, C_nH_m, and O_2 are the mole fraction of hydrogen, carbon monoxide, methane, higher hydrocarbons, and oxygen in the fuel gas, respectively. Note that "mol air/mol fuel" is equal to m^3 air/m^3 fuel at standard conditions, i.e., at 0°C and 101.3 kPa. It is assumed that no H_2S is present in the fuel gas.

Excess Air Ratio

To achieve complete combustion of the fuel, an excess of air over the theoretical amount must be supplied. The actual air-fuel ratio AF_a is often stated in terms of the excess air ratio or the percent of excess air. The excess air ratio (or dilution coefficient) λ is defined as the ratio of the actual air-fuel ratio AF_a to the theoretical (stoichiometric) air-fuel ratio AF_t:

$$\lambda = AF_a/AF_t \tag{2.32}$$

The percent of excess air is defined as follows:

$$\text{Percent excess air} = 100(\lambda - 1) \quad \% \tag{2.33}$$

The excess air ratio λ depends on the fuel characteristics, the furnace design, and the burner type. Thus for pulverized coal, oil, and gas burners, λ may be as low as 1.03–1.1, whereas for stoker coal furnaces it typically lies between 1.3 and 1.6. In fluidized bed furnaces, λ is above 2, whereas in gas turbine combustors it is even higher (about 4). The actual excess air ratio in a furnace may be determined from the flue gas analysis.

Actual Air Requirements

The actual dry air-fuel ratio is given by

$$AF_a = \lambda AF_t \quad \text{kg air/kg fuel} \tag{2.34}$$

The actual wet air-fuel ratio, which accounts for the water vapor content of the atmospheric air, is

$$AF_{a,w} = \lambda AF_t(1 + w) \quad \text{kg air/kg fuel} \tag{2.35}$$

where w is the humidity ratio or mass of water vapor per unit mass of dry air ($\sim$0.009 kg/kg in summer and 0.002 kg/kg in winter).

The volume of dry and humid air supplied to the furnace is given by

$$V_a = \lambda AF_t/\rho_{air} \quad m^3 \text{ dry air/kg fuel} \tag{2.36}$$

$$V_{a,w} = \lambda AF_t(1/\rho_{air} + w/\rho_{H_2O}) \quad m^3 \text{ humid air/kg fuel} \tag{2.37}$$

where ρ_{air} and ρ_{H_2O} are the density of air and water vapor (1.293 and 0.804 kg/m^3 at 0°C and 101.3 kPa), respectively.

Lean and Rich Fuel-Air Mixtures

For liquid and gaseous fuels the excess air ratio, AF_a/AF_t, may be used to differentiate lean fuel-air mixtures from rich ones. If this ratio is less than 1, the mixture is called rich; otherwise, it is lean.

Table 2.7. Masses and volumes of flue gas constituents for the stoichiometric combustion of a solid or liquid fuel

Ultimate analysis as-fired	Theoretical oxygen, kg/kg	Products of combustion	
		kg/kg	m^3/kg
C	2.67C	3.67C	1.867C
H	7.94H	8.94H	11.11H
S	S	2S	0.68
N	—	N	0.8N
O	O	—	—
Moisture (M)	—	M	1.24M

Products of Combustion

Mass and Volume of the Products of Combustion

The products of complete combustion of a fuel contain CO_2, H_2O (water vapor), SO_2, N_2, and O_2. When incomplete combustion occurs, the flue gas also contains CO and unburned hydrocarbons.

From the stoichiometric equations, Eqs. (2.21)–(2.27), the amount of flue gas constituents may be calculated. Thus the mass of each constituent for complete combustion of a solid or liquid fuel will be found from the stoichiometric equations, Eqs. (1.21)–(1.23). The volume is then found by dividing the mass by the density. Table 2.7 gives the masses and volumes of flue gas constituents for the stoichiometric combustion of a solid or liquid fuel.

Total nitrogen volume in the flue gas is

$$V_{N_2} = 0.8N + 0.79V_a \quad m^3/kg \tag{2.38}$$

Taking into account the humidity of the atmospheric air, the total volume of water vapor is

$$V_{H_2O} = 11.11H + 1.24M + 1.24w\lambda AF_t \quad m^3/kg \tag{2.39}$$

The volume of oxygen in the flue gas is

$$V_{O_2} = 0.21(\lambda - 1)V_a \quad m^3/kg \tag{2.40}$$

Here, λ is the excess air ratio (dilution coefficient), w is the humidity ratio of the combustion air (in kg/kg), and V_a is the actual air volume per kg fuel.

Alternatively, the number of moles of the flue gas constituents may be calculated from the number of moles n of elements C, H, S, O, and N per kg of fuel from the af ultimate analysis. The mole balances for each element in the fuel and in the flue gas yield the number of moles of flue gas.

By complete combustion of a gaseous fuel with an excess of air (for $\lambda > 1$), the following volumes of gas constituents are formed (in m^3/m^3 fuel gas, at standard conditions).

Carbon dioxide

$$V_{CO_2} = CO_2 + CO + CH_4 + nC_nH_m \tag{2.41}$$

Nitrogen

$$V_{N_2} = N_2 + 0.79\lambda AF_{t.m} \tag{2.42}$$

Oxygen

$$V_{O_2} = 0.21(\lambda - 1)AF_{t.m} \tag{2.43}$$

Water vapor volume (without and with air moisture)

$$V_{H_2O} = H_2 + 2CH_4 + m/2C_nH_m \tag{2.44}$$

$$V'_{H_2O} = V_{H_2O} + 1.6w\lambda AF_{t.m} \tag{2.45}$$

The wet flue gas volume is thus the sum of the volumes of all constituents:

$$V_g = V_{CO_2} + V_{SO_2} + V_{N_2} + V_{O_2} + V_{H_2O} \tag{2.46}$$

Partial Pressure of a Flue Gas Constituent

The partial pressure p_i of a flue gas constituent i may be calculated from its mole fraction r_i and the total flue gas pressure p:

$$p_i = r_i p \quad \text{Pa} \tag{2.47}$$

The mole fraction of a flue gas constituent (CO_2, H_2O, etc.) may be calculated from its volume V_i and the total volume of the wet flue gas V_g:

$$r_i = V_i / V_g \tag{2.48}$$

The combustion calculations for solid, liquid, and gaseous fuels are presented in the following examples.

Example 2.3

Bituminous coal with the following ultimate analysis (in mass fractions), $C = 0.705$, $H = 0.047$, $S = 0.034$, $O = 0.051$, $N = 0.013$, $A = 0.05$, and $M = 0.10$, is fired in a steam generator furnace with 25% excess air ($\lambda = 1.25$). Determine (a) the stoichiometric and actual air-fuel ratios and (b) the actual volume of the products of combustion. Assume that the humidity ratio of the humid air is 0.01 kg/kg. The air density at standard conditions is 1.293 kg/m³.

Solution

1. Stoichiometric (theoretical) oxygen and products of combustion

Ultimate analysis af, kg/kg	Theoretical oxygen, kg/kg	Stoichiometric products of combustion	
		kg/kg	m³/kg
$C = 0.705$	$2.67C = 1.88$	$m_{CO_2} = 3.67C = 2.59$	$V_{CO_2} = 1.867C = 1.32$
$H = 0.047$	$7.94H = 0.37$	$m_{H_2O} = 8.94H = 0.42$	$V_{H_2O} = 11.11H = 0.52$
$S = 0.034$	$S = 0.034$	$m_{SO_2} = 2S = 0.068$	$V_{SO_2} = 0.68S = 0.023$
$N = 0.013$	—	$m_{N_2} = N = 0.013$	$V_{N_2} = 0.8N = 0.01$
$O = 0.051$	$-O = -0.051$	—	—
$M = 0.10$	—	$m_{H_2O} = M = 0.10$	$V_{H_2O} = 1.24M = 0.124$
$A = 0.05$	—	0.05 (solid residue)	—
Total 1	$m_{O_2} = 2.233$	3.191	1.997

2. Theoretical and actual air-fuel ratios and volumes of the air are

$$AF_t = m_{O_2}/0.232 = 2.233/0.232 = 9.625 \text{ kg air/kg fuel or}$$

$$V_t = 9.625/1.293 = 7.444 \text{ m}^3/\text{kg}$$

$$AF_a = \lambda AF_t = 1.25 \times 9.625 = 12.03 \text{ kg air/kg fuel or } V_a = 12.03/1.293 = 9.305 \text{ m}^3/\text{kg}$$

3. For combustion with $\lambda = 1.25$, the masses and volumes (at 0°C and 101.3 kPa) of flue gas constituents are

$$m_{CO_2} = 2.59 \text{ kg/kg} \qquad V_{CO_2} = 1.32 \text{ m}^3/\text{kg}$$

$$m_{SO_2} = 0.068 \text{ kg/kg} \qquad V_{SO_2} = 0.023 \text{ m}^3/\text{kg}$$

$$m_{N_2} = N + 0.768AF_a = 0.013 + 0.768 \times 12.03 = 9.25 \text{ kg/kg}$$

$$V_{N_2} = 0.8N + 0.79V_a = 0.8 \times 0.013 + 0.79 \times 9.305 = 7.36 \text{ m}^3/\text{kg}$$

$$m_{O_2} = 0.21(\lambda - 1)AF_t = 0.21(1.25 - 1)9.625 = 0.50 \text{ kg/kg}$$

$$V_{O_2} = m_{O_2}/\rho_{O_2} = 0.50/1.43 = 0.35 \text{ m}^3/\text{kg}$$

4. Dry flue gas mass and volume

$$m_{g.d} = m_{CO_2} + m_{SO_2} + m_{N_2} + m_{O_2} = 2.59 + 0.068 + 9.25 + 0.50 = 12.41 \text{ kg/kg}$$

$$V_{g.d} = V_{CO_2} + V_{SO_2} + V_{N_2} + V_{O_2} = 1.32 + 0.023 + 7.36 + 0.35 = 9.05 \text{ m}^3/\text{kg}$$

5. Total moisture in the flue gas is

$$m_{H_2O} = 8.94H + M + wAF_a = 8.94 \times 0.047 + 0.1 + 0.01 \times 12.03 = 0.64 \text{ kg/kg}$$

$$V_{H_2O} = m_{H_2O}/\rho_{H_2O} = 0.64/0.804 = 0.80 \text{ m}^3/\text{kg}$$

6. Wet flue gas mass and volume are

$$m_g = m_{g.d} + m_{H_2O} = 12.41 + 0.640 = 13.05 \text{ kg/kg}$$

$$V_g = V_{g.d} + V_{H_2O} = 9.05 + 0.80 = 9.85 \text{ m}^3/\text{kg}$$

Example 2.4

For a fuel oil with $C = 0.86$ kg/kg, $H = 0.12$ kg/kg, and $S = 0.02$ kg/kg, find the actual wet air-fuel ratio AF_a and the wet flue gas volume V_g, when the excess air ratio λ is 1.1.

Solution

1. The respective number of moles of the constituents C, H, and S in the fuel oil, per kg fuel, is

$$n_C = (0.86 \text{ kg/kg})/[12 \text{ kg/(kg mol)}] = 0.0717 \text{ kg mol}$$

$$n_H = (0.12 \text{ kg/kg})/[2.016 \text{ kg/(kg mol)}] = 0.0595 \text{ kg mol}$$

$$n_S = (0.02 \text{ kg/kg})/[32 \text{ kg/(kg mol)}] = 0.0006 \text{ kg mol}$$

2. The combustion equation may be written on a mole basis as follows:

$$n_C + n_H + n_S + a\lambda(O_2 + 3.76N_2) = bCO_2 + cH_2O + dSO_2 + eO_2 + fN_2 \tag{E1}$$

3. Balances for C, H, S, O_2, and N_2, on a mole basis, give the unknown values a, b, c, d, e, and f:

For C: $n_C = b$, i.e., 0.0717 kg mol CO_2 are formed per kg fuel

For H: $n_H = c$, i.e., 0.0595 kg mol CO_2 are formed per kg fuel

For S: $n_S = d$, i.e., 0.0006 kg mol SO_2 are formed per kg fuel

For O_2 with $\lambda = 1.1$: $1.1a = b + c/2 + d + e$ $\tag{E2}$

For N_2: $3.76 \times 1.1a = f$ $\tag{E3}$

Equations (E2) and (E3) contain the three unknowns a, e, and f. An additional equation may be written for O_2 from the excess air:

$$0.1a = e \tag{E4}$$

From Eqs. (E2)–(E4),

$$a = b + c/2 + d = 0.0717 + 0.0595/2 + 0.0006 = 0.1020$$

$$e = 0.0102$$

$$f = 3.76 \times 1.1 \times 0.1020 = 0.4218$$

4. Actual dry air-fuel ratio

(i) on a mole basis:

$$a\lambda4.76 = 0.1020 \text{ kg mol/kg} \times 1.1 \times 4.76 = 0.534 \text{ kg mol air per kg fuel}$$

(ii) on a mass basis:

$$AF_{t,d} = a\lambda 4.76 M_{air} = 0.1020 \times 1.1 \times 4.76 \times 28.97 = 15.47 \text{ kg air/kg fuel}$$

(iii) on a volumetric basis:

$$AF_{t,v,d} = a\lambda 4.76 V_M = 0.1020 \text{ kg mol/kg} \times 1.1 \times 4.76 \times 22.4 \text{ m}^3/(\text{kg mol})$$
$$= 11.96 \text{ m}^3 \text{ air/kg fuel}$$

5. Wet flue gas volume

$$V_{g,w} = (b + c + d + e + f)V_M$$
$$= (0.0717 + 0.0595 + 0.0006 + 0.0102 + 0.4218)22.4 = 12.629 \text{ m}^3/\text{kg fuel}$$

Here V_M is the mole volume at standard conditions, i.e., $22.4 \text{ m}^3/(\text{kg mol})$ (mean value for ideal gases).

Example 2.5

A natural gas with 98% CH_4, 0.8% N_2, and 1.2% CO_2 (all % by volume) and with an HHV of 36,290 kJ/m^3 (at standard conditions of 0°C and 101.3 kPa) is burned in a gas turbine. The excess air ratio λ is 2.0.

Calculate the dry air-fuel ratio in kg/kg and in m^3/m^3, and the constituents of the wet flue gas in % by volume.

Solution

1. Numbers of kilogram-moles of C, H_2, N_2, and O_2 in the fuel:

$$n_C = 0.98 + 0.012 = 0.992$$
$$n_H = 2 \times 0.98 = 1.96$$
$$n_N = 0.008$$
$$n_O = 0.012$$

2. The combustion equation is given by

$$(0.98CH_4 + 0.008N_2 + 0.012CO_2) + a\lambda(O_2 + 3.76N_2) = bCO_2 + cH_2O + dN_2 + eO_2 \quad \text{(E5)}$$

3. Theoretical (stoichiometric) dry air-fuel ratio

$$AF_{t,d} = 4.76(n_C + n_H/2 - n_O) = 4.76(0.992 + 1.96/2 - 0.012) = 9.33 \text{ mol air/mol fuel gas}$$

or

9.33 m^3 air/m^3 gas

4. Molar mass of the fuel gas

$$M_g = CH_4 M_{CH_4} + n_{N_2}M_{N_2} + n_{CO_2}M_{CO_2} = 0.98 \times 16 + 0.008 \times 28 + 0.012 \times 44$$
$$= 16.43 \text{ kg/(kg mol)}$$

5. For $\lambda = 2.0$, the actual dry air-fuel ratio is

$$AF_{a,d} = \lambda AF_{t,d} = 2 \times 9.33 = 18.66 \text{ m}^3 \text{ air/m}^3 \text{ gas}$$

or

$18.66 \rho_{air}/\rho_g = 18.66 \times 1.293 \text{ kg/m}^3/(16.43/22.4) = 32.9 \text{ kg air/kg gas}$

6. For the wet flue gas, it follows from Eq. (E1), in mol constituent/mol fuel gas, dry air

balance for C: $b = 0.98 + 0.012 = 0.992$ mol CO_2

balance for H_2: $c = 2 \times 0.98 = 1.96$ mol H_2O

balance for N_2: $d = 0.008 + 2a3.76 = 0.008 + 7.52a$ (E6)

balance for O_2: $b + c/2 + e = 0.012 + 2a$

Consequently,

$$e = 0.012 + 2a - b - c/2 \tag{E7}$$

On the other hand,

$$e = \text{excess oxygen} = (\lambda - 1)a = (2 - 1)a = a \tag{E8}$$

Equations (E3) and (E4) yield $e = -0.012 + b + c/2 = -0.012 + 0.992 + 1.96/2 = 1.96$ and $a = 1.96$. Therefore $d = 0.008 + 7.52 \times 1.96 = 14.75$.

7. Total number of moles in the flue gas

$$n_{g.w} = 0.992 + 1.96 + 14.75 + 1.96 = 19.66 \text{ mol flue gas/mol fuel gas}$$

8. The wet flue gas analysis, in % by volume

$\% \, CO_2 = 100b/n_{g,w} = 100 \times 0.992/19.66 = 5.04\%$

$\% \, H_2O = 100c/n_{g,w} = 100 \times 1.96/19.66 = 9.97\%$

$\% \, N_2 = 100d/n_{g,w} = 100 \times 14.75/19.66 = 75.02\%$

$\% \, O_2 = 100e/n_{g,w} = 100 \times 1.96/19.66 = 9.97\%$

$\text{Total} = 100\%$

ADIABATIC COMBUSTION TEMPERATURE

The theoretical or adiabatic combustion temperature may be calculated from an energy balance for the adiabatic steady-flow combustion chamber. The adiabatic combustion temperature t_{ad} is calculated assuming complete combustion and no dissociation of products of combustion (Combustion Engineering, 1981). The energy balance for the adiabatic steady-flow combustion chamber per kg of fuel is given by

$$\text{LHV} + c_f t_f + AF_a c_{pa} t_a = m_g c_{pg} t_{ad} + m_{ash} c_{ash} t_{ash} \tag{2.49}$$

where LHV is in kJ/kg; AF_a is actual air/fuel ratio in kg air/kg fuel; m_g and m_{ash} are mass of flue gas and ash, respectively, in kg/kg fuel; c_f, c_{pa}, c_{pg} and c_{ash} are specific heat of fuel, air, flue gas, and ash, respectively, in kJ/kg K; t_f, t_a, and t_{ash} are temperatures of fuel and ash, respectively, in °C; t_{ad} is adiabatic combustion temperature in °C.

The first term is the LHV of fuel, the second term is the sensible heat of fuel, the third term is the sensible heat of combustion air, the fourth term is the enthalpy of products of combustion, and the fifth term is the sensible heat of solid residue (ash).

If the sensible heat of the fuel, combustion air, and ash can be ignored, then

$$t_{ad} = \text{LHV}/V_g \rho_g c_{pg} \tag{2.50}$$

where V_g is volume of the flue gas per kg fuel in m^3/kg, ρ_g is the density of the flue gas in kg/m^3, and c_{pg} is the mean isobaric specific heat of the flue gas in kJ/(kg K).

The actual furnace temperature takes into account the heat transfer in the furnace (combustor) and depends on the furnace cooling. Therefore it is much smaller than the adiabatic combustion temperature of the fuel. At temperatures above 1500°C, the dissociation of CO_2 and H_2O with heat absorption occurs so that the furnace temperature becomes even smaller (Combustion Engineering, 1981; Bartok and Sarofim, 1991; Smoot and Smith, 1985; Merrick, 1984).

Table 2.8. Typical values of adiabatic
combustion temperature

Fuel	Adiabatic combustion temperature, °C
Hard coal	2200–2300
Lignite	1400–1500
Fuel oil, natural gas	2000–2100
Producer gas	~1700

Typical values of adiabatic combustion temperature of various fuels are given in Table 2.8. The incomplete combustion reduces the temperature in the furnace. At very high combustion temperatures the endothermic dissociation reaction of CO_2 may occur:

$$2CO_2 + heat \rightarrow 2CO + N_2 \tag{2.51}$$

In addition, the following reaction takes place:

$$N_2 + O_2 + heat \rightarrow 2NO \tag{2.52}$$

This causes a reduction in the temperature. These reactions are undesirable because their products are pollutants of the atmospheric air.

Example 2.6

Determine the adiabatic combustion temperature for the bituminous coal of Example 2.3, ignoring the product dissociation. Assume that the specific heat of flue gas is $c'_p = 1.34$ kJ/(m³ K).

Solution

From Examples 2.2 and 2.3, the coal LHV = 28,800 kJ/kg, and the wet gas volume is $V_g = 9.85$ m³/kg.
Adiabatic combustion temperature is

$$t_{ad} = LHV/V_g c'_p = 28,800/9.85 \times 1.34 = 2182°C.$$

FLUE GAS ANALYSIS

To control the combustion process, an Orsat gas analysis of the flue gas will be carried out. Thereby contents O_2 and CO_2 in the dry flue gas will be determined. The CO_2 content of the flue gas is an important indicator of the quality of the combustion process.
    The maximum CO_2 content in the flue gas of a solid or liquid fuel is

$$CO_{2max} = 1.87C/V_g \quad vol.\% \tag{2.53}$$

where C is the carbon content (in kg/kg) of the fuel and V_g is the flue gas volume (in m³/kg).
    For pure carbon, $CO_{2max} = 21$. For various coal grades and lignites the value of CO_{2max} lies between 17.2 and 19.8; for fuel oil it is 16, and for natural gas 12.5 (all values are in vol.%).
    The actual excess air ratio λ may be determined from flue gas analysis on the basis of the following considerations. For the complete combustion of a fuel without the excess air, i.e., for $\lambda = 1$, the CO_2 content in the flue gas is equal to the maximum value CO_{2max} for the particular fuel under consideration.
    When the CO_2 content in flue gas is known from the Orsat gas analysis, the O_2 content in the flue gas is given by (Combustion Engineering, 1981; Smoot and Smith, 1985)

$$O_2 = 21(1 - CO_2/CO_{2max}) \quad vol.\% \tag{2.54}$$

Correspondingly, the excess air ratio is

$$\lambda = CO_{2max}/CO_2 \tag{2.55}$$

If the O_2 content in the flue gas is evaluated from the gas analysis, then

$$\lambda = 21/(21 - O_2) \tag{2.56}$$

The above formulae are valid only for complete combustion. The carbon monoxide content in products of incomplete combustion is

$$CO = (CO_2 + O_2)_{theor} - \varphi(CO_2 + O_2) \tag{2.57}$$

where

$$\varphi = 1/(21/CO_{2max} - 0.4) \tag{2.58}$$

If the flue gas temperature drops below its dew point, condensation of water vapor occurs in the flue gas and sulfurous and sulfuric acids, H_2SO_3 and H_2SO_4, are formed. They cause corrosion of economizers and air preheaters in boilers as well as of heat recovery steam generators in combined-cycle plants. The fuel sulfur content affects the corrosion behavior of these so-called cold-end components. To prevent corrosion, the flue gas temperature must be above the dew point. The dew point temperature t_{dp} of the flue gas is the saturation temperature of water vapor at its partial pressure in the flue gas. It can be found on the basis of the partial pressure of water vapor in the flue gas from the steam tables. The presence of sulfur oxides in the flue gas may dramatically increase the dew point temperature of the flue gas as compared to that of the pure water vapor with sulfur-free fuels.

COAL GASIFICATION

Chemistry of Coal Gasification

Coal gasification is the chemical conversion of coal with a gasifying medium to a fuel gas. Figure 2.2 shows the gasification process schematically. Generally, the gasification process can be presented as follows:

$$coal + gasifying\ medium + heat = product\ (fuel)\ gas + slag\ (ash) \tag{2.59}$$

The majority of coal gasification processes are based on the interaction of carbon in coal with steam H_2O as the gasifying medium (Smoot and Smith, 1985; Merrick, 1984; Smoot, 1993; Richards, 1994). The gasification process is endothermic, i.e., it occurs with heat absorption. The heat required for the coal gasification can be obtained through combustion (oxidation) of a relatively small quantity of coal with oxygen within the gasifier according to the following reactions of complete or partial oxidation of carbon with oxygen:

$$C + O_2 = CO_2 + 406\ kJ/mol \tag{2.60}$$

$$C + 1/2O_2 = CO + 123\ kJ/mol \tag{2.61}$$

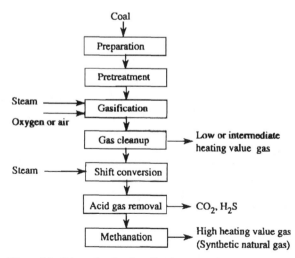

Figure 2.2. Schematic of coal gasification process.

The gasification process consists of two stages. The first stage is coal pyrolysis, i.e., decomposition by heat, which occurs at high temperatures in the absence of air (oxygen). Some gas and liquids are produced from the volatile matter of coal, but the main product is a solid residue, i.e., char or coke. The coke gas may contain CO, H_2, CH_4, C_nH_m, H_2O, and CO_2. The amounts of gas and liquids depend on the coal rank and the temperature of the process.

The pyrolysis process may be described as

$$\text{coal} + \text{heat} = \text{char (coke)} + \text{coke gas (CO, } H_2, H_2O, C_nH_m, \text{tar, } CO_2) \tag{2.62}$$

In the second stage of the process, the products of pyrolysis are converted to a product gas.

Fuel, e.g., coal or naphtha, and a gasifying medium are introduced into the gasifier. The product gas and the solid residual leave the gasifier. Usually, steam is used as a gasifying medium. Pure oxygen, oxygen, or air is supplied to the gasifier for partial oxidation of the original fuel. During gasification, certain heterogeneous (solid/gas) and homogeneous (gas/gas) reactions take place. They may be described as follows (Smoot and Smith, 1985; Merrick, 1984; Smoot, 1993; Richards, 1994).

The heterogeneous water-gas reaction is

$$C + H_2O \text{ (steam)} = CO + H_2 \ (\Delta H_r = +119 \text{ kJ/mol}) \tag{2.63}$$

where carbon is the char (coke). The reaction enthalpy ΔH_r is referred to 298 K and 1 bar. The heterogeneous Boudouard reaction is

$$C + CO_2 = 2CO(\Delta H_r = +162 \text{ kJ/mol}) \tag{2.64}$$

The homogeneous water-gas reaction is

$$CO + H_2O(\text{steam}) = H_2 + CO_2(\Delta H_r = -42 \text{ kJ/mol}) \tag{2.65}$$

The homogeneous methanation reaction is

$$CO + 3H_2 = CH_4 + H_2O(\Delta H_r = -206 \text{ kJ/mol}) \tag{2.66}$$

The total gasification process is

$$\text{coal} + \text{gasifying medium (steam} + \text{oxygen or air)} + \text{heat}$$
$$= \text{product gas (CO} + H_2 + CH_4 + CO_2 + N_2) + \text{refuse (slag, ash)} \tag{2.67}$$

Classification of Fuel Gases

The aim of coal gasification is to convert the carbon in the coal into fuel gas. When coal is burned as a fuel, the sulfur present appears as sulfur oxides, which must be removed from the stack gases before they are discharged to the atmosphere. In coal gasification, however, the sulfur is usually converted into hydrogen sulfide (H_2S), nearly all of which is removed before the gas is burned. The volume of fuel gas is considerably less than that of the stack gases to be treated for the same heat generation, and furthermore, desulfurization of fuel gases is more highly developed than the removal of sulfur from stack gases.

The different reactions in the gasification process take place to different extents, which depend largely on the coal rank and the gasification temperature and pressure. The composition and heating value of the product gas vary accordingly. However, the combustible constituents of the product gas are CO and H_2, together with a smaller proportion of CH_4. Some CO_2 is usually present as an inert component. If the gasification occurs in air, N_2 from the air is another inert diluent. If oxygen gas is used, the product gas contains little or no nitrogen.

The major combustible components of coal gas are H_2, CO, and CH_4. The heating value of CH_4 is about 37 MJ/m^3, and the values for H_2 and CO are both almost 12 MJ/m^3. The heating value of a product gas depends on the proportions of these fuel constituents and of the inert gases N_2 and CO_2.

Fuel gases are commonly classified into three categories: gases having a high heating value (35–39 MJ/m^3), an intermediate (medium) heating value (11–17 MJ/m^3), or a low heating value (4.8–7.4 MJ/m^3). Fuel gases of the first category consist largely of CH_4. Gases of intermediate heating value contain 65–70% of CO and H_2 in various proportions, and 5–15% of CH_4. The remainder is mostly CO_2.

A fuel gas of intermediate heating value can be converted into a high heating value syngas or substitute natural gas (SNG) by subjecting it to the water-gas shift reaction followed by methanation. The syngas is used for the production of liquid hydrocarbon fuels or methanol. Low heating value gas cannot be economically transported, and therefore it will normally be used on the production site. It may be burned in the gas turbine combustor of an advanced combined-cycle power plant or cogeneration plant with a high overall energy conversion efficiency.

Gasification Efficiency

The gasification efficiency, η_{gas}, is defined as the ratio of the chemically bound energy of the product gas to that of the fuel:

$$\eta_{gas} = LHV_g \times V_g / LHV \tag{2.68}$$

where LHV_g is the lower heating value of the product gas in MJ/m^3, V_g is the product gas output in m^3/kg fuel, and LHV is the lower heating value of the fuel in MJ/kg.

Carbon Conversion Efficiency

The carbon conversion efficiency, η_c, is defined as the ratio of the carbon content in the product gas to that of the coal per kg coal gasified. Thus

$$\eta_c = \text{carbon content of product gas/carbon content of fuel} \tag{2.69}$$

Thermal Efficiency

The thermal efficiency of the gasification process, η_{tg}, is defined as the ratio of the total energy content of the product gas to that of fuel and gasifying medium. This includes both the chemically bound energy and the sensible heat. Thus

$$\eta_{tg} = (LHV_g + c_{pg}t_g)V_g / (LHV + m_{gm}h_{gm}) \tag{2.70}$$

where c_{pg} is the isobaric specific heat of the product gas in $MJ/(m^3 \text{ K})$, t_g is the temperature of the product gas in °C, m_{gm} is the mass of the gasifying medium in kg/kg fuel, and h_{gm} is the specific enthalpy of the gasifying medium in MJ/kg. Since air or oxygen is supplied into the gasifier without preheating, h_{gm} is the specific enthalpy of steam, and m_{gm} is the mass of steam.

Gasification Processes and Gasifier Types

The three gasification systems are commercially applied at the present time (Smoot and Smith, 1985; Merrick, 1984; Smoot, 1993; Richards, 1994; Schellberg and Kuske, 1992; Uhde AG, 1993; Lurgi AG, 1992): fixed bed systems, fluidized bed systems, and entrained flow systems. They have been developed and realized in a number of coal gasification processes. The major processes, i.e., Lurgi dry ash, British Gas/Lurgi (BGL) slagging, Koppers-Totzek, Texaco, Winkler, and high-temperature Winkler processes, are outlined below. Operating conditions of these three gasification processes are compared in Tables 2.9 and 2.10.

Lurgi Dry Ash Gasification Process

This is one of the most highly developed commercial coal gasification processes. The Lurgi fixed bed dry ash gasifier was developed by Lurgi AG in Germany for the gasification of noncaking coals with gasifying medium steam and oxygen or air. It can also be modified for the gasification of noncaking coals.

The Lurgi gasifier operates in a counterflow mode (Figure 2.3). Thus the coal, entering the top of the gasifier, travels slowly downward, interacting with the rising hot gases. The gasifying

Table 2.9. Operating conditions of main gasification processes

Gasifier type	Fixed bed	Fluidized bed	Entrained flow
Process	BGL	KRW, Winkler	Texaco, Koppers-Totzek, Shell
Operating temperature, °C	800–1200	1000	1300–1800
Operating pressure, MPa	0.1–10	0.1–3	0.1–4
Gasifying medium	Steam + O_2	Steam + air or steam + O_2	O_2
O_2/steam ratio	1 : 8 to 1 : 4	1 : 2	2 : 1
Coal rank	Bituminous coal Anthracite Petrol coke	Lignite	Bituminous coal Lignite
Coal particle size, mm	3–30	1–10	< 0.1
Gas flow scheme	Counterflow	Parallel flow	Parallel flow
Residence time	30–60 min	1–10 min	< 0.1 s

Table 2.10. Performance comparison of main gasification processes

Process	Fixed bed	Fluidized bed	Entrained flow
Gas output, kg/kg	2.1	5.3	2.2
O_2 requirements, kg/kg	0.7	—	1.1
Air requirements, kg/kg	—	3.75	—
Steam requirements, kg/kg	0.5	0.6	0.15
Composition of dry fuel gas, % by volume			
CO	60	19	55
H_2	28	20	34
CH_4	9	1	—
N_2	1	50	1
CO_2	2	10	10
Heating value, MJ/kg	15.4	5.5	11.4
Gasification efficiency, %	89	73	79
Carbon conversion efficiency, %	99	95	99
Thermal efficiency, %	94	92	95

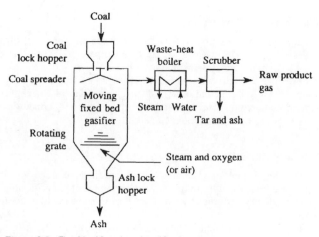

Figure 2.3. Fixed bed Lurgi coal gasification process.

medium is introduced at the bottom through spaces in a rotating grate. Rotation of the grate facilitates ash removal and favors uniform distribution of the gasifying medium throughout the coal bed. In oxygen-blown gasifiers an intermediate heating value fuel gas is produced. In air-blown gasifiers, a low heating value fuel gas with some 50% N_2 and only 20% CO and 20% H_2 is produced.

BGL Slagging Gasification Process

Unlike the original Lurgi, the BGL process is carried out at a temperature higher than the melting point of ash, which melts and forms a liquid slag in the slagging gasifier that is suited to operate with caking or noncaking coals without pretreatment. Crushed coal is fed through the lock hopper at the top of the gasifier, oxygen and steam are injected through nozzles near the bottom of the gasifier. The coal passes several zones in counterflow to the ascending gases. The coal is heated, dried, and gasified there, and part of it is burned with oxygen. Due to high temperatures, the carbon conversion efficiency of the slagging gasifier is higher than that of the dry ash Lurgi gasifier. The product gas from BGL gasifier contains about 60% CO, 28% H_2, and 9% CH_4 and has the heating value of around 15.4 MJ/m^3. This gas can be converted into SNG with methane as the main constituent by means of the water-gas shift reaction, followed by methanation (Smoot and Smith, 1985).

Koppers-Totzek Process

This commercial process for coal gasification in entrained flow is mostly used for the production of hydrogen or of a synthesis gas (Smoot and Smith, 1985; Richards, 1994). Nearly any kind of coal can be gasified with steam and oxygen by this process without pretreatment. Finely pulverized coal, entrained in a stream of oxygen gas and low-pressure steam in a mixing nozzle, is introduced into the gasifier (Figure 2.4). The entrained coal particles react very rapidly with the oxygen and steam in the gasifier at a temperature of about 1800–1900°C and a pressure of 1 to 40 bars. Such temperatures are achieved by the oxidation of a portion of the carbon in the coal with oxygen. The remaining carbon reacts with steam to produce carbon monoxide and hydrogen. The coal ash melts in the gasifier, and liquid slag is withdrawn from the bottom of the gasifier.

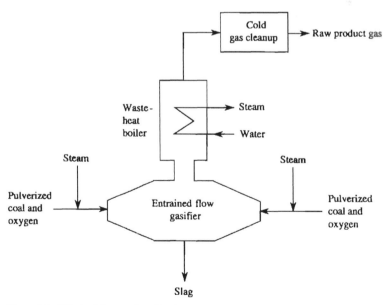

Figure 2.4. Entrained flow coal gasification process.

The product gas at a temperature of about 1500°C is used to raise steam in a waste-heat boiler, then cooled and scrubbed with water, and treated for hydrogen sulfide removal. The resulting clean fuel gas contains around 55% CO and 34% H_2. Most of the remainder is CO_2. The product gas has a heating value of about 11 MJ/m^3, and it can be converted into SNG.

Prenflo

The s.c. Prenflo process, i.e., the pressurized entrained flow oxygen-blown gasification process, was developed by Krupp-Koppers from the Koppers-Totzek process. At a pressure of 24–30 bars and temperature above 2000°C, the fuel gas with CO and H_2 as the main constituents and LHV = 11.5 MJ/m^3 is produced from bituminous coal containing up to 40% ash and up to 2–4% sulfur (Schellberg and Kuske, 1992). This process is well suited for integrated gasification combined-cycle (IGCC) plants (see Chapter 9).

Winkler Process

This process for fluidized bed coal gasification was developed in Germany in the 1920s. Most commercial plants use lignite and subbituminous noncaking coals. The gasifying medium is steam with oxygen or air. The product is an intermediate or low heating value fuel gas, depending on whether oxygen or air is used. The crushed coal fed to the bottom of the fluidized bed gasifier is fluidized by the upward flow of the gasifying medium (Figure 2.5). Due to thorough mixing of the coal particles with the gasifying medium in the fluidized bed, the coal is rapidly gasified. The temperature in the gasification zone ranges from 1000 to 1150°C. Lignite requires lower gasification temperatures than less reactive coals. The heat required for the reaction is provided by the partial combustion of carbon. Most of the carbon in the coal interacts with steam to produce carbon monoxide and hydrogen. The larger ash particles are discharged from the bottom of the gasifier. The fly ash and unreacted char, however, are carried over with the product gas. To prevent softening and depositing of fly ash particles in the gas exit duct at high temperatures, the gas is partially cooled by a steam boiler in the upper part of the gasifier. With reactive coals, the gasifier temperature is lower, and the boiler is not used. The raw product gas leaving the gasifier is passed

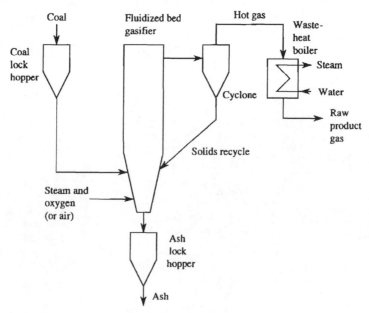

Figure 2.5. Fluidized bed Winkler coal gasification process.

through a waste-heat boiler, a cyclone separator, and a water scrubber to remove solid particles. The gas is treated for removal of hydrogen sulfide and contains about 40% by volume of CO, 35% H_2, and 3% CH_4. Most of the remainder is inert CO_2. In an oxygen-blown gasifier, the gas heating value is approximately 10 MJ/m^3. The product gas can be converted to an HHV gas. In the air-blown gasifier the gas heating value is around 4.5 MJ/m^3.

The high-temperature Winkler (HTW) gasification process was used by Rheinbraun AG (Germany) (Uhde AG, 1993). It features the following characteristics: pressurized high temperature gasification, the coal is supplied in the bottom part of the gasifier, the oxygen is supplied in two or more levels, the unburned char particles are separated from the product gas stream in a cyclone and returned into the fluzidized bed, and the raw gas is cooled in a waste-heat steam boiler. This process has several advantages such as high specific heat capacity of the gasifiers and high C conversion (98%). The application of the HTW process to IGCC plants is described in Chapter 9.

Texaco Process

This is an entrained flow process for coal gasification with steam and oxygen to produce an intermediate heating value (9–10 MJ/m^3) fuel gas (Smoot and Smith, 1985; Richards, 1994). The process is applicable to almost any type of coal and also to heavy residual oils. A slurry of pulverized coal in water is fed with pressurized oxygen into the top of the downflow gasifier. At an operating temperature of 1350–1500°C and pressure of 2–10 MPa, the carbon in coal reacts very rapidly with steam (from the slurry water) to produce gas mainly containing CO and H_2. The ash forms a slag, which after quenching with water, is removed from the bottom of the gasifier. The hot product gas is used to generate steam in a boiler. The gas is then treated for removal of sulfur and particulate matter. Table 2.11 shows the main data for the Texaco entrained flow coal gasification process.

In the alternative, direct-quench process, the gas is cooled with water before it leaves the gasifier. The product is then saturated with steam and is suitable for the water-gas shift reaction.

Claus Process

This is a process for oxidizing hydrogen sulfide (H_2S) with oxygen of the air into elemental sulfur according the following reaction:

$$2H_2S + O_2 \rightarrow 2H_2O + S \tag{2.71}$$

The oxidation of H_2S is carried out in two stages: in a high-temperature thermal stage and a low-temperature catalytic stage. In the first stage the mixture of product gas to be processed and air is fed into a furnace at a temperature of 600°C. The major portion of H_2S in the gas is oxidized there to give elemental sulfur in the form of vapor and water vapor. The gas is then cooled to condense the sulfur vapor. Then the gas is reheated to about 250°C, and the remaining H_2S is oxidized in a

Table 2.11. Main data of Texaco entrained flow coal gasification process

Item	Value
Temperature, °C	1200–1600
Pressure, bars	40
Coal concentration in suspension, % by mass	70
Syngas output, m^3/kg coal	1.85
Gas composition, dry, % by volume	
CO	54
H_2	34
CH_4	0.1
N_2	0.6
CO_2	11
$H_2S + COS$	0.3
Gasification efficiency, %	78
Carbon conversion efficiency, %	>99

Source: Uhde AG (1993).

catalytic reactor. The sulfur formed is removed by condensation, and the end product is solid sulfur. Small amounts of carbon oxysulfide (COS) and carbon disulfide (CS_2), together with residual H_2S and SO_2, are removed by further treatment, e.g., in the Wellman-Lord process (Richards, 1994). The gas to be treated in the Claus process should have at least 10% by volume of H_2S.

CLOSURE

The fundamentals of fuel combustion presented in this chapter are applied to all types of fossil fuel fired power plants. The equations for combustion calculations are used to determine the air requirements and flue gas volumes in combustors and burners of steam, and gas turbine combined-cycle and cogeneration plants. Modern coal gasification processes and proprietary gasifier types are also thoroughly discussed in this chapter. Designs of combustors and burners with low NO_x emissions are described in Chapters 3, 4, and 5, whereas the fluidized bed combustion technology with its intrinsic capability to reduce the sulfur and nitrogen oxide emissions is discussed in Chapter 9.

PROBLEMS

2.1. A bituminous coal has the following composition (in % by mass, dry, ash free): carbon 86.8, hydrogen 5.4, oxygen 5.0, nitrogen 1.6, and sulfur 1.2. The ash and moisture fractions in the proximate analysis are 9.7% and 4.3%, respectively. Calculate (a) the coal ultimate analysis, (b) the stoichiometric and actual air-fuel ratios for 18% excess air in kg/kg, (c) the volumes of the components of the products of combustion, and (d) the mole fractions of the products of combustion. Assume the air humidity ratio of 0.009 kg/kg.

2.2. A lignite has the following composition (in % by mass, dry, ash free): carbon 67, hydrogen 7, oxygen 22, nitrogen 2, and sulfur 2. The ash and moisture fractions in the proximate analysis are 5% and 50%, respectively. For combustion with 125% of the stoichiometric air, calculate (a) the coal ultimate analysis, (b) the stoichiometric and actual air-fuel ratios in kg/kg, (c) the volumes of the components of the products of combustion, and (d) the mole fractions of the products of combustion. Assume the air humidity ratio of 0.009 kg/kg.

2.3. Find the (a) stoichiometric and actual air-fuel ratios, in m^3/m^3, and (b) the mole fractions of the products of combustion for a complete combustion of methane CH_4 with a 15% excess air. Assume the air humidity ratio of 0.008 kg/kg.

2.4. A natural gas has the following composition (in % by volume): methane (CH_4) 90.3, ethane (C_2H_6) 3.2, propane (C_3H_8) 1.7, and nitrogen 4.8. For combustion in a gas turbine with 400% of the stoichiometric air, calculate (a) the actual air-fuel ratio in kg/kg and in m^3/m^3 and (b) the mole fractions of the products of combustion. Assume the air humidity ratio of 0.008 kg/kg.

2.5. A fuel oil has the following composition (in % by mass): carbon 86.3, hydrogen 13.4, and sulfur 0.3. Calculate (a) the stoichiometric and actual air-fuel ratios for the combustion with a 10% excess air in kg/kg, and (b) the mole fractions of the products of combustion. Assume the air humidity ratio of 0.008 kg/kg.

2.6. Find the adiabatic temperature of combustion of a coal with a heating value of 32.5 MJ/kg, if the wet flue gas volume is 10.75 m^3 (at standard conditions) per kg coal. The dissociation of the products of combustion may be ignored. Assume the specific heat of the products of combustion of 1.35 kJ/(m^3 K).

2.7. Determine the temperature of combustion of a fuel oil with a lower heating value of 46 MJ/kg, if the wet flue gas volume is 12.3 m^3 at standard conditions per kg fuel. The dissociation of the products of combustion may be ignored. Assume that the specific heat of the products of combustion is 1.35 kJ/(m^3 K).

REFERENCES

Bartok, W., and Sarofim, A. F., eds. 1991. *Fossil fuel combustion: A source book*. New York: Wiley.

Combustion Engineering. 1981. *Combustion: Fossil power systems*, 3rd ed. Windsor.

Lurgi AG. 1992. CFB power plants. Germany.

Merrick, D. 1984. *Coal combustion and conversion technology*. New York: Elsevier.

Richards, P. C. 1994. Gasification: The route to clean and efficient power generation with coal. *Can. Min. Metall. Bull. CIM Bull.* 87:135–140.

Schellberg, W., and Kuske, E. 1992. The qualities of Prenflo coal gas for use in high-efficiency gas turbines. ASME paper 92-GT-263, pp. 1–5.

Smoot, L. D., ed. 1993. *Fundamentals of coal combustion for clean and efficient use*. Amsterdam: Elsevier.

Smoot, L. D., and Smith, P. J. 1985. *Coal combustion and gasification*. New York: Plenum.

Uhde AG. 1993. Texaco Kohleverasung. Germany.

Chapter Three

REDUCTION OF ENERGY SYSTEMS' ENVIRONMENTAL IMPACT

Fossil fuel fired power plants exert a great adverse impact on the environment because of the emissions of the greenhouse gas carbon dioxide and such pollutants as dust (fly ash), sulfur, and nitrogen oxides. This chapter deals with the formation of pollutants in combustors and burners and with the abatement techniques applied to control pollutant emissions from power plants. Electrostatic precipitators (ESPs), bag filters, and ceramic filters are used to control particulate emissions.

The formation of sulfur dioxide in the fuel combustion process and the control of SO_2 emissions by using sorbents such as lime, limestone, and dolomite are discussed in detail. One of the most crucial issues is the control of NO_x emissions. First, the mechanisms of thermal, fuel, and prompt NO formation are discussed. Then the abatement techniques used to control NO_x emissions are described. Finally, advanced multipollutant control methods such as two-sorbent, urea injection, char-based, and Noxsol processes are presented.

POLLUTANT EMISSIONS FROM CONVENTIONAL POWER PLANTS

The task of reduction of environmental impact is one of the most important challenges of modern energy technology. There are no environmentally benign energy systems. The following pollutant emissions from thermal power plants directly impact the environment (Bald and Heusinger, 1996; Makansi, 1993; Elliott, 1989):

- gaseous pollutant emissions
- particulate pollutant emissions
- wastewater
- rejected heat
- noise emissions

The products of combustion of a fuel may contain gaseous constituents and solid residuals. Gaseous constituents consist of carbon oxides (CO_2 and CO), water vapor (H_2O), sulfur oxides (SO_2 and SO_3), nitrogen (N_2), oxygen (O_2), nitrogen oxides (NO and NO_2), nitrous oxide (N_2O), and unburned hydrocarbons (UHCs). Solid residuals consist of fly ash, slag, heavy metals, chlorides, etc. Only three of these substances are harmless to the environment, namely, H_2O, N_2, and O_2; all the rest impact adversely. The concentration of the substance in the flue gas depends upon the fuel type and composition, the type of combustion unit, and the peculiarities of the combustion process. The rate of pollutant emissions from a power plant is lower when its efficiency is higher. The higher the efficiency of a power plant, the better the fuel energy utilized and therefore the lower its fuel requirements per unit of electrical energy produced.

The control of CO and UHC emissions is achieved by increasing the efficiency of combustion through improved designs of combustors and burners, in particular. The control of particulate matter is accomplished by using efficient dust collection systems such as electrostatic precipitators

Table 3.1. Pollutant emissions (in mg/m^3) from advanced coal-fired
power plants

Power plant technology	SO$_x$	NO$_x$	Dust
Pulverized coal (reference technology)	2450	800	50
Pulverized coal +FGD +SCR	200	200	50
FBC	250	170	50
IGCC	10	50	5

FGD, flue gas desulfurization; SCR, selective catalytic reduction; IGCC, integrated
gasification combined cycle; FBC, fluidized bed combustion.

and fabric filter baghouses. It is worth mentioning that, in some cases, the nitrogen and sulfur oxide abatement techniques might exert an adverse impact on combustion efficiency.

In general, pollutant emissions from power plants can be reduced, first, by increasing the energy conversion efficiency. By comparing various types of fossil fuel fired power plants, one can conclude that combined-cycle power plants and cogeneration plants are beneficial, as they have a high degree of fuel energy utilization. The combustion conditions in boiler furnaces and in gas turbine combustors differ in the excess of air used. Owing to high excess air ratios (4–5) in gas turbine combustors, the combustion there is practically complete, and this results in very low concentrations of unburned components such as carbon monoxide and UHCs in the gas turbine exhaust gas. Because of large combustion air-fuel ratios, the pollutants are rather strongly diluted in the flue gas with an oxygen concentration of 13–17% (Makansi, 1993).

Pollutant emissions from advanced coal-fired power plants, in mg/m^3 gas, are given in Table 3.1. Combined-cycle power and cogeneration plants burning natural gas or light fuel oil have intrinsically higher efficiency of fuel energy utilization and therefore exert less harmful impact on the environment than conventional coal-fired power plants and heating plants. For this reason, they are better suited for use in heavily populated areas. While burning natural gas, the only pollutant emissions contained in the flue gas are nitrogen oxides NO and NO$_2$, which are commonly designated as NO$_x$. Thus the most important environmental problem of gas turbine based power plants is NO$_x$ emissions. Nitrogen oxides are very harmful environmental pollutants. In particular, sulfur and nitrogen oxides combine with water to form sulfuric and nitric acids, which considerably increase the acidity of rain. Acid rain with a pH below 5 causes lake acidification and results in damage to fisheries and forests and has a harmful impact on human health. Increased concentrations of SO$_2$, NO$_x$, ozone, and acid aerosols in the atmosphere also negatively affect human health.

There are a number of abatement techniques that are employed to reduce NO$_x$ emissions. The pollutant emissions from furnaces and combustors used in power plants can be drastically reduced by the fuel preparation. This is particularly applicable in gas turbines (Lefebvre, 1995). The most important issues of pollutant emissions control will be discussed below.

PARTICULATE EMISSIONS CONTROL

The amount of solid residuals (ash and slag) of a combustion process per MJ of input fuel energy and per kWh of generated electrical energy is given by

$$m_{sr} = m_{ash} + m_{slag} = A/LHV \quad kg/MJ \tag{3.1}$$

$$m'_{sr} = 3.6A/LHV \; \eta_o \quad kg/kWh \tag{3.2}$$

where A is the mass fraction of ash in the coal in kg/kg, LHV is the lower heating value of the coal in MJ/kg, and η_o is the overall efficiency of the power plant.

The amount of fly ash carried with the flue gas is given by

$$m_{fa} = Am_f(1 - A_r) \quad kg \tag{3.3}$$

where m_f is the mass of coal burned in kg and A_r is the mass fraction of ash in the refuse collected from the bottom of the furnace (per kg of coal) in kg/kg.

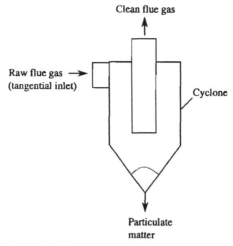

Figure 3.1. Principle of operation of the cyclone.

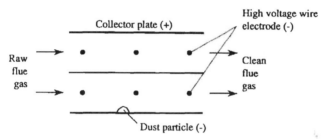

Figure 3.2. Principle of operation of the electrostatic precipitator.

The fly ash concentration in a cubic meter of the flue gas is given by

$$c_a = m_{fa}/(m_f V_g) \quad kg/m^3 \tag{3.4}$$

where V_g is the volume of wet flue gas in m^3 at standard conditions, i.e., at 0°C and 1.013 bars, per kg coal.

Particulate matter control is accomplished by means of the following dust collection systems (Elliott, 1989): cyclone, electrostatic precipitator, fabric filter, and wet scrubber. The dust (fly ash) collection efficiency is given by

$$\eta_{coll} = (1 - c_e/c_i) \times 100\% \tag{3.5}$$

where c_i and c_e are particulate matter concentration in the gas stream at the inlet and exit of the dust collection system, respectively, in kg/m^3.

A cyclone is shown schematically in Figure 3.1. Its operation is based on centrifugal force acting on the particles. The gas stream enters the cyclone through a tangential inlet and therein is brought into a swirling motion. The collection efficiency of a cyclone depends on the difference of densities of solid particles and gas. Cyclones are capable of removing over 99% of coarser particles with a diameter above 20 μm. With smaller particles they are less efficient, e.g., they remove only about 40% of particles smaller than 5 μm.

In an electrostatic precipitator (see Figure 3.2), high voltage (30–60 kV) is applied to a large number of wire electrodes (negatively charged) suspended in the precipitator across the gas flow and below the positively charged collector plates. The particles in the gas stream acquire a negative

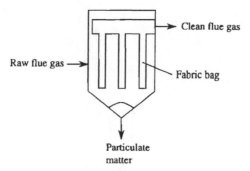

Figure 3.3. Fabric filter (baghouse).

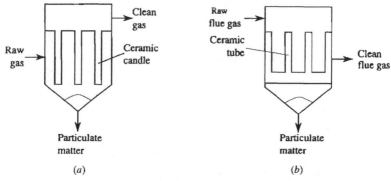

Figure 3.4. Hot gas cleanup with (*a*) ceramic candle filter or (*b*) ceramic tube filter.

charge and settle on the collector plates. Collection efficiency is in excess of 99.5% for particle sizes above 10 μm but is only about 90% for particles smaller than 5 μm.

Fabric filters (see Figure 3.3) are used for cleanup of particulate matter from the flue gas while it passes through a porous fabric structure. Bags of woven glass fiber and Teflon felt are employed for the final cleanup of particulates from the hot flue gas at temperatures up to $\sim$250°C. Fabric filters are capable of removing over 99.5% of the smallest particles of fly ash, in particular, from the flue gases of utility boilers. They are used as large-size baghouses.

Hot gas cleanup (HGCU) systems operating at temperatures above 550°C require ceramic materials that must be mechanically and thermally stable in the high-temperature environment and be resistant to oxidation and corrosion from the coal's inorganic constituents (ash), sulfur, and alkali metals (sodium and potassium). Figures 3.4*a* and 3.4*b* show the principle of operation of ceramic barrier filters, such as ceramic candle filters and ceramic tube filters, respectively. This technology is of utmost importance for the implemention of advanced integrated gasification combined cycle (IGCC) plants and will be discussed further in Chapter 9.

FORMATION AND RETENTION OF SULFUR OXIDES

Sulfur Content of Fuel

The sulfur content in fossil fuels varies widely. Fuels such as coal and fuel oil contain significant amounts of sulfur. There are three forms of sulfur in coal: (1) inorganic or pyritic sulfur (FeS_2), (2) organic sulfur compounds such as thiophenes and sulfides, and (3) calcium or iron sulfates. In most cases the sulfur content of coal is 1–3% by mass. By mechanical separation methods applied for coal beneficiation, a large portion of the pyrite can be removed. The sulfur content of fuel oils is between a minimum of $>0.1\%$ by mass for light fuel oils and a maximum value of 4% for heavy fuel oils. Natural gas is sulfur free except for very small amounts of mercaptan, the odor additive.

Formation of Sulfur Pollutants

Sulfur oxides are formed in combustors from the sulfur contained in the fuel. During combustion, virtually all of the sulfur contained in the fuel is oxidized to sulfur dioxide according to the following reaction:

$$S + O_2 \rightarrow SO_2 \tag{3.6}$$

Thereby, 2 kg of SO_2 are formed per kg of S in the fuel. Hence for a fuel with a sulfur content S kg per kg fuel, the quantity of SO_2 formed per MJ or kWh of input fuel energy is given by

$$m_{SO_2} = 2S/LHV \quad kg/MJ \tag{3.7}$$

where LHV is the lower heating value of the fuel in MJ/kg.

The concentrations of sulfur oxides in the flue gas mainly depend on the sulfur content of the fuel. Gas turbines generally use clean fuels; thus sulfur oxide is a minor problem in combined-cycle power plants in comparison with coal-fired power plants.

In coal gasification plants the sulfur remains predominantly in the form of hydrogen sulfide (H_2S), and such species as COS and CS_2 (Longwell et al., 1995; Smoot and Smith, 1985).

Technologies to Reduce Sulfur Pollutants

There are no benign gaseous substances to which sulfur-containing pollutants from combustion and gasification plants can be converted. Advanced power plant configurations including fluidized bed combustor (FBC) or IGCC schemes with HGCU can provide high reduction levels of sulfur pollutants. The introduction of sorbents directly into FBC or into the postcombustion chamber is one option that may help to control SO_2 emissions from existing and new combustors (Lurgi AG, 1992; Smoot, 1993).

Innovative coal desulfurization methods can be classified into four categories (Bartok and Sarofim, 1991; Smoot and Smith, 1985; Smoot, 1993): (1) precombustion coal cleaning, (2) advanced combustion technologies, (3) in situ sulfur retention, and (4) flue gas desulfurization (FGD). To remove sulfur from the coal prior to combustion, physical, chemical, or biological cleaning methods may be employed. About 30–50% of the pyritic sulfur, which comprises 10–30% of the total sulfur in the coal, can be removed by physical cleaning methods, which separate inert matter from the coal on the basis of different densities or surface properties. Advanced coal-cleaning methods are expected to remove an even greater amount of pyritic sulfur. Chemical and biological cleaning methods are capable of removing 90% of the total sulfur in the coal, including some chemically bound sulfur. Chemical coal cleaning involves leaching both the organic sulfur and the mineral matter from the coal using hot sodium-based or potassium-based caustic. By biological cleaning, the coal is exposed to bacteria or fungi that have an affinity for sulfur. Sulfur-digesting enzymes may be added.

Advanced Combustion Technologies for Sulfur Pollutant Reduction

There are a few combustion concepts that can be used to abate SO_2 emissions from coal combustors. New techniques that convert coal into clean gaseous or liquid fuels with minor environmental impact are now being developed and tested. Advanced or clean coal combustion technologies include FBC configurations, coal-water slurry combustion, as well as combustion of clean fuels produced by coal gasification integrated into combined-cycle power generation systems, or by coal liquefaction.

Clean fuel combustion technologies including IGCC systems are discussed in detail in Chapter 9. However, it is worthwhile to present here a brief description of an IGCC system, which consists of the following units:

- unit for high-temperature coal gasification with steam and oxygen (or air),
- hot gas cleanup unit,
- combined-cycle gas and steam turbine power plant.

Most of the sulfur in the coal is converted to H_2S during gasification. In the HGCU the raw gas constituents harmful to the gas turbine such as metals like vanadium and nickel, as well as sodium and atmospheric pollutants, must be removed from the raw fuel gas. This step requires some method of desulfurization. Several alternatives are available for efficient removal of H_2S. In some gasification combined-cycle systems, the hot gases are passed through a bed of zinc ferrite, which absorbs H_2S. Commercial-grade sulfur is produced when the ferrite bed is regenerated. In other applications, limestone sorbent is used for H_2S capture.

However, the main advantage of gasification combined-cycle systems is that pollutant reduction is accomplished before combustion takes place. The volume of gas to be cleaned is much less, and the pollutant concentrations are significantly higher; thus a higher cleanup efficiency can be reached.

In Situ Sulfur Capture

Sulfur pollutants can be efficiently removed during the combustion process by adding sorbents directly into the combustor, as is done in fluidized bed combustors (Lurgi AG, 1992) or by injecting sorbents into the postcombustion flue gas duct (Bald and Heusinger, 1996). Sorbents are used to react with sulfur pollutants. The most suitable are calcium- and magnesium-based sorbents, especially lime, limestone, and dolomite, and calcium hydroxide. Other sorbents include oxides of zinc, iron, and titanium. Calcium-based sorbents have low cost and combine with sulfur to form inert by-products such as calcium sulfate.

FBC technology is described in detail in Chapter 9.

Sulfur Retention by Means of Calcium-Based Sorbents

Both hydrated lime, i.e., calcium hydroxide [$Ca(OH)_2$], and limestone, i.e., calcium carbonate ($CaCO_3$), can be used as sorbents. In a fluidized bed combustor, for example, limestone will be calcinated in situ in the furnace. The calcination (dissociation) reaction is endothermic, and its rate depends mainly on the sorbent particle size and temperature. Thus

$$CaCO_3 \rightarrow CaO + CO_2 - 183 \quad kJ/mol \tag{3.8}$$

Similarly,

$$Ca(OH)_2 \rightarrow CaO + H_2O \tag{3.9}$$

Now the sulfur capture from the products of combustion occurs according to the following reactions:

$$CaO + 1/2O_2 + SO_2 \rightarrow CaSO_4(solid) \tag{3.10}$$

$$CaO + SO_3 \rightarrow CaSO_4(solid) \tag{3.11}$$

The overall desulfurization process involves diffusion of sulfur species and oxygen to the calcined sorbent surface and through the intergranular pores, as well as the solid-state diffusion through the $CaSO_4$ product layer, and finally, reaction with solid CaO. The molar volume of $CaSO_4$ is greater than that of CaO. Therefore formation of $CaSO_4$ decreases the pore radius and thus sets a limit to the extent of desulfurization conversion possible. Figure 3.5 shows schematically the lime/limestone FGD system.

Hydrogen sulfide must be removed from fuel gas produced in the coal gasification process. It occurs according to the following reaction:

$$CaO + H_2S \rightarrow CaS(solid) + H_2O \tag{3.12}$$

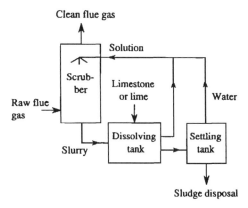

Figure 3.5. Flow diagram of the lime/limestone flue gas desulfurization system.

Flue Gas Desulfurization Systems

The sulfur abatement technologies employed for FGD can be classified either as wet, semidry, or dry, or as nonregenerative or regenerative FGD systems. More than 250 FGD processes have been proposed worldwide.

The wet FGD systems include (Longwell et al., 1995; Elliott, 1989) lime/limestone scrubbing, caustic scrubbing, ammonia scrubbing, and double-alkali scrubbing. The lime/spray dryer is a semidry system.

In the dry FGD systems the reactant (absorbent) such as lime, limestone, or bicarbonate is mixed with SO_2. The obtained by-products cannot be sold and must be disposed of near the plant. Various waste materials such as fly ash and kiln dust may be used as reagents in the FGD process to absorb SO_2.

Wet FGD systems are most frequently used at utility power stations. Figures 3.6–3.8 show schematically FGD plants that can reduce up to 90% of the sulfur oxide emissions from a coal-fired power plant.

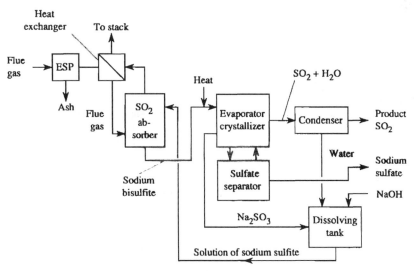

Figure 3.6. Flow diagram of the sodium sulfite nonregenerative FGD system (Wellman-Lord absorption system).

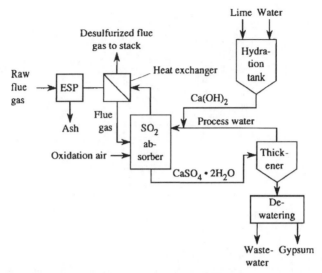

Figure 3.7. Flow diagram of the regenerative FGD system with gypsum by-product.

In nonregenerative FGD systems, SO_2 retention occurs using a reactant (absorbent) available locally at low cost. The by-products obtained in the process are not saleable and must be disposed of near the plant. A flow diagram of the sodium sulfite nonregenerative FGD system, the Wellman-Lord absorption system, is depicted in Figure 3.6.

In regenerative FGD systems, more expensive reactants (absorbents) are used that allow higher efficiency of the desulfurization process. Usually, reactants are at least partially recovered, and extracted by-products can be sold commercially.

Advanced regenerative FGD systems improve the already proven technology of conventional nonregenerative systems and produce by-products that are harmless to the environment and can be sold commercially. For example, the nonregenerative limestone process becomes an advanced regenerative process when the by-product is gypsum. The regenerative FGD system shown in Figure 3.7 uses lime or limestone as sorbent and produces gypsum as by-product.

A number of alternative saleable by-products, such as liquid sulfur dioxide, solid elemental sulfur, or sulfuric acid, may be produced. However, calcium sulfite-sulfate sludge is a nonusable waste that will be disposed in ponds, which is only a mid-term solution. The by-product quantity depends on the sulfur content of the coal and the required SO_2 removal efficiency according to current regulations, e.g., those established by the U.S. Environmental Protection Agency (EPA). Thus a 900-MWe power station that burns coal with a 2.7% sulfur content and operates with a capacity factor of 0.8 annually requires roughly 460×10^6 kg of lime or 580×10^6 kg of limestone for sulfur removal from the stack gases to meet emissions standards. Depending on the process used, the following quantities of alternative by-products (in millions of kilograms) can be produced annually: 80×10^6 kg of sulfur dioxide (100%), 40×10^6 kg of solid sulfur (elemental), or 130×10^6 kg of sulfuric acid (95%). It is seen that the amounts of by-products recovered from FGD plants are immense, and the problem of disposal of unusable by-products can be very serious. Therefore precombustion cleaning of coal with an aim to remove some 40–50% of sulfur content may be a proper near-term solution.

In wet regenerative FGD processes the absorber of SO_2 is usually an aqueous solution of either an alkali (sodium hydroxide or ammonia) or a sulfite (sodium, ammonium, or magnesium sulfite). The addition of lime (CaO) or limestone ($CaCO_3$) to the spent absorber solution results in the regeneration of the absorber and the precipitation of solid calcium sulfite ($CaSO_3$) and sulfate $CaSO_4$, which are disposed of as sludge. If the absorber is a solution of sodium hydroxide (NaOH)

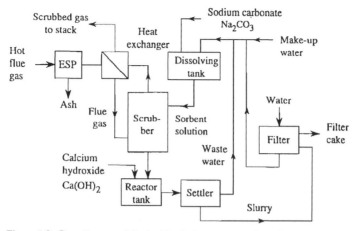

Figure 3.8. Flow diagram of the double-alkali nonregenerative FGD system.

or ammonia (NH_3), the interaction with sulfur dioxide in the flue gas leads to the formation of soluble sodium sulfite (Na_2SO_3) or ammonium sulfite [$(NH_4)_2SO_3$], respectively. Regeneration of the alkali absorber occurs through the addition of lime.

When the absorber is a sulfite solution, namely, sodium sulfite, ammonium sulfite, or magnesium sulfite ($MgSO_3$), the products are sodium bisulfite ($NaHSO_3$), ammonium bisulfite (NH_4HSO_3), or magnesium bisulfite [$Mg(HSO_3)_2$], respectively. Addition of limestone or lime causes precipitation of solid calcium sulfite (and some sulfate) and regeneration of the original sulfite absorber in solution.

The double-alkali flue gas scrubbing system shown in Figure 3.8 uses two absorbents, e.g., sodium carbonate (Na_2CO_3) and calcium hydroxide [$Ca(OH)_2$] or sodium hydroxide (NaOH), to remove the oxides of sulfur from the combustion gases. The double-alkali process includes three steps: absorption, chemical regeneration, and solids dewatering. During absorption, dirty flue gas enters the bottom part of the SO_2 absorber (scrubber). Then the flue gas flows into the reaction zone, which has packing trays, a venturi, or a spray device, where it is mixed with the sorbents.

NITROGEN OXIDE FORMATION

There are three mechanisms of nitrogen oxide formation in flames and combustors burning solid, liquid, or gaseous fuels: (1) fuel NO mechanism, (2) thermal NO mechanism, and (3) prompt NO mechanism. The first mechanism involves the oxidation of nitrogen contained in the fuel to NO, the second mechanism is the fixation of molecular nitrogen by oxygen atoms at high temperatures, and the third mechanism means the attack of hydrocarbon free radicals on molecular nitrogen, producing NO precursors.

Fuel Nitric Oxide Mechanism

The fuel NO accounts for 75–95% of total NO_x in coal flames and 50% in fuel oil flames. Hence fuel NO is by far the most significant source of nitric oxide formed during the combustion of nitrogen-containing fossil fuels. The dominance of fuel NO in coal systems is due to the moderately low temperatures (1200–1700°C) and the locally fuel-rich nature of most coal flames (Longwell et al., 1995; Smoot and Smith, 1985). Fuel NO is formed more readily than thermal NO because the N—H and N—C bonds common in fuel-bound nitrogen are weaker than the triple bond in molecular nitrogen, which must be dissociated to produce thermal NO.

First, the fuel nitrogen is evolved from heterocyclic compounds (pyridine, quinoline, etc.) as HCN and from amines as NH_3. Then, HCN decays rapidly to NH_i ($i = 0, 1, 2, 3$), which in turn, reacts with O_2, O, and OH to form NO and N_2 according to the following simplified scheme:

$$\text{fuel-N} \rightarrow \text{HCN} \rightarrow \text{NH}_i \rightarrow \text{NO (and } N_2) \tag{3.13}$$

During coal devolatilization, the fuel nitrogen is converted into two fractions: volatile nitrogen (mainly, HCN and NH_3) and char-bound nitrogen. The relative amount of the volatile nitrogen released from coal depends on the devolatilization temperature, heating rate, and coal type. Nitric oxide is generated during the combustion of both the volatiles and the char. Simultaneous with the production of NO, the reduction of NO to N_2 can occur in the combustion process. The NO can be reduced by gas-solid reactions, gas-solid catalytic reactions, and homogeneous gas phase reactions. Nitric oxide (NO) is reduced to nitrogen by the reaction on the char (C) surface:

$$C + 2NO \rightarrow N_2 + CO_2 \tag{3.14}$$

$$C + NO \rightarrow 1/2N_2 + CO \tag{3.15}$$

Volatile nitrogen compounds can react to either NO or molecular nitrogen N_2.
The most important and rate-controlling reaction for NO formation is

$$N + OH \rightarrow NO + H \tag{3.16}$$

In fuel-lean flames (excess air ratios higher than 1), the following reactions of NO formation and its reduction are also important:

$$N + O_2 \rightarrow NO + O \tag{3.17}$$

$$N + NO \rightarrow N_2 + O \tag{3.18}$$

Thermal NO Mechanism

The thermal nitric oxide formation mechanism is described by the following chemical reactions, which represent the modified Zel'dovic mechanism (Smoot, 1993):

$$N_2 + O \rightarrow NO + N \tag{3.19}$$

$$N + O_2 \rightarrow NO + O \tag{3.20}$$

For the first reaction, which is the rate-limiting step of the whole process, high temperatures are required because of the high activation energy of 314 kJ /mol. Under fuel-rich conditions (excess air ratios below 1), one more reaction should be included in the mechanism (Smoot, 1993):

$$N + OH \rightarrow NO + H \tag{3.21}$$

However, for most fuel-lean flames this reaction may be neglected. At relatively low temperatures, NO may form through a nitrous oxide (N_2O) intermediate. In practical combustors with high operating temperatures, this mechanism may be neglected.

When hydrocarbon fuels are burned, the results are influenced by prompt NO and NO_2 formation. The general conclusion is that although the nonequilibrium effects are important in describing the initial rate of thermal NO formation, the accelerated rates are still sufficiently low that very little thermal NO is formed in the combustion zone (Smoot and Smith, 1985; Smoot, 1993) and that the majority is formed in the postflame regime, where the residence time is longer. Thus the prerequisites for thermal NO formation are high temperatures above 1400°C, total or local excess oxygen, and large residence times in a high-temperature oxidizing environment.

Prompt Nitric Oxide Mechanism

Prompt NO formation occurs by the collision and fast reaction of hydrocarbon free radicals with molecular N_2. This mechanism is significant in fuel-rich hydrocarbon flames. The following reactions are suggested to be significant in prompt NO formation (Smoot, 1993):

$$CH + N_2 \rightarrow HCN + N \qquad\qquad (3.22)$$

$$C + N_2 \rightarrow CN + N \qquad\qquad (3.23)$$

The cyanide species (HCN and CN) are oxidized with oxygen to NO. The significance of the second reaction increases with increasing temperature.

Dominating NO_x Mechanism

No definite rules exist to determine which NO formation mechanism dominates for a given stationary combustor configuration because of the complex interactions between burner fluid dynamics and both fuel oxidation and nitrogen species chemistry. It has been shown that the fuel NO mechanism dominates in pulverized, coalfired boilers, although thermal NO is also important in the postflame regions, where overfire is used. However, the thermal NO contributions become significant at temperatures above 1400°C in coal flames (Smoot, 1993). Such high temperatures lead to the thermal NO formation in regions where local excess air ratios are near unity. The prompt NO formation is an important mechanism only in liquid fuel combustors.

The effect of temperature on the three mechanisms of NO formation is shown in Figure 3.9.

For FBC, at a typical temperature range of 850–900°C, the formation of thermal and prompt NO_x is less than 5% of the total NO_x emitted and is of only minor importance.

Total Nitrogen Oxide Emissions

Most of the nitrogen oxide emitted to the atmosphere by combustion systems is in the form of nitric oxide (NO), with only a small fraction appearing as nitrogen dioxide (NO_2). The other two oxides, i.e., N_2O and NO_2, are insignificant in oil and coal flames as well as in unquenched effluent gas of most gas combustors. Hence NO is normally the only significant nitrogen oxide species emitted from practical combustors.

With the exception of benign molecular nitrogen, nitrogenous species, such as ammonia, hydrogen cyanide, amine compounds, and nitrous oxide (N_2O), are air pollutants whose anthropogenic origins also include combustion systems.

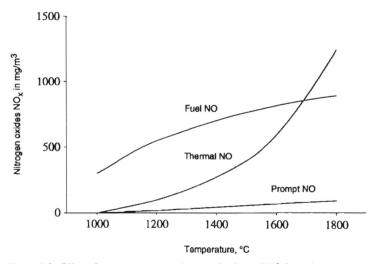

Figure 3.9. Effect of temperature on the three mechanisms of NO formation.

CONTROL OF COMBUSTION NO$_x$ EMISSIONS

In-Furnace NO$_x$ Control Methods

The following three parameters influence NO formation in boiler furnaces and gas turbine combustors (Longwell et al., 1995; Smoot and Smith, 1985): temperature, availability of oxygen, and residence time. NO$_x$ is produced in large quantities only at very high temperature levels. The situation in a gas turbine combustor is quite different, first, because combustion actually takes place and, second, because residence times at high temperatures are fairly limited.

All the in-furnace NO$_x$ control techniques are based on lowering the temperature and limiting the availability of oxygen to reduce both thermal and fuel NO formation. Implicit to the reaction rate controled formation of NO is the need to reduce the residence time available for NO formation.

The following methods are practically used for NO$_x$ control during combustion (Bald and Heusinger, 1996; Makansi, 1993):

- air staging
- fuel staging (reburning)
- low-NO$_x$ burners
- low excess air (LEA) operation
- flue gas recirculation

All these techniques seek to redistribute air and fuel flows in the furnace. This is done at the burner and/or within the combustor. A balance must be carefully established and maintained among coal particle size, carbon monoxide (CO), unburned hydrocarbons (UHCs), and unburned carbon (UC) in the bottom ash and fly ash and NO$_x$ emissions. It is very important to avoid a reducing environment near the water wall to prevent tube corrosion.

Air Staging

Air staging in boiler furnaces is also called overfire air (OFA), burners out of service (BOOS), and biased or off-stoichiometric firing. The principle of NO$_x$ control through air staging is illustrated in Figure 3.10. The total combustion air is divided into primary air and secondary air. To minimize NO$_x$ formation, the mixing of fuel and primary air in the furnace is delayed to enable the occurrence of reducing conditions (with excess air ratio below 1). The secondary air is then introduced to achieve burnout of the remaining combustibles at lower temperatures and thus to prevent additional NO$_x$ formation. However, it should be stressed that this method has limitations. It cannot be applied to boilers with slagging furnaces. In the dry bottom coal-fired boilers, evaporator tube corrosion may accelerate under reducing conditions in the furnace. Flame stability problems and furnace vibrations may occur in gas- and oil-fired boilers.

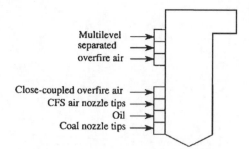

Figure 3.10. Air staging in boiler furnaces with concentric firing system (CFS) air nozzle tips, close-coupled overfire air, multilevel overfire air, burners out of service, and biased or off-stoichiometric firing.

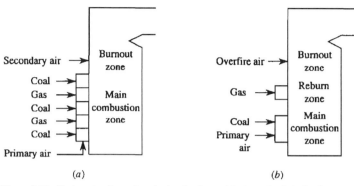

Figure 3.11. Fuel staging (gas reburning) technology with air staging in boiler furnaces: (a) coal-gas co-firing, (b) close-coupled gas reburning.

Advanced low-NO_x firing systems are based on the following control mechanisms: firing-zone stoichiometry control, pulverized coal fineness control, combustion process control, and concentric firing. The process of NO_x formation is controlled by staged introduction of air through coal nozzle tips and multiple levels of separated overfire air (SOFA) and close-coupled overfire air (CCOFA). Innovative preassembled wind boxes include some elevations of coal compartments with flame-attachment coal nozzle tips, several elevations of air/oil compartments with concentric firing system (CFS) air nozzle tips, and one or two elevations of CCOFA. Some of the secondary air is directed away from the fuel streams in the main firing zone by CFS air nozzle tips (Makansi, 1993). This increases the O_2 concentration near the furnace water wall and thus reduces the potential for tube corrosion from such coal constituents as sulfur and alkali metals. Postretrofit NO_x emissions as low as 0.1 kg per GJ of fuel heat input can be achieved.

Fuel Staging (Reburning)

Fuel supply to the furnace can be staged as shown in Figure 3.11. About 80% of the fuel (pulverized coal) is burned under fuel-lean (with excess air ratio above 1) conditions in the lower portion of the furnace. This is followed by the staged introduction of the remaining 20% of the fuel (pulverized coal, gas, or oil) at a higher elevation in the furnace to create reducing (fuel-rich) conditions with an excess air ratio below 1, under which the NO is converted into N_2. Through overfire air ports in the upper portion of the furnace, air is added to complete the combustion at lower temperatures. at which no additional NO_x is formed. By utilizing gas and coal reburning, NO_x reductions may exceed 50% (Makansi, 1993).

Low-NO_x Burners

The use of low-NO_x burners is the most important method of combustion NO_x control. Most low-NO_x burners, particularly those for pulverized coal, employ air staging. However, in some gas-fired burners, fuel staging is employed to reduce combustion temperatures and control flame stoichiometry. Under optimized conditions, NO_x reductions over 40% may be achieved. Adverse side effects may, in particular, include increased carbon in the fly ash, increased CO emissions, increased furnace exit temperature, and increased corrosion and slagging due to localized reducing conditions.

Low Excess Air Operation

LEA is based on limiting the availability of oxygen for NO formation, subject to constraints of flame stability and CO emissions. In general, it can lead to about 15–20% NO_x reduction, which may be utilized in conjunction with other control technologies. It is applicable to all combustion devices for NO_x control.

Flue Gas Recirculation

There are two modes of flue gas recirculation for NO_x control: either into the furnace or into the burner. The latter mode can be used with low NO_x burners to reduce flame temperature and oxygen concentration. Usually, 20–30% of the flue gas is recirculated, subject to operational constraints of flame stability and impingement as well as boiler vibrations.

NO_x Reduction in Boiler Firing

The two main combustion technologies in utility boilers are front-wall or opposed-wall firing and tangential firing. Front- or opposed-wall burners operate as individual burners. In tangential fired boilers, fuel and air are injected into the furnace tangentially from the four corners, using the furnace as a mixing chamber for production of a single-flame envelope. The relatively slow, even burning characteristics of tangential firing reduce highly turbulent hot spots, producing lower peak and bulk gas temperatures, hence producing less NO_x than front-wall firing.

Low-NO_x system technologies include staged combustion with gas mixing or gas recirculation and reburning (Bald and Heusinger, 1996). These technologies apply to both tangentially and front-wall and opposed-wall fired boilers.

Staged Combustion

Pulverized coal is burned under substoichiometric conditions to reduce flame temperature and, consequently, the formation of thermal NO_x. Burn-out over fire air (OFA) is added above the main combustion zone. In addition, combustion air is mixed with flue gas to reduce oxygen partial pressure and therefore the flame temperature.

In tangentially fired boilers, NO_x formation is reduced by the use of the OFA nozzles added above the tangential-firing wind boxes. This diverts a portion of the total combustion air away from the burners, resulting in a staged combustion which retards the fuel NO and thermal NO formation mechanisms.

Reburning in Boiler Furnace

This method removes NO_x from combustion products by using fuel as a reducing agent. The reburning fuel is injected over the primary combustion zone to form a fuel-rich second stage, where NO_x is reduced to molecular nitrogen. Compared with original emissions, NO_x reductions of up to 80% have been obtained. Recirculated flue gas (inert atmosphere) and reburning fuel are injected into the furnace. These technologies are applied to oil, gas, and coal firing.

FLUE GAS TREATMENT FOR NO_x EMISSIONS CONTROL

It is often impossible to achieve the low NO_x levels required through combustion controls alone. This is especially the case when required outlet NO_x levels are extremely low or when NO_x must be controlled over a wider range of system loads than low NO_x burners (LNBs) or other combustion controls can operate. Older boilers built for high efficiency and small plan area, in particular, generally have combustion systems that are not amenable to retrofit combustion controls for NO_x. In these cases, exhaust gas treatment methods may be used to reduce the amount of NO_x remaining in the gas stream after combustion.

The postcombustion exhaust gas treatment for NO_x control includes (Bald and Heusinger, 1996; Makansi, 1993; Smoot, 1993) selective catalytic reduction (SCR), selective noncatalytic reduction (SNCR), and hybrid or combined SNCR-SCR method.

Selective Catalytic Reduction

SCR is the only commercially available flue gas treatment technology that has the demonstrated capability to remove over 90% of the NO_x contained in exhaust gas. SCR is widely used in

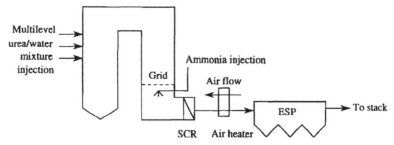

Figure 3.12. NO_x abatement system with multilevel urea/water mixture injection, ammonia injection, and selective catalytic reduction (SCR).

Germany and Japan on all types of combustion systems. Most of the U.S. SCR installations are used in gas turbine cogeneration plants. SCR technology operates by reacting ammonia (NH_3) with NO_x in the presence of oxygen and a catalyst to form nitrogen and water. The chemical reactions involved are as follows (Smoot, 1993):

$$4NO + 4NH_3 - O_2 \rightarrow 4N_2 + 6H_2O \tag{3.24}$$

$$6NO_2 + 8NH_3 \rightarrow 7N_2 + 12H_2O \tag{3.25}$$

By injecting ammonia into the exhaust gases prior to catalyst, ~30% of the NO_x can be removed. Although SCR is a well-proven technique, it encounters the following problems: high capital costs, e.g., about 20% of the cost of the gas turbine; high replacement costs because of the relatively short life expectancy of the catalyst; and danger of ammonia slipping into the atmosphere.

Figure 3.12 shows an NO_x abatement system with multilevel urea-water mixture injection, ammonia injection, and SCR.

As commercial catalysts, V_2O_5, TiO_2, and WO_3 are widely used. Zeolites offer, as a new catalyst base, improved performance at higher temperatures. The catalyst's ability to minimize oxidation of SO_2 to SO_3 is very important, since the unreacted NH_3 can react with SO_3 to form ammonium bisulfate, which in turn, can reduce catalyst activity by depositing on active surfaces and causing corrosion and plugging of downstream heat exchange surfaces of the boiler. Another problem with catalysts is poisoning by arsenic compounds and alkali metals.

Depending upon the catalyst used, it must be installed in the zone with a temperature from about 200°C to 550°C, most frequently from 300 to 400°C. At low oxygen concentration in the exhaust gas, the SCR reaction becomes much less effective.

Figure 3.13 presents an efficient hot gas high-dust SCR configuration used in boiler flue gas cleaning systems.

Selective Noncatalytic Reduction

Selective noncatalytic NO_x reduction (SNCR) uses ammonia or urea that reacts with NO in the presence of oxygen to reduce NO_x at temperatures between 870 and 1150°C (1600 and 2100°F)

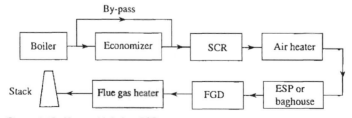

Figure 3.13. Hot gas high-dust SCR arrangement.

without the need for a catalyst. Levels of NO_x reduction from 30 to 75% have been achieved using SNCR techniques. The performance of the SNCR process is affected by the raw gas NO_x level, the temperature, the reactant distribution in the gas stream, and the residence time in the reaction zone. Poor reactant distribution results in low NO_x reduction and/or high ammonia escape.

Hybrid or Combined SNCR-SCR Process

It is possible to reduce the necessary SCR catalyst requirement by using a combined SNCR-SCR process consisting of an upstream SNCR unit followed by a SCR unit. For example, if 95% of total NO_x reduction is required and 50% of NO_x reduction is achieved by SNCR, then the SCR system needs to be designed for only 90% reduction.

Methods suitable for NO_x control in gas turbine based simple-cycle and combined-cycle power plants are described in Chapters 5 and 6.

ADVANCED SYSTEMS FOR POLLUTANT EMISSIONS CONTROL

High-performance emissions control systems include the following:

*   improved single-pollutant removal systems such as advanced FGD systems, low-cost SO_2 removal techniques with in-duct and/or in-furnace sorbent injection, or advanced in-furnace NO_x control systems, and
*   multipollutant removal processes.

Figures 3.14 and 3.15 show the flow diagrams of char-based and ammonia-based multipollutant removal processes.

SNO$_x$ Process

The SNO$_x$ process is the process for simultaneous SO_2/NO_x removal that features high efficiency of removal of both pollutants. The world's largest SNO$_x$ plant has been operated at a 300-MW unit in Denmark for some years. The 35-MWe-scale SNO$_x$ demonstration plant at Ohio Edison Co.'s Niles station consists of a fabric filter to remove particulates, a flue gas heater, an SCR unit to remove NO_x, a catalytic reactor to oxidize SO_2 to SO_3, and a sulfuric acid (93% purity) condenser. The heat released in the oxidation reactor is partially recovered in the flue gas heating step. The process is highly efficient, so that 96% SO_2 removal from coal with 3.4% sulfur and 94% NO_x removal are achieved. In addition, 99% of heavy metals are removed (Makansi, 1993).

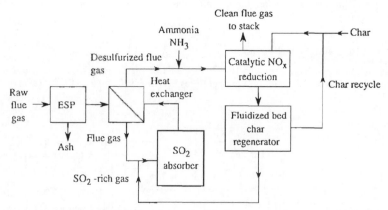

Figure 3.14. Flow diagram of the char-based multipollutant removal process with fluidized bed char regeneration and catalytic NO_x reduction.

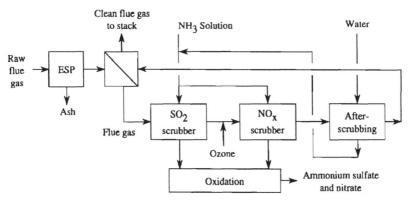

Figure 3.15. Flow diagram of the ammonia-based multipollutant removal process.

The integrated installation of an SNO_x plant will result in an increase in net steam production. Typically, each 1% of sulfur in fuel results in 1% additional steam production, due to the recovery of the heat of formation of sulfuric acid from SO_2 (2.4 kWh per kg sulfur recovered as sulfuric acid) and the heat of cooling the flue gas.

The SNO_x process includes five key steps: particulate control, NO_x reduction, SO_2 oxidation, sulfuric acid condensation, and acid conditioning.

The NO_x removal efficiency of this process is limited to less than 90%. The SO_3 in the gas leaving the SO_2 converter is hydrated and condensed in two steps to give sulfuric acid. Over 95% of the entering SO_2 is catalytically oxidized to SO_3. The flue gas then enters a tube-and-shell falling-film condenser with ambient air used as a cooling medium on the shell side. Condensation and capture of the sulfuric acid at concentrations of 94–97% (by mass) occur there.

SNRB Process

The SO_x-NO_x, Rox-Box (SNRB) process of Babcock and Wilcox Co. is capable of simultaneous removal of fly ash, SO_2, and NO_x in a compact device with a high-temperature catalytic fabric filter. Dry sorbents and ammonia (NH_3) are injected upstream. The air heater is located downstream of the fabric filter. Since the sulfur is removed from the flue gas upstream of the air heater, operation at lower flue gas exit temperatures, with associated heat-rate improvement, could result from SNRB application.

Removal efficiencies of over 80% for SO_2, over 90% for NO_x, and 99% for fly ash have been achieved. The fabric filter operating temperature is 430–460°C, and the calcium-to-sulfur ratio is 2 or lower with lime sorbent. Higher SO_2 removal rates can be achieved with sodium bicarbonate as sorbent (Makansi, 1993).

A very efficient method for combined NO_x-SO_x removal involves streamer corona discharge to produce radical species in the flue gas. In the presence of ammonia and water, nitrogen oxides and SO_2 are reduced by means of the radicals (Kumar et al., 1993). Sawtooth pulses having a 40-kV peak voltage and 0.36–0.8 kV/ns rising rates are applied. The peak corona currents exceed 10 A.

NO_xSO and Soxal Processes

Another multipollutant removal process is the NO_xSO process. It can remove over 95% SO_2 and 70% NO_x. Flue gas passes through a fluidized bed adsorber operated at 620°C, in which SO_2 and NO_x are adsorbed. The sorbent is regenerated with hot air. The waste-heat utilization improves the cost-effectiveness of the process (Makansi, 1993).

Table 3.2. Overall efficiency and specific CO_2 emissions g_{CO_2} (kg/MJ) and g'_{CO_2} (kg/kWh) of power plants

Fuel	η_0	LHV, MJ/kg	V_{CO_2}, m^3/kg	g_{CO_2}, kg/MJ	g'_{CO_2}, kg/kWh
Lignite	0.38	10.26	0.58	0.11	1.05
Bituminous coal	0.43	28.8	1.32	0.09	0.75
No. 6 fuel oil	0.44	42.33	1.61	0.07	0.57
Natural gas^a	0.44	36.73	1.02	0.05	0.41

LHV, lower heating value.
^{a} Values of LHV and V_{CO_2} for natural gas refer to 1 m^3 of fuel volume.

The Soxal process is capable of removing over 98% SO_2 and over 90% NO_x (Makansi, 1993). The prescrubbed flue gas passes an absorber that uses sodium sulfite as sorbent. A bipolar membrane cell stack regenerates the sodium sulfite for use in the absorber and produces sulfurous acid. Steam is used to strip SO_2 from the acid. NO_x is removed in a two-stage process using methanol or urea injection into the reactor. NO is oxidized to NO_2, which is then reduced to N_2 by reaction with sodium sulfite in the scrubbing solution.

Example 3.1

Calculate the CO_2 emissions in kg per 1 MJ of released heat and per kWh of power output of a conventional fossil fuel fired power plant with solid, liquid, or gaseous fuel. The lower heating values of fuels, the overall power plant efficiency η_0, and the carbon dioxide volume V_{CO_2} are given in Table 3.2.

Solution

The specific CO_2 emissions of the power plant are calculated as follows:

$$g_{CO_2} = V_{CO_2} \rho_{CO_2}/LHV \quad \text{kg/MJ}$$
$$g'_{CO_2} = 3.6 g_{CO_2}/\eta_0 \quad \text{kg/kWh}$$

where V_{CO_2} is carbon dioxide volume in m^3 per kg fuel, and ρ_{CO_2} is density of CO_2 at standard conditions (1.965 kg/m^3). The calculation results are presented in Table 3.2.

Example 3.2

An FGD power plant has a flue gas flow rate of 2.526×10^6 m^3/h. The SO_2 concentrations in the flue gas upstream and downstream of the FGD are $c_i = 9$ g/m^3 and $c_e = 200$ mg/m^3, respectively. Calculate the requirements in lime (CaO) per hour and the hourly gypsum (calcium sulfate dihydrate [$CaSO_4 \times 2H_2O$]) production.

Solution

Hourly amount of SO_2 to be removed from the flue gas in the FGD plant is

$$m_{SO_2} = V_g(c_i - c_e) = 2.526 \times 10^6 \text{ m}^3\text{/h } (9000 - 200) \text{ mg/m}^3 = 22,257 \text{ kg/h}$$

Lime requirements are

$$m_{CaO} = (56/64) \times 22,256.7 \text{ kg/h} = 19,450 \text{ kg/h}$$

Hourly gypsum production is

$$(172/64)m_{SO_2} = (172/64) \times 22,256.7 \text{ kg/h} = 59,740 \text{ kg/h}$$

Example 3.3

Calculate the hourly ammonia requirements for an NO_x reduction plant to remove NO_x from the flue gas flow of 2.526×10^6 m³/h operating on an SCR process if the NO_x concentration in the flue gas upstream and downstream of the NO_x reduction plant are 900 and 160 mg/m³, respectively.

Solution

Hourly ammonia requirements are

$$m_{NH_3} = (17/30) \times 2.526 \times 10^6 \text{ m}^3/\text{h} \times (900 - 160) \text{ mg/m}^3 = 1059.2 \text{ kg/h}$$

Example 3.4

For a coal-fired 930-MWe power plant, calculate the fuel rate, the annual CO_2 production rate, the cooling water rate, the sulfur dioxide release and collection rates, the mass of the FGD adsorbent limestone ($CaCO_3$) required, the mass of the FGD by-product $CaSO_3$ produced, mass of ash produced, and the annual particulate emissions under the following conditions:

Coal:
Carbon fraction	$C = 80\%$
Ash fraction	$A = 7\%$
Sulfur fraction	$S = 3.5\%$
Lower heating value	$LHV = 29.3$ MJ/kg
Overall cooling water temperature difference	$\Delta t_w = 9.6$ K
Overall plant thermal efficiency	$\eta_o = 0.4$
Overall boiler efficiency	$\eta_b = 0.91$
Overall sulfur dioxide allowable emissions	7%
Overall electrostatic precipitator efficiency	$\eta_{esp} = 0.99$

Solution

Fuel energy input rate

$$Q_f = P_{el}/\eta_o = 930 \text{ MW}/0.4 = 2325 \text{ MW}$$

Plant fuel rate

$$m_f = 3600 Q_f/LHV = 3600 \times 2325/29.3 = 285,665.5 \text{ kg/h}$$

Annual CO_2 production

$$m_{CO_2} = 3.67C \times m_f = 3.67 \times 0.8 \times 285,665.5 \text{ kg/h}$$
$$= 838,713.9 \text{ kg/h} \times 8766 \text{ h/yr} = 7.35 \times 10^9 \text{ kg/yr}$$

CO_2 production rate per kWh

$$m_{CO_2}/P_{el} = 838,713.9 \text{ kg/h}/(930 \times 10^3 \text{ kW}) = 0.902 \text{ kg/kWh}$$

Boiler heat loss rate

$$Q_{loss} = Q_f (1 - \eta_b) = 2325(1 - 0.91) = 209.25 \text{ MW}$$

Condenser heat release rate

$$Q_{cw} = Q_f (\eta_b - \eta_o) = 2325(0.91 - 0.4) = 1185.75 \text{ MW}$$

Cooling water rate

$$m_{cw} = Q_{cw}/c_{pw}\Delta t_w = 1185.75 \times 10^6 \text{ W}/[1.16 \text{ Wh}/(\text{kg K}) \times 9.6 \text{ K}]$$
$$= 208.78 \times 10^6 \text{ kg/h}$$

Sulfur dioxide production rate

$$m_{SO_2} = 2Sm_f = 2 \times 0.035 \times 285,665.5 = 19,996.6 \text{ kg/h}$$

Sulfur dioxide production rate per kWh

$$m_{SO_2}/P_{el} = 19,996.6 \text{ kg/h}/930 \times 10^3 \text{ kW} = 0.0215 \text{ kg/kWh}$$

Sulfur dioxide allowable release rate

$$m_{SO_{2,a}} = 0.07 m_{SO_2} = 0.07 \times 19,996.6 \text{ kg/h} = 1399.8 \text{ kg/h}$$

Total sulfur dioxide collection rate

$$m_{SO_{2,c}} = (1 - 0.07)m_{SO_2} = 0.93 \times 19,996.6 \text{ kg/h} = 18,596.8 \text{ kg/h}$$

Mass of limestone required

$$m_{CaCO_3} = m_{SO_{2,c}} \times M_{CaCO_3}/M_{SO_2} = 18,596.8 \times 100.09/64.06 = 29,056.4 \text{ kg/h}$$

where $M_{CaCO_3} = 100.09 \text{ kg/(kg mol)} = $ molar mass of $CaCO_3$

$M_{SO_2} = 64.06 \text{ kg/(kg mol)} = $ molar mass of SO_2

$M_{CaSO_3} = 120.14 \text{ kg/(kg mol)} = $ molar mass of $CaSO_3$

Mass of $CaSO_3$ produced

$$m_{SO_{2,c}} M_{CaSO_3}/M_{SO_2} = 18,596 \times 8 \times 120.14/64.06 = 34,877 \text{ kg/h}$$

Mass of ash produced

$$m_a = 8766 m_f A = 8766 \text{ h/yr} \times 285,665.5 \text{ kg/h} \times 0.07 = 175.3 \times 10^6 \text{ kg/yr}$$

Annual particulate emissions

$$m_{part} = m_a(1 - \eta_{ep}) = 175.3 \times 10^6 \text{ kg/yr } (1 - 0.99) = 1.753 \times 10^6 \text{ kg/yr}$$

Table 3.3 contains data about efficiency of different combustion-related emissions control systems being developed within the U.S. Department of Energy's clean coal technology program.

Table 3.3. Combustion-related emissions control systems in the U.S. Clean Coal Technology program

Technology	Pollutant removal efficiency, %	
	SO_2	NO_x
SO_2 removal		
Advanced FGD	95	
NO_x removal		
Low-NO_x burners + gas reburning		70
Selective catalytic reduction		80
Combined SO_x and NO_x control		
Low-NO_x burners + gas reburning	50	70
Limestone injection + multistage burners	70	50
NO_x SO dry regenerable flue gas cleanup	97	70
Dry NO_x/SO_2 flue gas cleanup	70	80
SNO_x catalytic advanced flue gas cleanup	96	94
SNRB combined SO_x and NO_x control	85	90

FGD, flue gas desulfurization; SNRB, SO_x-NO_x Rox Box.

CLOSURE

The state-of-the-art pollutant emissions abatement techniques are thoroughly discussed in this chapter with emphasis on SO_2 and NO_x emissions control including precombustion fuel preparation, a partial in situ pollutant retention, and postcombustion flue gas cleaning.

As far as the NO_x emissions are concerned, the staged combustion approach realized in advanced combustors and burners or such technologies as fluidized bed combustion allow significant reductions in NO_x emissions. All the possible combustion NO_x control processes such as air staging, fuel staging, flue gas recirculation, low-NO_x burners, and low excess air mode of operation are considered. For final control of NO_x emissions, SCR and SNCR methods are applied to remove NO_x from the flue gas. Some innovative and efficient techniques for pollutant emissions control, which allow simultaneous removal of both SO_2 and NO_x from the flue gas, are described. The most advanced emissions control techniques allow removal of up to 95–98% of SO_2 and up to 90% and more of NO_x.

Thus this chapter creates a basis for the evaluation of the environmental impact of fossil fuel fired power plants of all types. Special issues related to gas turbine power plants, particularly those of combustion NO_x control, are thoroughly discussed in Chapter 5. Fluidized bed combustion technology with its intrinsic capability of reducing the SO_2 and NO_x emissions in situ is considered in Chapter 9.

PROBLEMS

3.1. A 1000-MW pulverized coal power plant is fired with (1) a bituminous coal and (2) a lignite having lower heating values of 31.8 and 9.63 MJ/kg, respectively. The plant overall efficiency is 42% and 40%, respectively. Calculate the specific and annual CO_2 emissions in kg per kWh of power output and in kg/yr, if the power plant capacity factor is 0.8. Assume that the carbon dioxide yield is 1.6 and 0.65 m^3 per kg of bituminous coal and lignite, respectively.

3.2. A power plant burns 8.85×10^4 kg/h of lignite with an S content of 1% (by mass of as-fired coal). The flue gas yield is 4.4 m^3 per kg fuel. An FGD plant reduces the SO_2 concentration in the flue gas to 180 mg/m^3. Calculate (a) the requirements in lime (CaO) per hour and (b) the hourly production of calcium sulfate ($CaSO_4$).

3.3. An advanced catalytic flue gas cleanup system removes 92% of NO from the flue gas of a power plant that burns 88,500 kg of lignite per hour. The wet flue gas volume is 4.4 m^3 per kg of coal. Calculate the hourly ammonia requirements if the initial NO concentration prior to catalytic cleanup is 1200 mg per m^3 of flue gas.

REFERENCES

Bald, A., and Heusinger, K. 1996. Power generation in advanced steam power plants relieves burden on environment. *Siemens Power J.* 1:5–11.

Bartok, W., and Sarofim, A. F. eds. 1991. *Fossil fuel combustion: A source book.* New York: Wiley.

Elliott, T. C., ed. 1989. *Standard handbook of powerplant engineering.* New York: McGraw-Hill.

Kumar, K. S., Feldman, P. L., and Jacobus, P. L. 1993. Pulse energization. *Proceedings American Power Conference,* pp. 1181–1185.

Lefebvre, A. H. 1995. The role of fuel preparation in low-emission combustion. *Trans. ASME J. Eng. Gas Turbine Power* 117:617–653.

Longwell, J. P., Rubin, E. S., and Wilson, J. 1995. Coal: Energy for the future. *Prog. Energy Combust. Sci.* 21:269–360.

Lurgi AG. 1992. CFB power plants. Germany.

Makansi, J. 1993. Reducing NO_x emissions from today's power plants. *Power* 137(5):11–28.

Smoot, L. D., ed. 1993. *Fundamentals of coal combustion for clean and efficient use.* Amsterdam: Elsevier.

Smoot, L. D., and Smith, P. J. 1985. *Coal combustion and gasification.* New York: Plenum.

Chapter Four

STEAM POWER PLANT TECHNOLOGY

This chapter reviews major issues related to steam power generation technology. The performance of a power plant is mainly characterized by its efficiency, power output, and heat rate. Economics and environmental impact of power plants became particularly important characteristics of power generation technologies.

The emphasis in this chapter is on the methods used in advanced steam power plants for efficiency enhancement. The general principle of efficiency improvement is based on an increase in the average heat addition temperature and a decrease in the average heat rejection temperature. The implementation of this principle includes using advanced steam conditions at the turbine inlet, steam reheating, and multiple regenerative feedwater heating along with a low condenser pressure.

Advanced coal-fired steam power plants apply supercritical steam parameters (up to 300 bars and 600°C), low condenser pressure (about 0.03–0.04 bar), single or double reheat, and eight to nine regenerative feedwater heaters. Thermodynamic considerations are suppported by numerous solved problems, each illustrating the effect of the efficiency enhancement methods listed above. The most advanced steam power plants achieve efficiencies of about 43% while firing lignites and about 45% while firing bituminous coals.

The major components such as steam generators, steam turbines, and condensers, along with heat rejection systems, are outlined in the final part of this chapter.

CARNOT CYCLE AND RANKINE CYCLE

Carnot Cycle

Conventional power plants, such as steam turbine or gas turbine power plants, convert heat released by combustion of fuel into useful work and electrical energy. The heat supplied from fuel burned in the furnace of a boiler or in the combustor of a gas turbine cannot be completely converted into useful work, and a certain portion of heat input is rejected to the ambient. Hence

$$W_{net} = Q_{in} - Q_{out} \quad \text{kJ} \tag{4.1}$$

where W_{net} is the plant net useful work, Q_{in} is the heat input to the plant, and Q_{out} is the rejected heat.

The most important criterion of performance of a power plant is the thermal efficiency, η_{th}. It is the measure of energy conversion in the power plant and is defined as the ratio of the net work output to the heat input:

$$\eta_{th} = W_{net}/Q_{in} \tag{4.2}$$

Alternatively, the thermal efficiency may also be expressed in terms of the power output P, in kW, and rate of heat addition Q_{in}, in kJ/s.

The most effective cycle for conversion of heat into work is the ideal reversible Carnot cycle with an ideal gas as working fluid. The Carnot cycle consists of four reversible processes (see Figure 4.1): isothermal (constant temperature) expansion 1-2 with a heat supply q_{in} from a hot

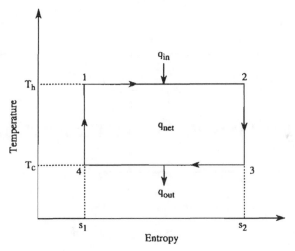

Figure 4.1. Carnot cycle.

heat source at a temperature T_h, isentropic (reversible adiabatic, constant entropy) expansion 2-3, isothermal compression 3-4 with a heat rejection q_{out} to cold surroundings at a temperature T_c, and isentropic compression 4-1.

The thermal efficiency of the Carnot cycle is given by

$$\eta_{th,C} = 1 - T_c / T_h \tag{4.3}$$

where T_h and T_c are the hot heat source and cold surrounding temperatures, respectively, in K. Equation (4.3) shows that in order to increase the thermal efficiency of the Carnot cycle, T_h must be increased and T_c decreased. For example, the Carnot thermal efficiency of an OTEC plant that uses ocean thermal energy to generate electricity with a temperature difference of only 20 K (surface water at $27°C = 300$ K and water in the ocean depth at $7°C = 280$ K) is only $1 - 280/300 = 0.067$. The Carnot thermal efficiency for a cycle using a heat source at a temperature of 1500 K and a heat sink at a temperature of 500 K is $1 - 500/1500 = 0.67$.

In a given temperature range, the Carnot cycle has the maximum thermal efficiency given by Eq. (4.3). Because of the irreversibilities, cycles of real heat engines, e.g., steam and gas turbines, internal combustion engines, have lower thermal efficiencies than the Carnot cycle efficiency in the same temperature range. Their efficiencies can be improved only by means of increasing the plant complexity and capital cost.

General Principle of Efficiency Enhancement

Figure 4.2 shows an arbitrary cycle ABCD that can be used for energy conversion. It represents in general the cycles of power plants such as steam and gas turbine power plants, or a diesel engine. The maximum temperature in the cycle ABCD is T_{max}, the minimum temperature is T_{min}, and the specific entropy change in the cycle is $\Delta s = s_2 - s_1$. The heat input to the cycle is q_{in}, and the rejected heat is q_{out}. This cycle has the same efficiency as an equivalent Carnot cycle 1234, which has the same quantities of added and rejected heat and the same entropy change as the cycle ABCD. The thermal efficiency of this cycle may be defined as the ratio of the net useful work, $w_{net} = q_{in} - q_{out}$, to the heat input, q_{in}. The heat quantities may be found as the integrals of the product of temperature T and specific entropy differential ds. They can also be found as the products of the corresponding mean temperature and specific entropy change. It is obvious that the thermal efficiency of the cycle ABCD, η_{th}, is equal to that of the equivalent Carnot cycle 1234.

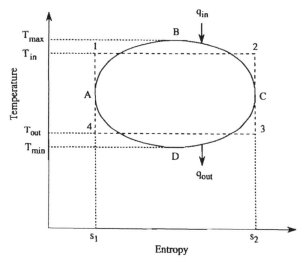

Figure 4.2. Arbitrary power cycle ABCD and equivalent Carnot cycle.

Thus

$$\eta_{th} = 1 - T_{out}/T_{in} \tag{4.4}$$

where T_{in} and T_{out} are mean temperatures of heat input to the cycle and heat rejection, in K.

Equation (4.4) represents the general principle of the thermal efficiency enhancement of power plants that employ heat engines as energy conversion devices. It may be formulated as follows. In order to increase the thermal efficiency of a heat engine cycle, the mean temperature of heat addition, T_{in}, must be raised, and the mean temperature of heat rejection, T_{out}, lowered. For example, an increase in T_{in} from 800 to 1200 K and a decrease in T_{out} from 350 to 290 K enhances the thermal efficiency of the ideal power cycle from 56.25% to 75%. Practical methods used to enhance the thermal efficiency depend on the power plant type and will be discussed in this chapter for steam power plants, in Chapter 5 for simple-cycle gas turbine plants, and in Chapter 6 for combined-cycle plants.

Rankine Cycle

The ideal Rankine cycle creates the thermodynamic basis for steam power plants [4.2, 4.4]. A schematic of the simple steam power plant depicted in Figure 4.3a operates on the Rankine cycle shown in Figure 4.3b on a T-s diagram. The plant consists of a steam generator, turbine/generator, condenser, and feed pump. It is seen that the Rankine cycle consists of the following processes: an isentropic steam expansion 1-2 in the turbine, isobaric-isothermal steam condensation 2-3 with heat rejection at a low pressure p_2, isentropic water compression 3-4 in the feed pump from p_2 to p_1, and isobaric heat addition 4-5-1 in a steam generator and superheater at a high pressure p_1. The steam expansion reversible 1-2s and irreversible 1-2 processes are shown on an h-s diagram in Figure 4.3c. Fuel is burned in the furnace of the steam generator, and thus steam is raised. The heat addition occurs at a constant pressure in the temperature range from the relatively low feedwater temperature to the highest cycle temperature, the live steam temperature. The conversion of thermal energy of steam into work occurs when the steam expands in a turbine that drives a generator delivering electrical power. The steam exiting from the turbine is condensed, and the condensate is returned by a feedwater pump into the boiler.

In order to find the thermal efficiency of the steam power plant, the net work and heat supplied to the cycle are required. The net specific work of the Rankine cycle per a unit mass of working

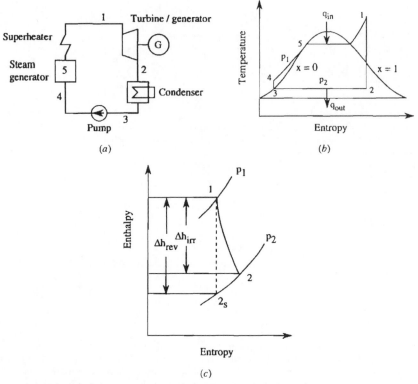

Figure 4.3. Rankine cycle power plant: (a) schematic diagram, (b) T-s diagram, and (c) reversible and irreversible steam expansion on an h-s diagram.

fluid steam is

$$w_{net} = w_t - w_p \approx h_1 - h_2 \quad \text{kJ/kg} \tag{4.5}$$

where w_t is specific turbine work, w_p is specific pump work, h_1 is live steam specific enthalpy, and h_2 is exhaust steam specific enthalpy. The value of w_p is much less than w_t, and thus it may be neglected in Eq. (4.5).

The heat addition (per a unit mass of working fluid steam) to the Rankine cycle is

$$q_{in} \approx h_1 - h_2' \quad \text{kJ/kg} \tag{4.6}$$

where h_2' is specific enthalpy of condensate (taken as the saturated liquid enthalpy at the condenser pressure p_2).

The thermal efficiency of the Rankine cycle is given by

$$\eta_{th} = w_{net}/q_{in} = (h_1 - h_2)/(h_1 - h_2') \tag{4.7}$$

The steam enthalpy values can be determined from the Mollier h-s diagram: h_1 at the live steam condition (pressure p_1 and temperature t_1), and at $s_2 = s_1$ on the $p_2 = \text{const}$ curve.

EFFICIENCY ENHANCEMENT METHODS FOR STEAM POWER PLANTS

As stated above, the efficiency of a power plant can be improved by raising the mean temperature of heat addition, T_{in}, and by lowering the mean temperature of heat rejection, T_{out} (Haywood, 1991; Kolle, 1994; Wark, 1994). In application to steam power plants, the mean temperature of heat addition can be increased by (1) raising live (main) steam pressure and temperature at the inlet of turbine, (2) regenerative feedwater heating, and (3) employing steam reheat.

The temperature of heat rejection in steam power plants can be lowered by reducing the condenser pressure, which in turn, depends on the cooling medium temperature. The condenser pressure is usually 0.04 bar in water-cooled condensers and 0.07–0.1 bar in air-cooled condensers. In advanced steam power plants, the condenser pressure can be as low as 0.03 bar or even lower, provided that cooling water with a temperature below 20°C is available (Author, 1995).

Thus the thermal efficiency of the simple Rankine cycle can be improved by increasing the live (main) steam temperature and pressure, by decreasing the condenser pressure, and by using one or two reheaters and several feedwater heaters. As explained above, the maximum temperature of live steam is limited by the materials used in the hot part of the steam generator (superheater, live steam supply line). The attainment of higher efficiencies by implementation of these methods requires additional components, such as multiple turbine-cylinders, additional heat transferring surfaces, and utilization of improved materials. Therefore advanced steam power plants are more complex and have higher capital costs. These additional investments should be assessed and balanced against resulting savings in fuel costs achieved through the improved efficiency.

REHEAT CYCLE

The live steam parameters used in advanced steam generators are as follows: pressure up to 250–300 bars and temperatures up to 580–650°C (Author, 1995). The effect of the live steam pressure and temperature on the thermal efficiency of the steam power plant is illustrated in Figure 4.4. At high live steam pressures, however, the quality x of the turbine exhaust steam deteriorates. It means that the moisture content, $1 - x$, of the wet exhaust steam increases, and this can shorten the turbine life because of turbine blade erosion.

To improve steam power plant performance, reheat can be used. Thereby, the quality x of the turbine exhaust steam improves, and the cycle thermal efficiency increases.

Figure 4.5a shows the steam turbine portion of a single-reheat steam power cycle. There are two turbine cylinders—high pressure (HP) and low pressure (LP). A steam reheater is placed between these two turbine cylinders. Consider a subcritical, single-reheat steam cycle shown on a T-s diagram in Figure 4.5b and on an h-s diagram in Figure 4.5c. The following energy quantities per kg of live steam may be determined.

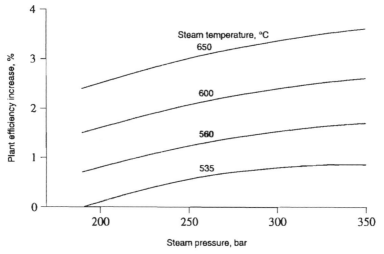

Figure 4.4. Enhancing the efficiency of the steam power plant by raising the main steam condition.

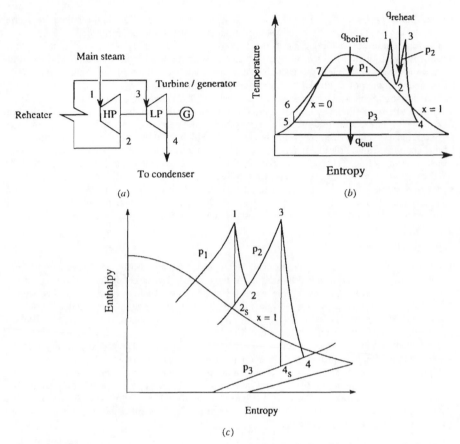

Figure 4.5. Single-reheat steam power plant: (*a*) schematic diagram (only turbine/generator and reheater are shown), (*b*) *T-s* diagram, and (*c*) *h-s* diagram.

Heat added to the cycle

$$q_{in} = h_1 - h_6 + h_3 - h_2 \quad \text{kJ/kg} \tag{4.8}$$

Turbine work

$$w_t = h_1 - h_2 + h_3 - h_4 \quad \text{kJ/kg} \tag{4.9}$$

Pump work

$$w_p = v_5(p_1 - p_3)/\eta_{ip} \quad \text{kJ/kg} \tag{4.10}$$

where v_5 is specific volume of saturated liquid at pressure p_3 and η_{ip} is pump isentropic efficiency (typically, 0.7).

Heat rejected

$$q_{out} = h_4 - h_5 \quad \text{kJ/kg} \tag{4.11}$$

Plant net specific work

$$w_{net} = w_t - w_p \quad \text{kJ/kg} \tag{4.12}$$

Cycle thermal efficiency

$$\eta_{th} = w_{net}/q_{in} \tag{4.13}$$

The actual specific enthalpies of the steam at states 2 and 4 are

$$h_2 = h_1 - (h_1 - h_{2s})\eta_{it} \quad kJ/kg \tag{4.14}$$

$$h_4 = h_3 - (h_3 - h_{4s})\eta_{it} \quad kJ/kg \tag{4.15}$$

where h_{2s} and h_{4s} are specific enthalpy (from h-s diagram) of steam after the isentropic expansion in HP and LP turbines, respectively, and η_{it} is isentropic efficiency of the steam turbine (typically, 0.9).

Figure 4.6a shows a schematic diagram of the steam turbine/generator portion of the double-reheat steam power plant. The turbine consists of three sections: HP, intermediate pressure (IP), and LP. Figure 4.6b presents the subcritical double-reheat steam power cycle on an h-s diagram. The analysis of the double-reheat cycle is similar to that of the single-reheat cycle.

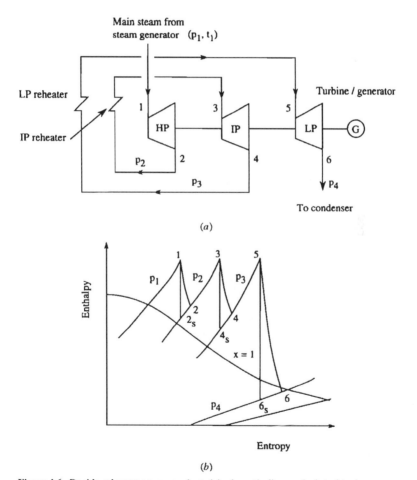

(a)

(b)

Figure 4.6. Double-reheat steam power plant: (a) schematic diagram (only turbine/generator with intermediate-pressure (IP) and low-pressure (LP) reheaters are shown) and (b) h-s diagram.

Ignoring the pump work, the heat added to the double-heat cycle and the cycle net specific work are given by

$$q_{in} = h_1 - h_4' + h_3 - h_2 + h_5 - h_4 \quad kJ/kg \tag{4.16}$$

$$w_{net} = h_1 - h_2 + h_3 - h_4 + h_5 - h_6 \quad kJ/kg \tag{4.17}$$

where h_1, h_2, h_3, h_4, h_5, and h_6 are specific enthalpy of steam at actual states 1, 2, 3, 4, 5, and 6, respectively, and h_4' is specific enthalpy of saturated water at the condenser pressure p_4. The cycle thermal efficiency is given by Eq. (4.13).

The following example presents an analysis of the double-reheat supercritical steam cycle.

Example 4.1

Calculate the net work, heat added, heat rejected, efficiency, and heat rate of a supercritical double-reheat power plant. The live steam pressure and temperature are 250 bars and 560°C, respectively. Two reheats occur at 40 bars to 560°C and at 6 bars to 580°C, respectively. Condenser pressure is 0.04 bar. Isentropic efficiencies of the HP turbine, IP turbine, LP turbine, and feed pump are 0.93, 0.93, 0.9, and 0.7, respectively.

Solution

Referring to Figures 4.6a and 4.6b, the specific enthalpies of steam at all states of the cycle are found from the h-s diagram of steam.

State 1 (HP turbine inlet):

$p_1 = 250$ bars $t_1 = 560°C$ $h_1 = 3370$ kJ/kg $s_1 = 6.22$ kJ/(kg K)

State 2 (HP turbine exit):

$p_2 = 40$ bars $h_{2s} = 2880$ kJ/kg $s_{2s} = s_1$

$h_2 = h_1 - (h_1 - h_{2s})\eta_{it} = 3370 - (3370 - 2880)0.93 = 2914.3$ kJ/kg

State 3 (IP turbine inlet):

$p_2 = 40$ bars $t_3 = 560°C$ $h_3 = 3580$ kJ/kg $s_3 = 7.26$ kJ/(kg K)

State 4 (IP turbine exit):

$p_3 = 6$ bars $h_{4s} = 3000$ kJ/kg $s_{4s} = s_3$ $p_{4s} = p_5 = 6$ bars

$h_4 = h_3 - (h_3 - h_{4s})\eta_{it} = 3580 - (3580 - 3000)0.93 = 3040.6$ kJ/kg

State 5 (LP turbine inlet):

$p_3 = 6$ bars $t_5 = 580°C$ $h_5 = 3655$ kJ/kg $s_5 = 8.22$ kJ/(kg K)

State 6 (LP turbine exit):

$p_4 = 0.04$ bars $h_{6s} = 2478$ kJ/kg $s_{6s} = s_5$ $p_{6s} = p_7 = 0.04$ bar

$h_6 = h_5 - (h_5 - h_{6s})\eta_{it} = 3655 - (3655 - 2478)0.9 = 2595.7$ kJ/kg

State 7 (pump inlet):

$p_7 = 0.04$ bar $h_7 = 137.77$ kJ/kg $v_7 = 0.00100395$ m³/kg (lookup in water/steam table)

State 8 (pump exit):

$p_1 = 250$ bars $h_8 = h_7 + w_{pump} = h_7 + v_7(p_1 - p_4)/h_{ip}$

$h_8 = 137.77 + 0.0010395(250 - 0.04) \times 10^2/0.7 = 137.77 + 37.12 = 174.89$ kJ/kg

Plant net specific work

$$w_{net} = w_{HP} + w_{IP} + w_{LP} - w_p = h_1 - h_2 + h_3 - h_4 + h_5 - h_6 - w_p$$

$$w_{net} = 3370 - 2914.3 + 3580 - 3040.6 + 3655 - 2595.7 - 37.12 = 2017.28 \text{ kJ/kg}$$

Heat addition

$$q_{in} = h_1 - h_8 + h_3 - h_2 + h_5 - h_4$$

$$q_{in} = 3370 - 174.89 + 3580 - 2914.3 + 3655 - 3040.6 = 4475.21 \text{ kJ/kg}$$

Thermal efficiency

$$\eta_{th} = w_{net}/q_{in} = 2017.28/4475.21 = 0.4508 = 45.08\%$$

STEAM POWER CYCLE WITH FEEDWATER HEATING

Open Feedwater Heaters

The feedwater is preheated by the heat of condensation of steam bled from the turbine and led to the feedwater heater. In the open-type feedwater heater, water is mixed with condensing steam, while in the closed-type, heat is transferred from condensing steam to water by means of a separating surface.

In the open-type or direct-contact feedwater heater, the feedwater leaves the heater at the saturation temperature corresponding to the extraction steam pressure. Figure 4.7a shows a schematic diagram of the turbine portion of the steam power plant with an open-type feedwater heater. Figures 4.7b and 4.7c show the supercritical reheat steam power cycle on T-s and h-s diagrams.

The condensate leaves the condenser as saturated liquid at pressure p_3 (state 4) and is pressurized to the pressure p_2 of the extraction steam at state 2 and pumped into the feedwater heater (state 5). The steam and water mix together in the feedwater heater and produce saturated water (state 6).

Figure 4.7d shows a schematic diagram of an open-type feedwater heater. The energy balance for the open-type feedwater heater is given by

$$ah_2 + (1 - a)h_5 = h_6 \tag{4.18}$$

where a is mass fraction of steam extracted from the turbine per kg of steam entering the turbine, h_2 is specific enthalpy of the steam, and h_5 and h_6 are specific enthalpy of the feedwater at the heater inlet and exit, respectively. The value of h_6 is equal to the specific enthalpy of the saturated liquid at the pressure in the feedwater heater, i.e., at the pressure of extracted steam.

From Eq. (4.18), the mass fraction of steam bled from the turbine is

$$a = (h_6 - h_5)/(h_2 - h_5) \tag{4.19}$$

Referring to Figure 4.7b, the specific energy quantities per kg steam at the turbine inlet (state 1) may be determined as follows.

Heat added to the cycle

$$q_{in} = h_1 - h_7 \quad \text{kJ/kg} \tag{4.20}$$

Turbine work

$$w_t = (h_1 - h_2) + (1 - a)(h_2 - h_3) \quad \text{kJ/kg} \tag{4.21}$$

Pump work

$$w_p = (1 - a)(h_5 - h_4) + (h_7 - h_6) = (1 - a)v_4(p_2 - p_3) + v_6(p_1 - p_2) \quad \text{kJ/kg} \tag{4.22}$$

where v is specific volume of saturated liquid at the corresponding pressure (p_3 for v_4 and p_2 for v_6).

Heat rejected

$$q_{out} = (1 - a)(h_3 - h_4) \quad \text{kJ/kg} \tag{4.23}$$

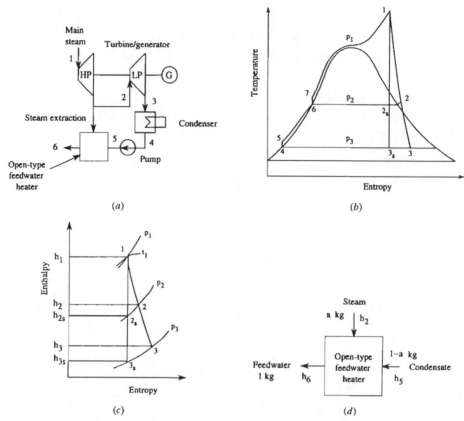

Figure 4.7. Regenerative steam power plant cycle: (*a*) schematic diagram (only turbine/generator and open-type feedwater heater are shown), (*b*) supercritical regenerative steam cycle on a *T*-*s* diagram, (*c*) steam expansion process on a *h*-*s* diagram, and (*d*) flow diagram of open-type feedwater heater.

Plant net specific work

$$w_{\text{net}} = w_t - w_p \quad \text{kJ/kg} \tag{4.24}$$

Cycle thermal efficiency

$$\eta_{\text{th}} = w_{\text{net}}/q_{\text{in}} \tag{4.25}$$

The actual steam specific enthalpies at states 2 and 3 must be calculated to account for the irreversibilities of steam expansion in the turbine. Thus

$$h_2 = h_1 - (h_1 - h_{2s})\eta_{\text{it}} \quad \text{kJ/kg} \tag{4.26}$$

$$h_3 = h_1 - (h_1 - h_{3s})\eta_{\text{it}} \quad \text{kJ/kg} \tag{4.27}$$

where h_{2s} and h_{3s} are specific enthalpy (from h-s diagram) of steam after the isentropic expansion from live steam state 1 to the extraction pressure p_2 and condenser pressure p_3, respectively, and η_{it} is steam turbine isentropic efficiency.

If two or more feedwater heaters are used, then the energy balances will be written for all the feedwater heaters, beginning with the feedwater heater having the highest steam pressure.

For any number of feedwater heaters there will be as many equations as there are unknowns. Each open feedwater heater requires a pump. Thus, for a plant with n open-type feedwater heaters, n pumps are needed in addition to the condensate pump, and a significant parasitic energy for the pump drive is required. Therefore, normally, one open-type feedwater heater and five to eight closed-type heaters are used in modern large power plants. The feedwater is preheated up to 280–325°C. The open-type feedwater heater serves as a deaerator, which removes the dissolved gases (air containing O_2) from the feedwater to prevent corrosion of boiler (Weston, 1992; Author, 1995).

Closed-Type Feedwater Heaters

Closed-type feedwater heaters are made as vertical or horizontal shell-and-tube heat exchangers with feedwater flow passing through the tubes, and the steam on the shell side. The steam is condensed on the outer surface of the tube bundle, and the heat of condensation heats the feedwater.

Usually, several closed-type feedwater heaters are employed in a steam turbine unit. The energy balance equation should be written for each closed-type feedwater heater in succession, starting with the feedwater heater having the highest steam pressure. There will be as many equations as there are unknown mass fractions of steam bled from the turbine. Consider the ith closed-type feedwater heater that receives the steam bled at a pressure p_i with an enthalpy h_i and increases the feedwater enthalpy from h_{wi} to h_{we} (see Figure 4.8a). This feedwater heater also receives condensate drained from the $(i + 1)$th closed-type feedwater heater, downstream of the ith heater, at a higher pressure. The mass flow rate of condensate leaving the ith heater is the sum of the mass flow rates of steam extracted for the two feedwater heaters, $a_i + a_{i+1}$.

Referring to Figure 4.8a, an energy balance on the ith feedwater heater per kg of steam entering the turbine may be written as follows:

$$h_{we} - h_{wi} = a_i(h_i - h_{ci}) + a_{i+1}(h_{c,i+1} - h_{ci}) \quad \text{kJ/kg} \tag{4.28}$$

where h_{wi} and h_{we} are feedwater enthalpy at the inlet and exit of the ith heater, respectively, in

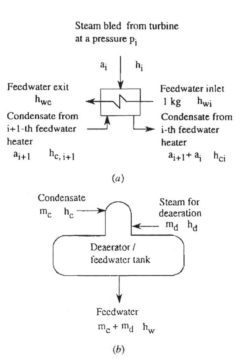

Steam bled from turbine
at a pressure p_i

a_i | h_i

Feedwater exit
h_{we}

Feedwater inlet
1 kg h_{wi}

Condensate from
i+1-th feedwater
heater
a_{i+1} $h_{c,i+1}$

Condensate from
i-th feedwater
heater
$a_{i+1} + a_i$ h_{ci}

(a)

Condensate
m_c h_c

Steam for
deaeration
m_d h_d

Deaerator /
feedwater tank

Feedwater

$m_c + m_d$ h_w

(b)

Figure 4.8. Flow diagrams of (a) closed-type feedwater heater and (b) deaerator.

kJ/kg; a_i and a_{i+1} are mass fractions of steam bled from turbine for the ith and $(i + 1)$th feed-water heaters, respectively, in kg per kg live steam; h_i is enthalpy of the bled steam in kJ/kg; and h_{ci} and $h_{c,i+1}$ are condensate enthalpy at the pressure in the ith and $(i + 1)$th feedwater heaters, respectively, in kJ/kg.

Provided that a_{i+1} and $h_{c,i+1}$ are known from the energy balance for the $(i + 1)$th feedwater heater, the mass fraction of steam bled at p_i is given by

$$a_i = [h_{wc} - h_{wi} - a_{i+1}(h_{c,i+1} - h_{ci})]/(h_i - h_{ci}) \tag{4.29a}$$

Figure 4.8b shows a deaerator/feedwater tank with corresponding enthalpies and mass flows of condensate, steam, and feedwater. The deaerator is an open-type feedwater heater, and its energy balance is given by

$$h_c m_c + h_d m_d = h_w(m_c + m_d) \tag{4.29b}$$

It follows for the steam mass flow rate:

$$m_d = m_c(h_w - h_c)/(h_d - h_w) \tag{4.29c}$$

Combination of Closed-Type and Open-Type Feedwater Heaters

Open-type feedwater heaters are thermodynamically more effective than closed-type feedwater heaters, since there is no temperature difference between the saturation temperature and exit temperature of feedwater. However, the open-type feedwater heaters need as many pumps as there are feedwater heaters plus one, and for a large number of feedwater heater stages (7–9) the energy consumption for feedwater pumping is therefore much higher than that for the utilization of closed-type feedwater heaters.

Except for the feedwater heater-deaerator, all other feedwater heaters are of the closed type. They are placed upstream and downstream of the deaerator and are classified as LP and HP feedwater heaters (see Figure 4.9).

Consider a steam power plant with two feedwater heaters, one of which is an HP closed-type feedwater heater and the second an LP open-type feedwater heater (see Figure 4.10a). Figure 4.10b shows the supercritical reheat cycle with the two feedwater heaters on a T-s diagram. Referring

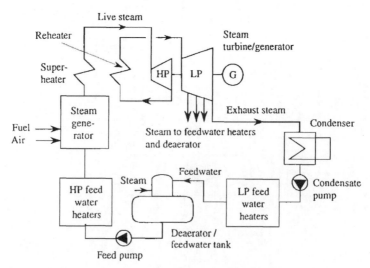

Figure 4.9. Regenerative steam power plant cycle with a block of low-pressure (LP) feedwater heaters, deaerator-feedwater tank, and block of high-pressure (HP) feedwater heaters.

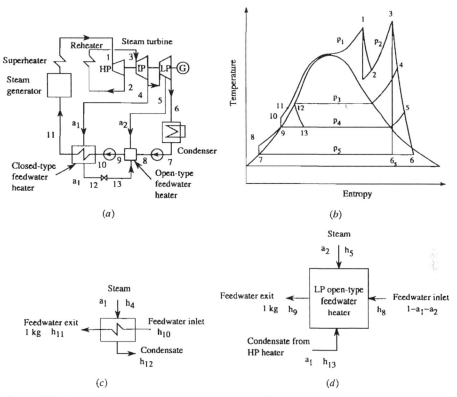

Figure 4.10. Supercritical reheat steam power plant cycle with one open-type feedwater heater and one closed-type feedwater heater: (a) plant schematic diagram, (b) cycle on a T-s diagram, (c) flow diagram of high-pressure (HP) closed-type heater, and (d) flow diagram of low-pressure (LP) open-type heater.

to Figure 4.10c, the energy balance on the HP closed-type feedwater heater is given by

$$a_1(h_4 - h_{12}) = h_{11} - h_{10} \tag{4.30}$$

where a_1 is mass fraction of steam bled from turbine at pressure p_3 for the HP feedwater heater, h_4 is specific enthalpy of the steam, and h_{10} and h_{11} are specific enthalpy of the feedwater at the inlet and exit of the HP heater, respectively.

The mass fraction of the steam bled for the HP feedwater heater is

$$a_1 = (h_{11} - h_{10})/(h_4 - h_{12}) \tag{4.31}$$

Referring to Figure 4.10d, the energy balance on the LP open-type feedwater heater is given by

$$a_2 h_5 + (1 - a_1 - a_2)h_8 + a_1 h_{13} = h_9 \tag{4.32}$$

where a_2 is mass fraction of steam bled from the turbine for the LP feedwater heater, h_5 is specific enthalpy of the steam, h_8 and h_9 are specific enthalpy of the feedwater at the inlet and exit of the LP heater, respectively, and h_{13} is specific enthalpy of the condensate from the HP heater.

Thus the mass fraction of the steam bled from the turbine for the LP feedwater heater is

$$a_2 = [h_9 - a_1 h_{13} - (1 - a_1)h_8]/(h_5 - h_8) \tag{4.33}$$

The feedwater exit temperature in closed-type heaters cannot reach the bled steam saturation temperature. The minimum difference between the saturation temperature of bled steam and the feedwater temperature is called the pinch point (PP). The value of PP is an important design

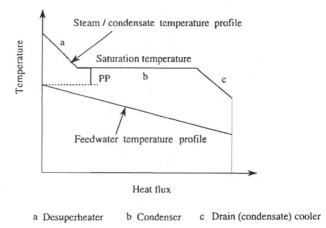

a Desuperheater   b Condenser   c Drain (condensate) cooler

Figure 4.11. Temperature–heat flux diagram of a high-pressure feedwater heater (PP is pinch point).

parameter, which affects the size and cost of feedwater heaters as well as cycle efficiency. At a very small value, the heater surface area is larger, and thus the heater is more expensive. At a higher value of PP, the exit temperature is lower, and therefore cycle efficiency is decreased.

For LP feedwater heaters that receive wet or dry saturated steam, the PP is 2–3 K (positive). In an HP feedwater heater receiving superheated steam bled from the turbine, the exit water temperature can be higher than the saturation temperature of the steam, and the PP can thus be negative (between 0 and −3 K). To cool the steam to the saturation temperature, a desuperheater is employed. The condensate leaving this heater can be slightly cooled below the saturation temperature in a drain cooler. All this enables a more effective utilization of energy in the closed-type feedwater heater. Thus the HP heater consists of a desuperheater, condenser, and drain (condensate) cooler. Figure 4.11 shows the temperature profiles of steam/condensate and feedwater.

ADVANCED SUPERCRITICAL REHEAT CYCLES

Advanced steam power plants, especially supercritical-pressure cycles, invariably use single reheat and often double reheat (Author, 1995; Pruschek and Oeljeklaus, 1996). An example of a supercritical reheat steam power plant cycle with blocks of HP and LP feedwater heaters and a deaerator is shown in Figure 4.9. Figure 4.10b shows a T-s diagram for a plant with only two feedwater heaters. In the supercritical reheat cycle the feedwater is pressurized beyond the critical pressure of 221.2 bars, and the heat is added by burning fuel in the boiler furnace. When the critical temperature of 374.15°C is attained by the heat addition in the boiler, the phase change from water to steam occurs without changing the density of working fluid, and the steam is then superheated to a live steam temperature. The current live steam parameters reach 560–600°C and 250–300 bars (Author, 1995; Kehr, 1994; Kather et al., 1994; Bald and Heusinger, 1996; Klein et al., 1996; Pruschek and Oeljeklaus, 1996). Only once-through steam generators (see below) can be employed in supercritical-pressure power plants.

At a high live steam pressure, the turbine exhaust steam quality x may be poor. To increase the value of x and to thus prevent blade erosion in the last stages of the LP steam turbine, reheat is used. In order to improve the cycle efficiency by applying reheat, a proper choice of reheat pressure is required. The reheat temperature is usually equal to the live steam temperature t_1. Since the reheat pressure is much lower than the live steam pressure, the reheat temperature may also be 20–30°C higher than t_1.

Analysis of a real power plant with double reheat and from seven to nine stages of feedwater heating is rather cumbersome. However, the analysis methodology is evident from the following example, which considers a power plant with one-stage reheat and two stages of feedwater heating in an HP closed-type feedwater heater and an LP open-type feedwater heater.

Referring again to Figure 4.10b, the energy quantities per kg steam at the turbine inlet (state 1) may be written as follows.

Heat added in boiler and reheater

$$q_{in} = h_1 - h_{11} + h_3 - h_2 \quad \text{kJ/kg} \tag{4.34}$$

Turbine work output

$$w_t = (h_1 - h_2) + (h_3 - h_4) + (1 - a_1)(h_4 - h_5) + (1 - a_1 - a_2)(h_5 - h_6) \quad \text{kJ/kg} \tag{4.35}$$

or

$$w_t = h_1 - h_2 + h_3 - h_6 - a_1(h_4 - h_5) - a_2(h_5 - h_6) \quad \text{kJ/kg} \tag{4.36}$$

Pump work consumption

$$w_p = (1 - a_1 - a_2)(h_8 - h_7) + (h_{10} - h_9) \quad \text{kJ/kg} \tag{4.37}$$

or

$$w_p = (1 - a_1 - a_2)v_7(p_4 - p_5) + v_9(p_1 - p_4) \quad \text{kJ/kg} \tag{4.38}$$

where v is specific volume of saturated liquid at the corresponding pressure (p_5 for v_7 and p_4 for v_9).

Heat rejected

$$q_{out} = (1 - a_1 - a_2)(h_6 - h_7) \quad \text{kJ/kg} \tag{4.39}$$

Cycle net specific work

$$w_{net} = w_t - w_p \quad \text{kJ/kg} \tag{4.40}$$

Cycle thermal efficiency

$$\eta_{th} = w_{net}/q_{in} \tag{4.41}$$

To attain the maximum cycle efficiency, an optimum pressure choice for extracted steam is required. This may be found by dividing into equal points the total enthalpy difference between the HP feedwater heater aimed to achieve the feedwater temperature of approximately 280–325°C and the condenser pressure. It is convenient to do this by means of the h-s diagram.

In some cases the condensate from the feedwater heater drain may be pumped forward into the downstream feedwater instead of being drained via a throttling valve to an LP feedwater heater or to the turbine condenser.

Example 4.2

Calculate the specific net work and the thermal efficiency for a power plant with an electric power output P_{el} of 900 MW. The plant operates on a single-reheat supercritical cycle with two feedwater heaters. Compare the turbine exhaust steam quality for a simple Rankine cycle and for the assigned plant.

The operation conditions are

Turbine inlet steam	250 bars and 560°C
Reheat	40 bars and 560°C
HP closed-type feedwater heater pressure	10 bars
LP open-type feedwater heater pressure	1.2 bars
Condenser pressure	0.05 bar
Isentropic efficiency of steam turbine η_{it}	0.9
Isentropic efficiency of feed pump η_{ip}	0.7
Energy losses in the mechanical drive and generator η_{mg}	0.97

Solution

The schematic and T-s diagrams are given in Figures 4.10a and 4.10b; the feedwater heater schematics are shown in Figures 4.10c and 4.10d.

The HP turbine steam enthalpies at states 1 and 2 (from h-s diagram and steam/water table) are

$$h_1 = 3370 \text{ kJ/kg}, h_{2s} = 2880 \text{ kJ/kg}$$

(from h-s diagram for isentropic expansion) and

$$h_2 = h_1 - (h_1 - h_{2s})\eta_{it} = 2929 \text{ kJ/kg}$$

(adiabatic expansion).

Similarly, the actual steam enthalpies at states 4, 5, and 6 are

$$h_4 = h_3 - (h_3 - h_{4s})\eta_{it} = 3175.2 \text{ kJ/kg}$$
$$h_5 = h_3 - (h_3 - h_{5s})\eta_{it} = 2759.4 \text{ kJ/kg}$$
$$h_6 = h_3 - (h_3 - h_{6s})\eta_{it} = 2349 \text{ kJ/kg}$$

Water enthalpies are based on the parameters of saturated liquid taken from the steam/water table. The respective water enthalpies at the condensate pump exit after isentropic and adiabatic compression are

$$h_{8s} = h_7 + v_7(p_4 - p_5) = 137.9 \text{ kJ/kg}$$
$$h_8 = h_7 + (h_{8s} - h_7)/\eta_{ip} = 137.96 \text{ kJ/kg}$$

Similarly, for the water enthalpy at the feed pump exit,

$$h_{10s} = h_9 + v_9(p_1 - p_4) = 465.42 \text{ kJ/kg}$$
$$h_{10} = h_9 + (h_{10s} - h_9)/\eta_{ip} = 476.59 \text{ kJ/kg}$$

The enthalpy of saturated liquid at 10 bars is $h_{12} = 762.61$ kJ/kg, and hence $h_{13} = h_{12}$ (due to throttling).

From the energy balance for the closed feedwater heater, the following relation for the steam extraction fraction is obtained:

$$a_1 = (h_{11} - h_{10})/(h_4 - h_{12}) = 0.124$$

The feed pump exit water temperature and enthalpy (at $p_{11} = p_1 = 250$ bars) are

$$t_{11} = t_{12} = 179.88°C$$

(saturation temperature at $p_{12} = 10$ bars, since approach is equal to zero) and

$$h_{11} = 775.8 \text{ kJ/kg}$$

(liquid at 250 bars and 179.88°C; look up in the water/steam table).

From the energy balance for the open-type feedwater heater (see Figure 4.10d),

$$a_2h_5 + (1 - a_1 - a_2)h_8 + a_1h_{13} = h_9$$

the steam extraction fraction is

$$a_2 = [h_9 - a_1h_{13} - (1 - a_1)h_8]/(h_5 - h_8) = 0.0854$$

Then the specific turbine work and the specific pump work, are

$$w_t = h_1 - h_2 + h_3 - h_6 - a_1(h_4 - h_6) - a_2(h_5 - h_6) = 1536.5 \text{ kJ/kg}$$
$$w_p = [(1 - a_1 - a_2)(h_8 - h_7) + (h_{10} - h_9)] = 37.38 \text{ kJ/kg}$$

Table 4.1. Thermodynamic properties of steam and water

State	Pressure p, bars	Temperature t, °C	Specific volume v, m³/kg	Enthalpy h, kJ/kg	Entropy s, kJ/(kg K)
1	250	560	0.013	3370	6.225
2s	40	275	0.055	2880	6.225
2	40	290	0.057	2929	
3	40	560	0.095	3582	7.260
4s	10	337	0.28	3130	
4	10	358	0.285	3175.2	7.260
5s	1.2	104.81	1.4	2668	7.260
5	1.2	142	1.6	2759.4	
6s	0.05	32.9	25	2212	7.260
6	0.05	32.9	25	2349	
7	0.05	32.9	0.0010052	137.77	
8s	1.2			137.9	
8	1.2			137.92	
9	1.2	104.8	0.0010476	439.36	
10s	250			465.42	
10	250			470.02	
11	250	179.88		775.8	
12	10	179.88	0.0011274	762.61	
13	1.2	104.81		762.61	

Therefore the plant net specific work is

$$w_{net} = w_t - w_p = 1497.03 \text{ kJ/kg}$$

The heat input per kilogram steam is

$$q_{in} = h_1 - h_{10} + h_3 - h_2 = 3546.41 \text{ kJ/kg}$$

The plant thermal efficiency is

$$\eta_{th} = w_{net}/q_{in} = 1497.03/3546.41 = 0.422$$

The mass flow rate of steam required to achieve the plant power output is

$$m_s = P_{el}/(w_{net}\eta_{m,el}) = 900 \times 10^3 \text{ kW}/(1497.03 \text{ kJ/kg} \times 0.97) = 619.78 \text{ kg/s}$$

The thermodynamic properties of steam/water at all states of the reversible and irreversible cycles are summarized in Table 4.1. Comparing the exhaust steam quality (condenser pressure is 0.05 bar) yields the following.

No reheat (isentropic expansion)

$$h_{2s} = 1902 \text{ kJ/kg} \qquad h_f = 137.77 \text{ kJ/kg} \qquad h_{fg} = 2423.8 \text{ kJ/kg}$$

$$x_{2s} = (1902 - 137.77)/2423.8 = 0.728$$

Reheat (isentropic expansion)

$$h_{6s} = 2212 \text{ kJ/kg} \qquad x_{6s} = (h_{6s} - h_f)/h_{fg} = (2212 - 137.74)/2423.8 = 0.856$$

Reheat (irreversible expansion)

$$h_6 = 2349 \text{ kJ/kg}$$

$$x_6 = (h_6 - h_f)/h_{fg} = (2349 - 137.77)/2423.8 = 0.912$$

POWER PLANT PERFORMANCE CRITERIA

Plant Net Power Output and Efficiency

Power plant performance is characterized by power output, efficiency, and heat rate. The plant gross (mechanical) power output is given by

$$P_{gross} = m w_{net} \quad kW \tag{4.42}$$

where m is the steam mass flow rate in kg/s and w_{net} is the cycle net specific work in kJ/kg.

The plant net (electrical) power output is less than the gross power output because of the energy losses in the boiler and in the turbine/generator train and because of the auxiliary energy consumption of pumps, induced-draft and forced-draft fans, coal crushers and pulverizers, and other motor-driven components (Author, 1972). Thus

$$P_{net} = P_{gross} - P_{loss} - P_{aux} \quad kW \tag{4.43}$$

Thus the overall efficiency of the steam power plant is given by

$$\eta_o = P_{net}/m_f HV \tag{4.44}$$

where m_f is the fuel mass flow rate in kg/s and HV is the heating value of the fuel.

The overall efficiency may be related either to HHV or LHV of the fuel. In the latter case it will be about 2% higher. Alternatively,

$$\eta_o = \eta_b \eta_{th} \eta_{it} \eta_m \eta_g \eta_{aux} \tag{4.45}$$

where η_b is the efficiency of boiler or combustor, η_{th} is the thermal efficiency of the reversible cycle, η_{it} is the isentropic efficiency of the turbine, η_m is the mechanical efficiency of the turbine/generator train, η_g is the electrical generator efficiency, and η_{aux} is the fraction of the auxiliary energy consumption (parasitic or service energy).

The overall efficiency of a power plant must be maximized in order to minimize the specific fuel energy consumption per a unit of the net work output:

$$SFC = 3600 m_f/P_{net} \quad kg/kWh \tag{4.46}$$

The major energy losses are the conversion losses in the cycle, which are accounted for by the thermal efficiency.

Plant Net Heat Rate

In addition to the thermal efficiency η_{th}, the performance of power plant is expressed in terms of plant net heat rate HR (Elliott, 1989; Li Kam and Priddy, 1985; Ordys et al., 1994; El-Wakil, 1984). It is defined as the ratio of heat input rate, Q_{in}, to plant net power output, P_{net}:

$$HR = 3600 Q_{in}/P_{net} \quad kJ/kWh \tag{4.47}$$

The heat rate is related to the thermal efficiency by

$$HR = 3600/\eta_{th} \quad kJ/kWh \tag{4.48}$$

or

$$HR = 3412/\eta_{th} \quad BTU/kWh \tag{4.49}$$

The heat rate as a measure of the plant performance is applied to all types of power plants, to both simple-cycle and combined-cycle power plants.

As the power plant load decreases, the heat rate increases. For a part-load operation the incremental heat rate (IHR) is used (Li Kam and Priddy, 1985). It is defined as the amount of heat required to generate an additional unit of power output at any given load P_L:

$$IHR = 3600 d(Q_{in})/dP_L = d(P_L HR)/dP_L \quad kJ/kWh \tag{4.50}$$

Table 4.2. Plant net heat rate and maximum efficiency of advanced power plants

Power plant type	Unit capacity, MW	Heat rate, kJ/kWh	Maximum efficiency, %
Steam power plants (lignite, bituminous coal)	300–1000	8370–9000	40–43
Steam power plants (gas, oil)	200–800	8180–8570	42–44
Simple-cycle gas turbine power plants	50–240	9110–9470	38–39.5
Combined-cycle power plants (STAG)	50–1000	6210–6545	55–60
Gas/diesel engine power plants	10–50	9230–10,600	34–39

or

$$IHR = HR + P_L dHR/dP_L \qquad (4.51)$$

The IHR is the slope of the input-output curve. As the plant net power output decreases, the heat rate increases and the IHR decreases. Table 4.2 indicates the highest values of plant net heat rate and thermal efficiency (based on LHV) of most advanced power plants of various types (1995 state-of-the-art power plants).

Advanced coal-fired steam power plants employ all the above discussed techniques for enhancement of performance. The following efficiency improvements can be achieved (Author, 1995; Kehr, 1994):

- Enhancement of main steam pressure from 165 bars (2400 psia) to 310 bars (4500 psia) results in a heat rate improvement of 2.8%.
- An increase in the main and reheat steam temperature from 535°C (1000°F) to 600°C (1110°F) at a pressure of 310 bars (4500 psig) provides a further heat rate improvement of 2.4%.
- A decrease in the condenser pressure from 0.065 bar (1.9 in Hg abs.) to 0.03 bar (0.9 in Hg abs.) improves the heat rate by 2.5%.
- A single reheat improves efficiency by 1.5%, and a double reheat by 1.2% as compared with the single-reheat plant.

The following example presents a 900 MWe lignite-fired steam power plant planned for construction by RWE Energie, Neurath, Germany. The net overall efficiency would be 42.8% at the main steam condition of 260 bars and 550°C, reheat at 46.5 bars to 580°C, and condenser pressure of 0.034 bar. The plant net heat rate would reach 8415 kJ/kWh. The main design parameters of the power plant are as follows: steam rate 2300 t/h, furnace capacity 1730 MW, boiler thermal capacity 1600 MW, boiler efficiency 92.6%, turbine efficiency 91% (HP) and 93.3% (LP), electrical power output (gross/net) 925/880 MW, heat rejection system—two wet cooling towers 162 m high, cooling water rate ~28,000 kg/s, cooling zone 9 K (Kehr, 1994).

Although it is still possible to improve the performance of conventional coal-fired power plants by raising steam parameters and by reducing the cold-end losses, only changing over to combined-cycle plants with either integrated gasification or pressurized combustion can radically improve the conversion efficiency of fuel energy to electric power.

Example 4.3

Calculate the gross and net overall efficiency and the net heat rate of a steam power plant with a net electrical power output of 900 MW. Assume that the boiler efficiency η_b is 0.93, the cycle thermal efficiency η_{th} is 0.5, the isentropic turbine efficiency η_{it} is 0.94, the mechanical efficiency η_m is 0.99, and the generator efficiency η_g is 0.99.

The auxiliary power consumption P_{aux} of the power plant is 54 MW, or 6% of the electrical power output.

Solution

The plant gross overall efficiency is

$$\eta_{gross} = \eta_b \eta_{th} \eta_{it} \eta_m \eta_g = 0.93 \times 0.5 \times 0.94 \times 0.99 \times 0.99 = 0.43$$

The plant net overall efficiency is

$$\eta_{net} = \eta_{gross}(1 - \eta_{aux}) = 0.43 \times (1 - 0.06) = 0.404$$

The plant net heat rate is

$$HR = 3600/\eta_{net} = 8902 \text{ kJ/kWh}$$

or

$$HR = 3412/\eta_{net} = 8437 \text{ BTU/kWh}$$

Example 4.4

Calculate the specific fuel consumption, the heat rate, and the flue gas flow rate for a coal-fired power plant with an electrical power output P_{el} of 900 MW if the plant net overall efficiency η_{net} of the power plant is 42%, the wet flue gas volume V_g is 9.85 m³/kg coal, and the LHV of bituminous coal is 30.11 MJ/kg.

Solution

Heat rate of the power plant is

$$HR = 3600/\eta_{net} = 3600/0.42 = 8571 \text{ kJ/kWh}$$

Fuel consumption of the power plant is

$$m_f = P_{el}/(LHV\eta_{net}) = 900/(30.11 \times 0.42) = 71.17 \text{ kg/s} = 71.17 \times 3.6 = 256.2 \text{ t/h}$$

Specific fuel consumption of the power plant is

$$SFC = 3600 m_f/P_{el} = 3600 \times 71.17/900 = 0.285 \text{ kg/kWh}$$

Hourly flue gas flow is

$$V_{g,h} = V_g \times SFC \times P_{el} = 9.85 \text{ m}^3/\text{kg} \times 0.285 \text{ kg/kWh} \times 0.9 \times 10^6 \text{ kW} = 2.526 \times 10^6 \text{ m}^3/\text{h}$$

Plant Capacity Factor

Capacity factor (CF) of a power-generating plant is defined as the ratio of the average power output of a power plant over a year to its rated power output. It can also be calculated as the ratio of the annual power generation in kWh to the product of the plant rated power output in kW and the duration of the year, i.e., 8760 hours. This means that the plant capacity factor is the ratio of the duration (in hours) of the annual plant operation at a rated power to 8760 hours.

According to the value of the plant CF, power plants are classified as follows.

1. Base-load power plants (over 6000 full-power hours per year, CF > 68%) usually have high rated power output, high capital costs, and low operating costs. Large coal-fired and nuclear power plants are base-load power plants.
2. Intermediate-load power plants (more than 2500 and less than 6000 full-power hours per year, CF between 28 and 68%) include some older less efficient plants.
3. Peaking units (less than 2500 full-power hours per year, CF < 28%) usually have low capital costs and high operating costs. Gas turbine, diesel engine, and pumped-storage power plants are employed as peaking units.

In recent years, highly efficient combined-cycle plants based on gas turbines have been developed and built in many countries to be employed as intermediate-load power plants and even as base-load power plants.

STEAM GENERATORS, STEAM TURBINES, AND HEAT REJECTION SYSTEMS

Boilers and Steam Generators

There are two general categories of boilers: fire tube and water tube. Modern high-pressure, high-temperature steam generators are all water-tube boilers, where the working fluid (water and steam) flows inside the tubes. Water-tube boilers are classified as natural circulation and forced circulation, and once-through or universal pressure (UP) boilers (Elliott, 1989; El-Wakil, 1984) (see Figures 4.12a–4.12c). The natural-circulation circuit consists of riser and downcomer tubes, a boiler drum, and a header (mud drum) (see Figure 4.13). The water flows from the boiler drum through several unheated downcomer tubes into the mud drum. Then it flows through the

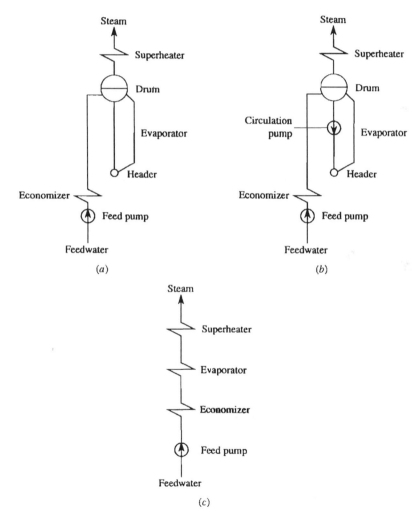

Figure 4.12. Simplified schematic diagrams of modern steam generator types: (a) natural-circulation boiler (pressure up to 18 MPa), (b) combined-circulation boiler (La Mont type, pressure up to 20 MPa), and (c) once-through boiler (Benson type, supercritical and subcritical pressure).

Boiler drum

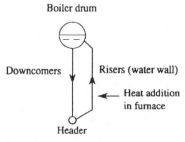

Downcomers   Risers (water wall)

                   ← Heat addition
                     in furnace

Header

Figure 4.13. Natural-circulation loop.

riser tubes arranged as tube walls of the boiler furnace. Inside these tubes the partial evaporation of water occurs, so that a water-steam mixture is formed, which has a lower density than the density of water in the downcomer tubes. The driving force that overcomes all the friction losses of the natural-circulation circuit is proportional to the density difference, $\rho_d - \rho_r$, between the fluids inside the downcomer and riser tubes and to the height H of the natural-circulation circuit:

$$\Delta p = g(\rho_d - \rho_r)H \quad \text{Pa} \tag{4.52}$$

The maximum heat flux in the evaporator of a natural-circulation boiler is about 0.4 MW/m². To maintain adequate cooling capacity of the natural-circulation boiler, a mass flow rate in the riser tubes of ∼600 kg/(m² s) is required. Natural-circulation boilers operate with a fixed vaporization end point. They are used only for subcritical steam parameters. The maximum pressure of the live steam is 170–180 bars in the drum. The pressure drop in a natural-circulation boiler is about 5–10% of the live steam pressure. The steam capacity of natural-circulation boilers varies from 100 to 3000 t/h. The live steam temperature reaches 540–560°C.

At higher steam pressures, forced-circulation boilers must be used to ensure the proper circulation of working fluid as the density difference becomes too small for the natural-circulation to be effective. Therefore a pump is used to assist the natural circulation at pressures above 165 bars. The water mass flow rate of 1000–2000 kg/(m² s) is required, and thus these boilers can be employed up to the live steam pressure of 185–215 bars in the drum. At the critical point of water (374.15°C and 221.2 bars), the density difference is zero; therefore the natural-circulation driving force is also zero. Only once-through (Benson) steam generators can be used for both the supercritical and subcritical steam conditions (Kather et al., 1994; Klein et al., 1996). They are also called the universal pressure boilers, as they can be used both for supercritical and subcritical steam parameters. The maximum pressure and temperature of live steam are limited only by the materials used to construct the superheater and other high-temperature components. They now reach 250–300 bars and 560–600°C. A modern pulverized coal fired once-through boiler of 930 MW thermal capacity with a steam rate of 2420 t/h has a total height of 163 m and a furnace height of 103 m (Kather et al., 1994; Bald and Heusinger, 1996).

Boilers comprise the following heat transfer surfaces: evaporator, superheater, reheater, attemperator, economizer, and air heater. Design of these heat transfer surfaces is based on the common heat transfer equations applied to heat exchangers. The required heat transfer surface area is given by

$$A = Q/U \, \Delta t_m \quad \text{m}^2 \tag{4.53}$$

where Q is rate of heat transfer in W, U is overall heat transfer coefficient in W/(m² K), and Δt_m is mean logarithmic temperature difference in K. The overall heat transfer coefficient U is given by

$$U = 1 \Big/ \Big[1/h_i + \sum (\delta/k)_i + 1/h_a\Big] \quad \text{W/(m}^2 \text{ K)} \tag{4.54}$$

where h_i is the film heat transfer coefficient for the working fluid inside the heat transfer surface, h_a is the film heat transfer coefficient for the flue gas outside the heat transfer surface, $(\delta/k)_i$ is the thermal resistance of the wall and contaminants, δ_i is the thickness of the wall and contaminant layer, and k_i is the thermal conductivity of the wall and contaminant layer. Thus in order to calculate U, the film heat transfer coefficients are required. Empirical equations that can be used to calculate the film heat transfer coefficients are given in the literature on heat transfer applications (e.g., Minkowitz, 1989).

Boiler Performance Criteria

The boiler performance is characterized by the thermal capacity Q_b, the steam capacity m_s, the fuel consumption rate m_f, and the efficiency η_b (Li Kam and Priddy, 1985; Ordys et al., 1994). The boiler thermal capacity is defined as the useful heat output and is measured in kilowatts:

$$Q_b = m_s\Delta h_s + m_{rh}\Delta h_{rh} + m_{at}\Delta h_{at} + m_{bd}\Delta h_{bd} \tag{4.55}$$

where m is the mass flow rate of steam or water through a component of the boiler and Δh is the enthalpy change of steam or water. Subscripts s, rh, at, and bd designate the plant parts, namely, steam generator, reheater, attemperator, and blowdown, respectively. The steam capacity m_s is the rate of steam production in the steam generator in kg/h.

The boiler fuel consumption is given by

$$m_f = Q_b/(HHV\eta_b) \quad \text{kg/s} \tag{4.56}$$

where HHV is the higher heating value of fuel and η_b is boiler efficiency. The boiler efficiency based on the HHV of fuel is given by

$$\eta_b = 100\,\text{(Total heat added to the working fluid/Total fuel input energy)} \tag{4.57a}$$

or

$$\eta_b = 100\,(1 - \text{Total heat losses/Total fuel input energy}) \quad \% \tag{4.57b}$$

or

$$\eta_b = 100\,(HHV - q_{loss})/HHV \quad \% \tag{4.58}$$

Total boiler heat losses include the dry-gas loss (DGL), the moisture loss (ML), the moisture in combustion air loss (MCAL), the incomplete combustion loss (ICL), the unburned carbon loss (UCL), and the radiation and unaccounted-for loss (RUL) (Elliott, 1989; Li Kam and Priddy, 1985).

The dry-gas loss is given by

$$DGL = m_{dg}c_{pg}(t_g - t_a) \quad \text{kJ/kg} \tag{4.59}$$

where $m_{dg} = AF_{a,d} + 1 - R - M - 9\,H$, kg dry flue gas/kg fuel; $AF_{a,d}$ is actual, dry-air fuel ratio in kg air/kg fuel; c_{pg} is specific heat of dry flue gas = 1.05 kJ/(kg K); R is refuse mass fraction per kg fuel; M is mass fraction of moisture in the fuel; H is mass fraction of hydrogen in the fuel; t_g is temperature of flue gas leaving the boiler; and t_a is temperature of air entering the boiler. The R (coal) introduced into the furnace is given by

$$R = \text{Mass of refuse/Mass of fuel} = A/A_f \tag{4.60}$$

where A and A_f are the ash mass fraction in coal and in refuse (from refuse analysis), respectively.
The moisture loss is given by

$$ML = (M + 9H)\Delta h_m \quad \text{kJ/kg} \tag{4.61}$$

where Δh_m is specific enthalpy change of the moisture in the boiler system. If $t_g > 300°C$,

$$\Delta h_m = 2442 + 2.093t_g - 4.187t_f \quad \text{kJ/kg} \tag{4.62}$$

If $t_g \leq 300°C$,

$$\Delta h_m = 2493 + 1.926 t_g - 4.187 t_f \quad \text{kJ/kg} \tag{4.63}$$

where t_f is temperature of the fuel entering the boiler system.

The moisture in combustion air loss is given by

$$MCAL = AF_{a,d} w c_{pw}(t_g - t_a) \quad \text{kJ/kg} \tag{4.64}$$

where w is humidity ratio of air entering the boiler and c_{pw} is specific heat of water vapor = 1.926 kJ/(kg K).

The incomplete combustion loss is given by

$$ICL = 23,630 C_b \, CO/(CO + CO_2) \quad \text{kJ/kg} \tag{4.65}$$

where $C_b = C - C_r$ is the mass of carbon burned per mass of fuel, C is kg of carbon in kg fuel (ultimate analysis), C_r is kg of unburned carbon in refuse per kg of fuel added to furnace, and CO and CO_2 are the contents of CO and CO_2 in the dry flue gas Orsat analysis in %.

The unburned carbon loss is given by

$$UCL = C_r \, HHV_{carbon} = 32,778 C_r \quad \text{kJ/kg} \tag{4.66}$$

The RUL occurs due to radiation and convection from the outside surface of the boiler to the surroundings as well as to the transient losses, including the startup and shut-down losses. The RUL is relatively small.

Once all the losses have been found, the boiler efficiency will be determined from Eq. (4.58).

In addition to major parts of the boiler plant (furnace and heat transfer surfaces of evaporator, superheater, reheater, attemperator, economizer, and air heater), there are also auxiliaries of the boiler plants. They include facilities for fuel preparation and transport, refuse removal and treatment, make-up water treatment and supply, forced-draft and induced-draft fans, emissions control equipment (electrostatic precipitator, FGD, SCR), and the automatic control system.

The size of the furnace required for a boiler of a particular thermal capacity is based on empirical values of the heat release rates per unit cross-sectional area (q_a) and per unit volume (q_v). The volume, the cross-sectional area, and the height of the boiler furnace are given by

$$V = Q_f/q_v \quad \text{m}^3 \tag{4.67}$$

$$A = Q_f/q_a \quad \text{m}^2 \tag{4.68}$$

$$H = V/A \quad \text{m} \tag{4.69}$$

Typical values of q_a and q_v for pulverized coal fired (PC) and heavy fuel oil fired (HFO) boilers are given in Table 4.3 and in Figures 4.14a and 4.14b.

Figures 4.15a and 4.15b show typical coal burner arrangements for front-wall and tangentially fired boilers whereas Figure 4.15c shows a typical coal burner with low pollutant emissions. Figure 4.16 shows the schematic of a bubbling fluidized combustor, which is characterized by low sulfur and nitrogen oxide emissions.

Table 4.3. Typical heat release rates per unit cross-sectional area (q_a) and per unit volume (q_v) for pulverized coal fired (PC) and heavy fuel oil fired (HFO) boilers

Boiler thermal capacity, MW	q_v, MW/m^3		q_a, MW/m^2	
	PC	HFO	PC	HFO
800	0.16	0.28	5.2	6.9
1600	0.13	0.24	6.3	8.4

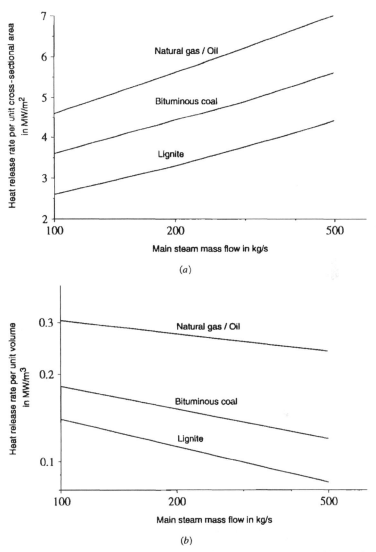

Figure 4.14. Furnace heat release rate (*a*) per unit cross-sectional area and (*b*) per unit volume as a function of steam mass flow in steam generators with various fuel firing.

Steam Turbines

Steam turbines can be classified as condensing, back-pressure, extraction condensing, and extraction back-pressure turbines (Elliott, 1989; Ordys et al., 1994; Author, 1972). A condensing turbine discharges steam to a condenser, where a low absolute pressure is maintained and the steam is condensed due to heat rejection to a cooling medium, usually water, sometimes ambient air. In water-cooled condensers the pressure is 0.03–0.06 bar; in the air-cooled it is ~0.07–0.1 bar. The level of pressure is determined through the temperature of the cooling medium. In extraction steam turbines, certain amounts of steam are bled from the turbine at various pressures and used for feedwater heating and/or for process steam delivery. In reheat steam power plants, reheat turbines are employed.

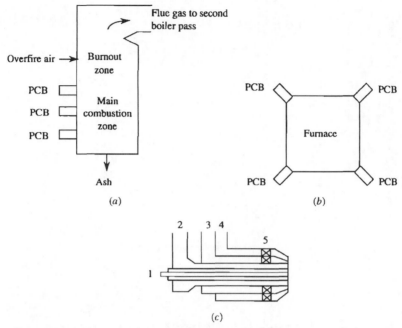

Figure 4.15. Typical coal burner arrangements for (*a*) front-wall and (*b*) tangentially fired boilers. (*c*) Typical coal burner with low pollutant emissions: 1, oil lance; 2, pulverized coal + primary air; 3, secondary air; 4, tertiary air; 5, air swirler.

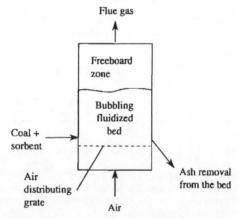

Figure 4.16. Schematic of a bubbling fluidized bed combustor.

Steam turbines are axial-flow machines and comprise a number of stages for steam expansion that can be either impulse or reaction stages (Elliott, 1989; El-Wakil, 1984). In an impulse stage, the steam first passes through nozzles, where the thermal energy of the steam is converted into kinetic energy in an adiabatic expansion process. If the inlet steam velocity is ignored, the velocity of steam leaving the nozzle is given by

$$c_1 = 2\sqrt{\Delta h} = 44.72\sqrt{\Delta h} \quad \text{m/s} \tag{4.70}$$

where Δh is the steam enthalpy drop in the nozzle in kJ/kg.

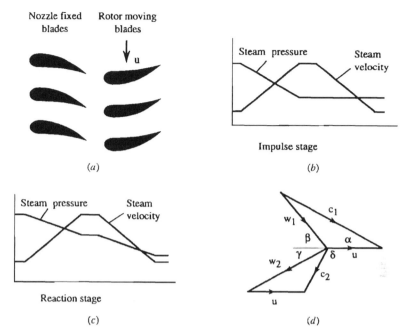

Figure 4.17. Schematic of (a) steam turbine stage, steam pressure and velocity profiles of (b) an impulse stage and (c) a reaction stage, and (d) velocity diagram.

For superheated steam the nozzle exit-to-inlet pressure ratio is 0.528. The kinetic energy flux of the steam leaving the nozzle is given by

$$P_{kin} = mc_1^2/2 \quad \text{W} \tag{4.71}$$

where m is the mass flow rate of steam through the blades in kg/s. Part of this energy flux is converted to mechanical work (power) while the steam flow passes through the moving blades. The power extracted by the blade is equal to the product of the force acting on the blade and the blade velocity. Thus

$$P_b = mu(c_1 \cos \alpha - c_2 \cos \delta) \quad \text{W} \tag{4.72}$$

Figure 4.17 shows a turbine stage, steam pressure and velocity in both an impulse stage and a reaction stage of a steam turbine, and a velocity diagram. Reaction blades are similar to airfoils so that a steam flow passage is shaped like a converging-diverging nozzle. The stationary and moving blades have approximately the same shape.

The following designations are used in Figure 4.17: c_1 and c_2 are absolute velocity of steam leaving nozzle and leaving blade, respectively; u is blade velocity; w_1 and w_2 are velocity of steam relative to the blade (all c, u, and w are in m/s); α is nozzle angle; δ is fluid exit angle; β is blade entrance angle; and γ is blade exit angle (all angles are in degrees).

The general equation for the blade power of both an impulse and a reaction stage is

$$P_b = m\left[\left(c_1^2 - c_2^2\right) - \left(w_1^2 - w_2^2\right)\right]/2 \quad \text{W} \tag{4.73}$$

For frictionless flow of steam through an impulse blade,

$$P_b = m\left(c_1^2 - c_2^2\right)/2 \quad \text{W} \tag{4.74}$$

The blade efficiency is defined as the fraction of the kinetic power of the inlet steam that is extracted by the blade, i.e., the ratio of P_b to $mc_1^2/2$. Thus for an impulse stage,

$$\eta_h = 2u(c_1 \cos \alpha - c_2 \cos \delta)/c_1^2 \tag{4.75}$$

The power and efficiency of an impulse blade are maximum when the velocity of steam leaving the blade is a minimum. For an impulse stage the optimum blade velocity is given by

$$u_{opt} = c_1 \cos \alpha/2 \tag{4.76}$$

The maximum blade power is given by

$$P_{b,max} = 2mu_{opt}^2 \tag{4.77}$$

and the maximum blade efficiency is

$$\eta_{b,max} = 1 - (c_2/c_1)^2 \tag{4.78}$$

For a reaction stage the blade power is given by

$$P_b = mu(c_1 \cos \alpha - u + w_2 \cos \gamma) \quad W \tag{4.79}$$

The optimum blade velocity is given by

$$u_{opt} = c_1 \cos \alpha \tag{4.80}$$

and the maximum blade power by

$$P_{b,max} = mu_{opt}^2 = m(c_1 \cos \alpha)^2 \quad W \tag{4.81}$$

The stage efficiency is defined as the ratio of the blade power to the enthalpy drop in the stage. Thus

$$\eta_{stage} = P_b/\Delta h_{stage} \tag{4.82}$$

In the high-pressure part of the turbine, impulse staging is commonly used, whereas reaction blading is employed in the intermediate- and low-pressure part of the turbine because the reaction stage is more efficient (Elliott, 1989).

Heat Rejection Systems

The amount of heat rejected from a steam power plant is given by

$$Q_{rej} = P_{el}(1/\eta_o - 1) \quad MJ/s \tag{4.83}$$

where P_{el} is the net electric power output in MW and η_o is the plant overall efficiency. Thus for a 900-MWe power plant with a net efficiency of 0.43,

$$Q_{rej} = 900(1/0.43 - 1) = 1193 \text{ MJ/s}$$

The cooling medium for the condenser is usually water. However, if a suitable cooling water source is not available, e.g., in arid regions, air is used to cool the condenser. The cooling medium requirements of a power plant are

$$m_{cm} = 1000 Q_{rej}/c_p \Delta t_{cm} \quad kg/s \tag{4.84}$$

where c_p is the specific heat of the cooling medium in kJ/(kg K) and Δt_{cm} is the temperature difference of the cooling medium in K (it is also called the cooling range). With water as the cooling medium and a typical value of 9 K for the cooling range,

$$m_{cm} = 1000 \times 1193 \text{ kW}/[4.19 \text{ kJ/(kg K)} \times 9 \text{ K}] = 31,636.8 \text{ kg/s}$$

If air were used as the cooling medium, then with $\Delta t_{cm} = 15$ K

$$m_{cm} = 1000 \times 1193 \text{ kW}/[1.005 \text{ kJ/(kg K)} \times 15 \text{ K}] = 79,137.6 \text{ kg/s}$$

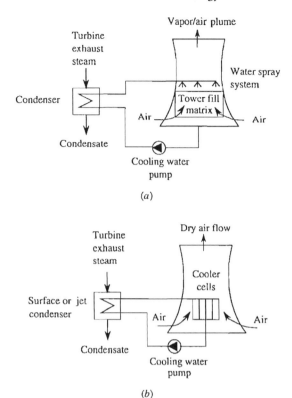

Figure 4.18. Heat rejection systems of a steam power plant: (*a*) wet cooling tower and (*b*) dry cooling tower.

In steam power plants, heat is rejected at a temperature of the cooling medium in the condenser. Lowering the heat rejection temperature is possible only by lowering the condenser pressure, which depends on the cooling medium temperature.

Cooling systems are classified as wet and dry systems (Elliott, 1989; El-Wakil, 1984; Author, 1972) (see Figure 4.18). Wet cooling systems are generally more efficient than dry cooling systems, as water has more favorable thermophysical properties than air. The volumetric specific heat of water is 3200 times that of air. Therefore the amount of cooling water required is much smaller than that of air. Owing to high specific heat of water and high heat transfer rates, the cooling range, i.e., the increase in the temperature of the cooling medium in the condenser, for water is small (about 9 K). The lowest values of condenser pressure of about 0.03–0.04 bar are attained by employing direct cooling systems.

There are two types of wet cooling systems in modern steam power plants: once-through and evaporative. Once-through cooling systems dispose the waste heat by dumping it directly into a river, lake, or sea. They are simpler, more efficient, and less costly than the evaporative cooling systems. With once-through water cooling, the lowest pressure in the condenser is attained. However, they are suitable primarily for power plants with small capacity, as the cooling water requirements are very large. Also, the source of cooling water is subjected to thermal pollution.

Wet cooling systems employ evaporative cooling towers (Elliott, 1989; Li Kam and Priddy, 1985; Author, 1972). The cooling of the turbine condenser cooling water is done mainly by evaporation of a portion of cooling water and partly by direct convective heat transfer from the cooling water to the air. In evaporative cooling systems the warmed water from the condenser is broken up into droplets and film, which falls down as it flows over a fill (or packing) of the cooling tower. The air flows upward and contacts directly with the water on a large exposed surface. The water is cooled mainly due to evaporation of a portion of it. There are two types of wet evaporative cooling systems: mechanical-draft and natural-draft cooling towers.

One of the largest natural-draft cooling towers is used at the Isar 2 nuclear power station in Germany. The tower is 165 m high, its base and top diameters are 153 and 85 m, respectively. The heat rejection system of the 900 MWe lignite-fired steam power plant to be constructed in Neurath, Germany, includes two cooling towers 162 m high, cooling water rate is about 28,000 kg/s, and cooling zone is 9 K (Kehr, 1994).

On sites where there is no cooling water available, e.g. in arid areas, direct or indirect dry cooling systems with air cooled condensers are employed. In a direct system, turbine exhaust steam is ducted through finned tubes of an air cooled condenser positioned over forced-draft fans. In an indirect cooling system (Heller system), exhaust steam is condensed using a surface or a direct-contact jet condenser. The cooling water or condensate is pumped through finned-tube conduits and recirculated back to condenser. At high temperatures of the ambient air, the condenser pressure is 0.07 bar or higher. This reduces the efficiency of power plants with air cooled condensers. The Trakya 1200-MW combined-cycle plant in Turkey uses two dry separate indirect Heller systems to cool four 100-MW steam turbines. The preheater/peak cooler cells of the fan-free systems are installed inside two 135-m high hyperbolic towers. A dry cooling system of the 614-MW combined-cycle cogeneration plant in Linden, USA, uses a 60 cell 30-m high structure, each with a 18-m diameter low-noise fan.

Hybrid wet/dry cooling systems are particularly used in urban sites that must be protected from plumes such as near highways and airports. As a rule, the wet and dry sections can be operated separately or in a series configuration. In the latter case, the cooling water first flows through finned-tube air cooled heat exchangers and then it is cooled in a tower of the wet section. Typically the dry section rejects about 20% of the power plant waste heat. The plume formation is prevented through the mixing of dry air from the air cooled heat exchangers with the wet air of the cooling tower. An example presents the cooling system of the West cogeneration plant in Frankfurt, Germany. Initially, the plant had a once-through cooling system. Now, with two additional units with 60-MW (electric capacity) and 100 MW (thermal capacity) each, the plant will use a hybrid wet/dry cooling system.

CLOSURE

Net efficiencies of state-of-the-art steam power plants of about 45% for hard coal fired power plants and about 43% for lignite-fired power plants have been currently achieved. Additional improvements of the net efficiency by 1–2% are predicted within the next 5 years, when even higher main steam parameters will be employed. Efficiencies up to 47–48% are predicted with main steam conditions of 350 bars and 700/720°C (Bald and Heusinger, 1996; Klein et al., 1996; Pruschek et al., 1996). This would be possible with new advanced materials for the hottest components, namely, the steam generator, steam conduits, and steam turbine. Austenitic steels and Inconel alloy are suitable for these components under the steam conditions mentioned above (Klein et al., 1996; Pruschek et al., 1996). Enhanced turbine efficiency due to optimized blade profiles, general process optimization, and reduced cold-end losses along with regenerative feedwater heating to over 300°C significantly contribute to the enhanced efficiency of advanced steam power plants.

A futher improvement of the efficiency of fuel energy conversion to power can be achieved in combined-cycle power plants, which are based on highly efficient large-frame gas turbines with an efficiency of about 40%. Combined-cycle power plants that can attain an overall efficiency of nearly 60% are discussed in Chapter 6.

PROBLEMS

4.1. In a Rankine cycle steam power plant the turbine receives 700 kg/s of steam at 200 bars and 550°C, and the condenser pressure is 0.04 bar. The turbine and pump isentropic efficiencies are 0.92 and 0.83, respectively. Calculate (a) the plant power output, in MW, (b) the plant thermal efficiency, and (c) the plant heat rate, in kJ/kWh.

4.2. In the steam power plant, the turbine receives 700 kg/s of steam at (i) 250 bars and 560°C, (ii) 250 bars and 600°C, (iii) 300 bars and 600°C. The condenser pressure is 0.04 bar. The turbine isentropic efficiency is 0.92. The work of the feedwater pump may be ignored. Calculate (a) the plant net power output, in MW, (b) the plant thermal efficiency, and (c) the plant net heat rate, in kJ/kWh.

4.3. In a Rankine cycle steam power plant with the main steam condition of 250 bars and 560°C, the condenser pressure is (i) 0.03 bar, (ii) 0.06 bar, (iii) 0.1 bar. The steam mass flow rate is 700 kg/s. The turbine isentropic efficiency is 0.92. The pump work may be ignored. Calculate (a) the plant power output, in MW, (b) the plant thermal efficiency, and (c) the plant heat rate, in kJ/kWh.

4.4. A 900-MW supercritical single-reheat steam power plant operates with the main steam condition of 250 bars and 560°C and the condenser pressure of 0.04 bar. The steam is reheated at a pressure of 30 bars to a temperature of 560°C. The turbine isentropic efficiency is 0.92. The feed pump work consumption may be ignored. Calculate (a) the plant net specific work, in kJ/kg, (b) the plant thermal efficiency, (c) the plant steam rate, in kg/kWh, and (d) the plant heat rate, in kJ/kWh.

4.5. A 900-MW supercritical double-reheat steam power plant operates with the main steam condition of 250 bars and 560°C and the condenser pressure of 0.03 bar. The steam is reheated first at 30 bars to 560°C and then at 4 bars to 560°C. The turbine efficiency is 0.92. The feed pump work may be ignored. Calculate (a) the plant net specific work, in kJ/kg, (b) the plant thermal efficiency, (c) the plant steam rate, in kg/kWh, and (d) the plant heat rate, in kJ/kWh.

4.6. A 1000-MW supercritical single-reheat steam power plant operates with the main steam condition of 300 bars and 600°C and the condenser pressure of 0.03 bar. The steam is reheated at 30 bars to 600°C. A certain quantity of steam is bled from the turbine at 2 bars for an open-type feedwater heater. The turbine efficiency is 0.92. The work of pumps may be ignored. Calculate (a) the plant net specific work, in kJ/kg, (b) the plant thermal efficiency, (c) the plant heat rate, in kJ/kWh, (d) the plant annual electricity production, in MWh/yr, if the plant capacity factor is 0.81.

4.7. A 960-MW steam power plant operates on a supercritical double-reheat cycle with two feedwater heaters. The turbine inlet steam condition is 250 bars/560°C, and the reheats are at 30 bars/560°C and 4 bars/560°C, respectively. The closed-type feedwater heater pressure is 3.0 bars, and the open-type feedwater heater pressure is 1.0 bar, the condenser pressure is 0.03 bar. The approach in the closed-type feedwater heater is 6 K. The steam turbine and generator efficiencies are 0.92 and 0.98, respectively. The pump work may be ignored. Calculate (a) the specific net work, (b) the efficiency in %, (c) the heat rate in kJ/kWh, and (d) the annual electricity generation in MWh/yr, if the capacity factor of the plant is 0.82.

REFERENCES

Bald, A., and Heusinger, K. 1996. Power generation in advanced steam power plants relieves burden on environment. *Siemens Power J.* 1:5–11.

Elliott, T. C., ed. 1989. *Standard handbook of powerplant engineering*. New York: McGraw-Hill.

El-Wakil, M. M. 1984. *Powerplant technology*. New York: McGraw-Hill.

Haywood, R. W. 1991. *Analysis of engineering cycles*. New York: Pergamon.

Kather, A., et al. 1994. Steam generators with advanced steam parameters. Presented at ASME Joint Int. PowerGen Conf.

Kehr, M. 1994. Kraftwerksprojekte der 90er Jahre. *VGB Kraftwerkstechnik*, 74(8):705–710.

Klein, M., Kral, R., and Wittchow, E. 1996. BENSON boilers—Experience in nearly 1000 plants and innovative design promise continuing success. *Siemens Power J.* 1:26–30.

Li Kam, W., and Priddy, A. P. 1985. *Power plant system design*. New York: Wiley.

Minkowitz, W. J., ed. 1989. *Handbook on numerical heat transfer*. New York: Wiley.

Ordys, A. W., Pike, A. W., Johnson, M. A., et al. 1994. *Modeling and simulation of power generation plants*. London-Berlin: Springer.

Pruschek, R., Oeljeklaus, G., and Brand, V. 1996. Zukuenftige Kohlekraftwerksysteme. *VGB Kraftwerkstechnik*, 76:441–448.

Rolle, K. C. 1994. *Thermodynamics and heat power*, 4th ed. New York: Macmillan Publ.

Wark, K. 1994. *Advanced thermodynamics for engineers*. New York: McGraw-Hill.

Weston, K. C. 1992. *Energy conversion*. New York: West Publishing.

Chapter Five

GAS TURBINE POWER GENERATION
TECHNOLOGY

Impressive progress has been made in the last decade in the area of power generation with gas turbines. This chapter deals with all major issues of modern gas turbine power generation technology.
First, the thermodynamic fundamentals are considered, beginning with the Joule cycle through
all the sophisticated techniques aimed at improving the efficiency, economic, and environmental
characteristics of gas turbine power plants. These include application of high gas turbine temperatures and compressor pressure ratios, intercooling, and reheating and conditioning of compressor
intake air. The most advanced industrial and, in particular, aeroderivative gas turbines achieve
efficiencies of about 40% and more. One of the crucial issues in gas turbine technology is the
control of NO_x emissions. This chapter describes major methods for reducing NO_x emissions such
as utilization of dry low-NO_x combustors and burners with precise control of fuel and air addition
and mixing, or water or steam injection. The most effective techniques include the simultaneous
utilization of most of these methods mentioned above. They are used in such advanced concepts
as recuperated water injected (RWI) and humid air turbine (HAT) cycles.

It is worthwhile noting that advanced gas turbine power plants can attain 40% efficiency and
more. The critical technological prerequisites for this achievement are advanced materials for
turbine blades and advanced blade-cooling techniques. All these issues are discussed in detail in
this chapter. Theoretical considerations are supported by numerous solved problems.

AIR–STANDARD JOULE CYCLE

Open-Cycle Gas Turbine Power Plant

There are basically two types of stationary gas turbine power plants that are used to generate
electric energy: open-cycle and closed-cycle (Cohen et al., 1987). They may also be designated
as internal and external combustion gas turbines, respectively. Figure 5.1 shows a flow diagram
of an open-cycle gas turbine power plant. It consists of a compressor, combustor, and gas turbine
with electric generator. In a closed-cycle gas turbine power plant, heat is added to a working fluid,
e.g., helium, from an external heat source by means of a heat exchanger—a heater that replaces
the combustor of the open-cycle plant. After expansion in the turbine, the working fluid is cooled
down to the compressor inlet temperature.

The ideal cycle for both the open-cycle and closed-cycle gas turbine power plants is the Air–
Standard Joule cycle (also called the Brayton cycle) (Cohen et al., 1987; Eastop and McConkey,
1993; Rogers and Mayhew, 1992). The Joule cycle is shown in Figure 5.2 on a T-s diagram. It
consists of four internally reversible processes: adiabatic (isentropic) compression (1-2) and
expansion (3-4) processes, as well as isobaric processes (2-3) and (4-1) of heat addition and
rejection. The working fluid in the Joule cycle is assumed to be an ideal gas with the properties of air.

Hence, in the Joule cycle, atmospheric air is first adiabatically compressed in the compressor
(1-2). Thereby, the air pressure rises from p_1 to p_2, and its temperature rises from T_1 to T_2. The

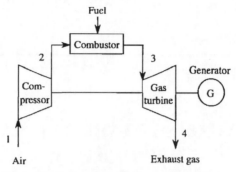

Figure 5.1. Open-cycle gas turbine power plant (schematic diagram).

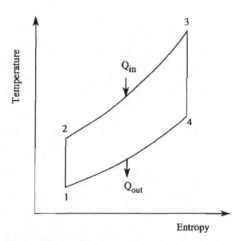

Figure 5.2. Air–Standard Joule cycle on a T-s diagram.

air temperature at the compressor discharge is thus

$$T_2 = T_1 \beta^{(k-1)/k} \quad \text{K} \tag{5.1}$$

where T_1 is the air intake temperature in K, $\beta = p_2/p_1$ is the compressor pressure ratio, and k is the isentropic exponent. The work input to the compressor per unit mass of air is

$$w_c = h_2 - h_1 = c_p(T_2 - T_1) \quad \text{kJ/kg} \tag{5.2}$$

where h_2 and h_1 are the air specific enthalpies at the compressor discharge and inlet in kJ/kg, respectively, and c_p is the constant pressure specific heat in kJ/(kg K). For the working fluid with the properties of air, $k = 1.4$ and $c_p = 1.005$ kJ/(kg K).

The fuel is next burned with compressed air in the combustor. The heat supplied to the combustor per unit mass of working fluid is

$$q_{in} = h_3 - h_2 = c_p(T_3 - T_2) \quad \text{kJ/kg} \tag{5.3}$$

where h_2 and h_3 are the specific enthalpies of the working fluid at the combustor inlet and outlet in kJ/kg and T_2 and T_3 are the corresponding temperatures in K.

In the gas turbine of an ideal cycle, an isentropic expansion of the working fluid (process 3-4 in Figure 5.2) takes place. The turbine work output per unit mass of working fluid is

$$w_t = h_3 - h_4 = c_p(T_3 - T_4) \quad \text{kJ/kg} \tag{5.4}$$

where h_3 and h_4 are the specific enthalpies of the working fluid at the turbine inlet and outlet in kJ/kg and T_3 and T_4 are the corresponding temperatures in K.

The turbine inlet and outlet temperatures are related to the pressure ratio β as follows:

$$T_3/T_4 = \beta^{(k-1)/k} \tag{5.5}$$

The heat rejected from the gas turbine per unit mass of exhaust gas is

$$q_{out} = h_4 - h_1 = c_p(T_4 - T_1) \quad \text{kJ/kg} \tag{5.6}$$

where h_4 is the specific enthalpy of the gas turbine exhaust gas in kJ/kg, T_4 is the exhaust gas temperature in K, and T_1 is the temperature of ambient air in K.

Once the specific compressor and turbine works are calculated, the net cycle work output is

$$w_{net} = w_t - w_c \quad \text{kJ/kg} \tag{5.7}$$

Substituting Eqs. (5.2) and (5.4) into Eq. (5.7) gives the net cycle work output per unit mass of air:

$$w_{net} = c_p[(T_3 - T_4) - (T_2 - T_1)] \quad \text{kJ/kg} \tag{5.8}$$

Joule Cycle Thermal Efficiency

The thermal efficiency η_{th} of the Joule cycle is defined as the ratio of the net work output of the cycle to the heat supplied to the cycle:

$$\eta_{th} = w_{net}/q_{in} \tag{5.9}$$

Substituting the expressions for w_{net} and q_{in} from Eqs. (5.8) and (5.3) into Eq. (5.9) yields

$$\eta_{th} = 1 - T_1/T_2 = 1 - 1/\beta^{(k-1)/k} \tag{5.10}$$

The thermal efficiency of the Joule cycle depends only on the compressor pressure ratio β. Compressor pressure ratios of 15–30 are employed in modern large-frame heavy-duty gas turbines. For light aeroderivative gas turbines, even higher values of β are typical. The calculated values of η_{th} for the reversible gas turbine cycle are given in Table 5.1. The relationship between the thermal efficiency of the Joule cycle and the compressor pressure ratio is plotted in Figure 5.3.

An important characteristic of the gas turbine cycle is the work ratio, i.e., the ratio of the gas turbine work output to the compressor work input:

$$w_t/w_c = (T_3/T_1)/\beta^{(k-1)/k} \tag{5.11}$$

The work ratio should be as large as possible. This can be achieved by increasing the ratio of the gas turbine and compressor inlet temperatures, T_3/T_1, and/or by decreasing the pressure ratio β.

Table 5.1. Joule cycle
thermal efficiency versus
pressure ratio

Pressure ratio	Thermal efficiency,%
1	0
5	36.9
10	48.2
15	53.9
20	57.5
25	60.1
30	62.2
40	65.1
60	69
80	71.4

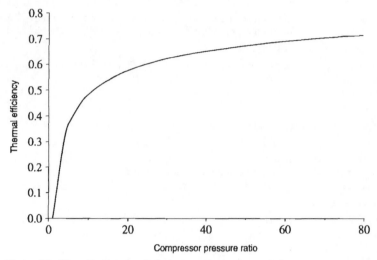

Figure 5.3. Thermal efficiency of the Joule cycle as a function of compressor pressure ratio.

PERFORMANCE OF SIMPLE-CYCLE GAS TURBINE PLANT

Effect of Irreversibilities on Simple-Cycle Performance

Contrary to the Air–Standard Joule cycle described above, irreversibilities in realistic turbomachinery, as well as pressure losses in combustor and flow inlet and exit channels, must be considered in the analysis of simple-cycle gas turbine power plants. For the isentropic exponent k and specific heat c_p of air and gas, different values should be adopted. Thus for air, $k = 1.4$ and $c_p = 1.005$ kJ/(kg K). However, for the working fluid in the gas turbine, c_p should be taken equal to the mean constant pressure specific heat of air at high temperatures (see Table 5.2).

Figure 5.4 shows the actual gas turbine power plant cycle on a T-s diagram. The irreversibilities in turbomachinery are accounted for by corresponding isentropic efficiencies. The isentropic efficiency of the compressor is defined as the ratio of the theoretical (reversible) to the actual compressor work input, $w_{c,rev}$ and w_c, respectively:

$$\eta_{ic} = w_{c,rev}/w_c \tag{5.12}$$

It can also be written in terms of air enthalpies or temperature at the initial and end states by isentropic (reversible) and actual (irreversible) compression:

$$\eta_{ic} = (h_{2s} - h_1)/(h_2 - h_1) = (T_{2s} - T_1)/(T_2 - T_1) \tag{5.13}$$

Typical values of η_{ic} are 0.88–0.9.

Table 5.2. Mean constant pressure specific heat c_p of air at high temperatures

Temperature, °C	c_p, kJ/(kg K)
500	1.039
1000	1.091
1500	1.132
2000	1.171

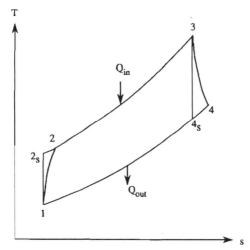

Figure 5.4. Actual (irreversible) gas turbine cycle on a
T-s diagram.

The actual compressor discharge temperature T_2 is higher than that in the isentropic compression, T_{2s}. It follows from Eq. (5.13):

$$T_2 = T_1 + (T_{2s} - T_1)/\eta_{ic} \tag{5.14}$$

The isentropic efficiency of the gas turbine is defined as the ratio of the actual (irreversible) and the theoretical (reversible) turbine work outputs, w_t and $w_{t,rev}$, respectively:

$$\eta_{it} = w_t/w_{t,rev} \tag{5.15}$$

In terms of gas enthalpies or temperatures at the initial and end states by isentropic (reversible) and actual (irreversible) expansion:

$$\eta_{it} = (h_3 - h_4)/(h_3 - h_{4s}) = (T_3 - T_4)/(T_3 - T_{4s}) \tag{5.16}$$

Typical values of η_{it} lie between 0.92 at $\beta = 5$ and 0.9 at $\beta = 30$.

The actual turbine exhaust gas temperature T_4 is higher than that in the isentropic (reversible) expansion, T_{4s}. From Eq. (5.16), it follows that

$$T_4 = T_3 - (T_3 - T_{4s})\eta_{it} \tag{5.17}$$

The actual specific compressor work, i.e., the work per unit mass of air, is

$$w_c = c_p(T_2 - T_1) = c_p(T_{2s} - T_1)/\eta_{ic} \tag{5.18}$$

The power input to the compressor is

$$P_c = m_a w_c \quad \text{kW} \tag{5.19}$$

where m_a is the air mass flow rate in kg/s. The rate of heat addition in the gas turbine combustor may be determined from

$$Q_{in} = m_a(1 + FA)c_{p.g}(T_3 - T_2) \quad \text{kW} \tag{5.20}$$

where FA is the actual fuel-air ratio in the combustor. Usually, FA is much less than the stoichiometric fuel-air ratio and often may be neglected with respect to 1.

Taking into account the pressure loss from the compressor to the gas turbine by means of a fractional pressure loss, f_i, the gas turbine inlet pressure may be expressed as

$$p_3 = p_2(1 - f_i) \quad \text{bars} \tag{5.21}$$

where p_2 is the compressor discharge pressure in bars. Similarly, the gas turbine exit pressure may be determined as

$$p_4 = p_1(1 + f_e) \quad \text{bars} \tag{5.22}$$

where p_1 is the ambient air pressure in bars and f_e is a fractional pressure loss at the gas turbine exit. The actual gas turbine pressure ratio is then

$$\beta_t = p_3/p_4 \tag{5.23}$$

For isentropic expansion, the gas turbine exit temperature is

$$T_{4s} = T_3/\beta_t^{(k_g-1)/k_g} \tag{5.24}$$

where k_g is the gas isentropic exponent that may be set equal to 1.33. The actual gas turbine exit temperature T_4 may be determined from Eq. (5.17). The actual gas turbine work per unit mass of gas is

$$w_t = c_{p.g}(T_3 - T_4) = c_{p.g}(T_3 - T_{4s})\eta_{it} \quad \text{kJ/kg} \tag{5.25}$$

Net Specific Work Output

The net specific work output of the gas turbine power plant, based on the unit mass of air, is

$$w_{net} = (1 + FA)w_t - w_c \quad \text{kJ/kg} \tag{5.26}$$

As was mentioned above, the fuel mass is much less than that of air, so that we assume that the fuel-air ratio FA may be neglected with respect to 1. Then

$$w_{net} = w_t - w_c = c_{p.g}(T_3 - T_{4s})\eta_{it} - c_p(T_{2s} - T_1)/\eta_{ic} \quad \text{kJ/kg} \tag{5.27}$$

General expressions for the cycle net work and thermal efficiency in terms of dimensionless parameters can be now derived. The following simplifying assumptions will be taken into consideration: fixed turbomachine isentropic efficiencies, η_{it} and η_{ic}, negligible pressure losses, temperature independent specific heat, and the fixed cycle maximum to minimum temperature ratio, T_3/T_1.

Rearranging and simplifying Eq. (5.18) yields (Eastop and McConkey, 1993; Rogers and Mayhew, 1992)

$$w_{net} = (c_p T_1/\eta_{ic})[(T_3/T_1)(1 - T_{4s}/T_3)\eta_{it}\eta_{ic} - (T_{2s}/T_1 - 1)] \quad \text{kJ/kg} \tag{5.28}$$

It follows from Eqs. (5.1) and (5.5),

$$T_{2s}/T_1 = T_3/T_{4s} = \beta^{(k-1)/k} \tag{5.29}$$

Designating $\beta^{(k-1)/k}$ as r and T_3/T_1 as n and rearranging Eq. (5.28) yields

$$w_{net} = (c_p T_1/\eta_{ic})(r - 1)(n\eta_{it}\eta_{ic}/r - 1) \quad \text{kJ/kg} \tag{5.30}$$

The heat addition to the combustor per unit mass of air is given by

$$q_{in} = (c_p T_1/\eta_{ic})[(n - 1)\eta_{ic} - (r - 1)] \quad \text{kJ/kg} \tag{5.31}$$

Thermal Efficiency of Simple-Cycle Gas Turbine

The thermal efficiency of the simple-cycle gas turbine power plant is given by

$$\eta_{th} = w_{net}/q_{in} = (r - 1)(n\eta_{it}\eta_{ic}/r - 1)/[(n - 1)\eta_{ic} - (r - 1)] \tag{5.32}$$

Effect of Pressure Ratio and Gas Turbine Inlet Temperature
on Simple-Cycle Performance

Equations (5.30) and (5.32) indicate that the major parameters affecting the thermal efficiency η_{th} and the net work output w_{net} of a simple-cycle gas turbine power plant with fixed compressor and gas turbine isentropic efficiencies are the maximum-to-minimum temperature ratio, T_3/T_1, and the pressure ratio β. It is obvious that the specific net work output is the power output per unit mass flow of working fluid:

$$w_{net} = P_{net}/m \quad kW/(kg/s) \tag{5.33}$$

The choice of the gas turbine inlet temperature (TIT) is based on the requirements of high efficiency, low cost, high reliability, and long life of gas turbines. Gas turbines with higher gas turbine inlet temperatures always have higher specific power outputs, w_{net}. There is an optimum gas turbine inlet temperature that corresponds to the gas turbine maximum thermal efficiency for a given plant configuration, cooling technology, and pressure ratio. Similarly, for a given gas turbine inlet temperature, there is an optimum pressure ratio. At higher gas turbine inlet temperatures, larger coolant flows are required, and therefore more heat should be spent to heat the coolant up to the gas temperature at the location of the coolant introduction. Thus a corresponding efficiency penalty follows. At present, the highest gas turbine inlet temperature is about 1250–1340°C. A further growth of the gas turbine inlet temperature up to about 1500°C is anticipated in more advanced gas turbines of the next generation.

To achieve efficient cooling, air is bled from both low- and high-pressure compressor stages. The compressor discharge air can also be cooled in an aftercooler. Although all these additional components make the plants more complicated and expensive, these configurations bring about substantial improvements in plant efficiency and power output.

Efficient cooling with increased coolant flow rates can reduce the maximum blade metal temperatures to about 800–830°C (ABB AG, 1991–1996; Becker and Finckh, 1995; Swanekamp, 1995). Increasing the gas turbine inlet temperature and using reheat increases the specific turbine work, thus providing an increase in the net specific work output of the gas turbine plant. The net power output P_{net} is the product of the net specific work output w_{net}, and the mass flow m of working fluid in the gas turbine. The steam/water injection increases the mass flow, thus increasing net power output of the gas turbine plant.

Figures 5.5 and 5.6 show the thermal efficiency and the net specific work output of the gas turbine power plant as a function of compressor pressure ratio and turbine inlet temperature (TIT). Owing to an increase in the gas turbine firing temperature from 800°C to over 1250°C, the gas turbine efficiency had increased from 22–25% in 1960 to 37–39% in 1995. For a given TIT or for a given ratio of TIT T_3 to compressor intake temperature T_1, both the thermal efficiency and

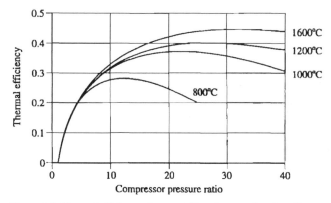

Figure 5.5. Thermal efficiency of a gas turbine plant as a function of compressor pressure ratio and turbine inlet temperature.

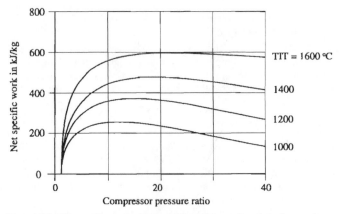

Figure 5.6. Net specific work of a gas turbine plant as a function of compressor pressure ratio and turbine inlet temperature (TIT).

the specific net work output of a simple-cycle gas turbine power plant attain a maximum value at a specified pressure ratio. It is seen that the optimum pressure ratio for the maximum efficiency is approximately twice as high as the optimum pressure ratio for the maximum specific power output when TIT is the same. The lowest specific cost of a simple-cycle gas turbine power plant as well as of the combined-cycle power plant is achieved when the plant design is optimized for the maximum specific power output per unit mass flow of air.

It should be pointed out that the TIT is limited by the maximum allowable temperature for hot gas path materials. Raising the TIT greatly increases the efficiency of the gas turbine power plant. At the present time the maximum turbine inlet temperature for gas turbines of the third generation lies around 1350°C. In the near future it may rise further to 1500°C (Kano et al., 1991). The compressor intake temperature T_1 is the site-dependent ambient air temperature. At a fixed value of the ratio of TIT to T_1, the work ratio w_t/w_c can be increased by decreasing the pressure ratio. However, this causes a decrease in the thermal efficiency of the plant. Therefore an optimum value of pressure ratio should be found that maximizes the net work output for a given gas turbine inlet temperature and mass flow rate.

Optimum Pressure Ratio

Let us now derive an expression for the optimum pressure ratio, β_{opt}, at which the specific power output per unit mass flow of air, w_{net}, attains its maximum value. Differentiate w_{net}—from Eq. (5.33)—with respect to β, and set the result equal to zero. All parameters except β are assumed to have fixed values. The resulting optimum pressure ratio is (Cohen et al., 1987)

$$\beta_{opt} = [\eta_{it}\eta_{ic}(T_3/T_1)]^{k/2(k-1)} \tag{5.34}$$

However, the maximum cycle efficiency is attained at a higher pressure ratio.

Power Output

The performance of a simple-cycle gas turbine power plant can be expressed in terms of power output, thermal efficiency, and plant net heat rate. The power output of the simple-cycle gas turbine power plant is

$$P_{net} = m_g w_t - m_a w_c \quad \text{kW} \tag{5.35}$$

where m_g and m_a are the gas and air flow rates in kg/s and w_t and w_c are the specific work of the gas turbine and compressor in kJ/kg, respectively.

As was mentioned above, the fuel mass is much less than that of air. Therefore m_g in Eq. (5.35) may be approximately replaced by the air mass flow m_a. The specific net work can be calculated by Eq. (5.29) or (5.31). Hence the power output of a simple-cycle gas turbine power plant is

$$P_{net} \approx m_a w_{net} = m_a c_p[(T_3 - T_{4s})\eta_{it} - (T_{2s} - T_1)/\eta_{ic}] \quad \text{kW} \tag{5.36}$$

or

$$P_{net} \approx m_a(c_p T_1/\eta_{ic})(r - 1)(n\eta_{it}\eta_{ic}/r - 1) \quad \text{kW} \tag{5.37}$$

If the power output is known, the air mass flow m_a required is given by

$$m_a = P_{net}/w_{net} \quad \text{kg/s} \tag{5.38}$$

Plant Net Heat Rate

Another important performance characteristic of a gas turbine power plant is the plant net heat rate. It is defined as the ratio of the heat input rate to the plant net power output:

$$HR = Q_{in}/P_{net} \quad \text{kJ/kWh} \tag{5.39}$$

Obviously, the plant net heat rate is related to the thermal efficiency as follows:

$$HR = 3600/\eta_{th} \quad \text{kJ/kWh} \tag{5.40}$$

The plant net heat rate as well as the plant thermal efficiency may be related to either the higher or the lower heating value of fuel, HHV or LHV, respectively. HR increases as the load decreases. With a decrease in load, the gas turbine inlet and exhaust temperature will decrease.

Examples 5.1–5.3 illustrate the calculation of performance parameters of the simple-cycle gas turbine plant.

Example 5.1

Consider a simple-cycle gas turbine power plant with the following parameters:

Compressor intake temperature	$T_1 = 17°C = 290$ K
Cycle maximum-to-minimum temperature ratio	$T_3/T_1 = 4$
Gas turbine isentropic efficiency	$\eta_{it} = 0.92$
Compressor isentropic efficiency	$\eta_{ic} = 0.89$

Assuming the isentropic exponent $k = 1.4$ and the mean isobaric specific heat $c_p = 1.005$ kJ/(kg K), estimate the specific power output per unit mass flow of air and the cycle thermal efficiency under the conditions of the optimum pressure ratio.

Solution

Optimum pressure ratio

$$\beta_{opt} = [\eta_{it}\eta_{ic}(T_3/T_1)]^{k/2(k-1)} = [0.92 \times 0.89 \times 4]^{1.4/2(1.4-1)} = 7.97$$

Parameter

$$r = \beta_{opt}^{(k-1)/k} = 7.97^{0.2857} = 1.809$$

Specific power output per unit mass flow of air

$$\begin{aligned} w_{net} &= (c_p T_1/\eta_{ic})(r - 1)(n\eta_{it}\eta_{ic}/r - 1) \\ &= (1.005 \times 290/0.89)(1.809 - 1)(4 \times 0.92 \times 0.89/1.809 - 1) = 214.72 \text{ kW/(kg/s)} \end{aligned}$$

Cycle thermal efficiency

$$\begin{aligned} \eta_{th} &= (r - 1)(n\eta_{it}\eta_{ic}/r - 1)/[(n - 1)\eta_{ic} - (r - 1)] \\ &= (1.809 - 1)(4 \times 0.92 \times 0.89/1.809 - 1)/[(4 - 1)0.89 - (1.809 - 1)] = 0.352 \end{aligned}$$

Table 5.3. Optimum pressure ratio β_{opt}, specific power output per unit mass flow of air w_{net}, and cycle thermal efficiency η_{th} versus maximum temperature ratio T_3/T_1

	T_3/T_1					
	4	4.5	5	5.5	6	6.5
β_{opt}	7.97	9.80	11.78	13.92	16.21	18.65
r	1.809	1.920	2.023	2.122	2.216	2.307
w_{net}, kJ/kg	214.72	276.89	342.95	412.34	484.60	559.39
η_{th}	0.352	0.385	0.413	0.437	0.458	0.476

Example 5.2

Evaluate the effect of the maximum temperature ratio T_3/T_1 on the specific power output and the thermal efficiency for the simple-cycle gas turbine power plant. Assume the following values of T_3/T_1 : 4, 4.5, 5, 5.5, 6, and 6.5. All the other parameters are the same as in Example 5.1.

Solution

The calculation results including those of Example 5.1 are given in Table 5.3.

Example 5.3

For the simple-cycle gas turbine power plant of Example 5.2, calculate the air mass flow m_a required for the net power output P_{net} of 170 MW and the plant net heat rate. The maximum temperature ratio T_3/T_1 is 4.5.

Solution

From Example 5.2, at $T_3/T_1 = 4.5$, the specific power output per unit mass flow of air w_{net} is 276.89 kW/(kg/s), and the cycle thermal efficiency η_{th} is 0.385.

Air mass flow required

$$m_a = P_{net}/w_{net} = 170,000 \text{ kW} /276.89 \text{ kJ/kg} = 614 \text{ kg/s}$$

Plant net heat rate

$$HR = 3600/\eta_{th} = 3600/0.385 = 9350.6 \text{ kJ/kWh}$$

Effect of Ambient Conditions on Gas Turbine Performance

The performance of the open-cycle gas turbine power plant described in terms of the power output and thermal efficiency is affected by plant site conditions, i.e., by temperature, pressure, and relative humidity of ambient air (Cohen et al., 1987).

The ambient air temperature is the major factor influencing the open-cycle gas turbine performance. An increase in the ambient air temperature T at the gas turbine site at a certain elevation, i.e., at a constant air pressure, causes a decrease in the air density ρ. At higher elevations, the atmospheric pressure p is lower than at sea level, and thereby, the air density is lower than at sea level. The relation between air density ρ at any temperature T and pressure p and that (ρ_0) at International Organization for Standardization (ISO) standard conditions ($T_0 = 273.15$ K and $p_0 = 1.013$ bars, sea level) is given by

$$\rho = \rho_0 \frac{T_0 p}{T p_0} \quad \text{kg/m}^3 \tag{5.41}$$

The change in ρ causes a corresponding change in the mass flow rate of the compressor intake air

$$m = \frac{m_0 \rho}{\rho_0} \quad \text{kg/s} \tag{5.42}$$

where m_0 is the mass flow of air at ISO standard conditions.

The power required to drive the compressor increases with increasing ambient air temperature. The mass flow through the gas turbine decreases as the air temperature rises. This, in turn, reduces

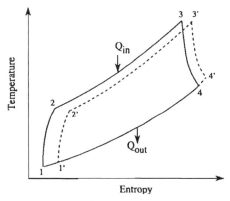

Figure 5.7. Effect of compressor intake condition on gas turbine cycle.

the pressure ratio in the turbine and, consequently, the power output of the gas turbine. Thus the thermal efficiency and the net power output of a gas turbine plant become lower at higher compressor intake temperature of the air.

Figure 5.7 shows the gas turbine cycle at two different ambient air temperatures on a temperature-entropy (T-s) diagram. As the air temperature increases, the pressure ratio decreases and the gas turbine exhaust gas temperature becomes higher at a constant turbine inlet temperature. Hence the thermal efficiency η_{th} of a simple-cycle gas turbine plant decreases at higher ambient air temperatures and increases when the ambient air temperature is lower. This causes a corresponding change in the plant heat rate that is determined from Eq. (5.29).

Figure 5.8 shows the effect of ambient air condition on power output of the performance parameters of simple-cycle gas turbine plant. Thus at 40°C the heat rate is about 5% higher and the power output about 17% lower than the design values at an optimum temperature of 15°C.

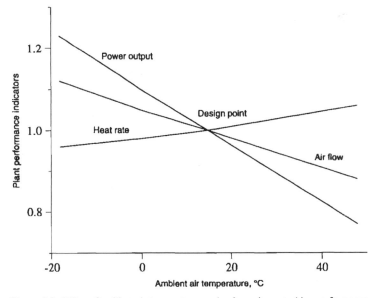

Figure 5.8. Effect of ambient air temperature on simple-cycle gas turbine performance parameters.

REGENERATIVE GAS TURBINE CYCLE

The general principle of thermal efficiency enhancement states: raise the average temperature of heat addition and lower the average temperature of heat rejection. There are several possibilities to implement this principle. One is to employ regenerative preheating of the working fluid. For the gas turbine cycle, it means the use of a heat exchanger, called the recuperator, for preheating the compressor discharge air with the gas turbine exhaust heat. Owing to the air preheating in the regenerative cycle, the fuel consumption in the combustor required to attain the gas turbine inlet temperature, T_5, is reduced as compared to a reference power plant without regeneration. Figures 5.9 and 5.10 show the schematic layout of the regenerative gas turbine power plant and its cycle on a T-s diagram. An increase in the thermal efficiency is achieved as the result of increased average temperature of heat addition and decreased average temperature of heat rejection. In the ideal case, the air temperature aft of the recuperator, T_4, is equal to the gas turbine outlet temperature T_6, and the gas outlet temperature T_8 is equal to the temperature of the compressor

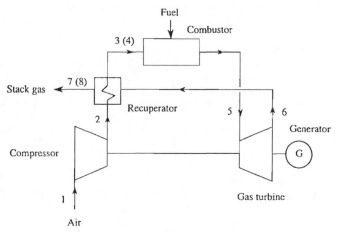

Figure 5.9. Regenerative gas turbine power plant.

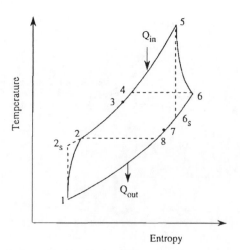

Figure 5.10. Regenerative gas turbine cycle on a T-s diagram.

discharge T_2. The actual recuperator has an effectiveness ε, typically between 0.6 and 0.8, and the actual air and gas outlet temperatures T_3 and T_7 differ from T_2 and T_6. The recuperator effectiveness ε is defined as the ratio of the actual to theoretical heat transfer rates of the recuperator:

$$\varepsilon = (T_3 - T_2)/(T_6 - T_2) \tag{5.43}$$

Temperatures T_2 and T_6 may be evaluated from Eqs. (5.14) and (5.17). With known values of ε, T_2, and T_6, the actual air temperature at the recuperator outlet, T_3, is

$$T_3 = T_2 + \varepsilon(T_6 - T_2) \tag{5.44}$$

The heat addition to the gas turbine combustor per unit mass of working fluid is

$$q_{in} = c_p(T_5 - T_2) \quad \text{kJ/kg} \tag{5.45}$$

in a nonregenerative cycle and

$$q_{in.reg} = c_p(T_5 - T_3) \quad \text{kJ/kg} \tag{5.46}$$

in a regenerative cycle. Thus the heat addition in the regenerative cycle is less than that in the nonregenerative cycle by

$$\Delta q_{in.reg} = c_p(T_3 - T_2) \quad \text{kJ/kg} \tag{5.47}$$

The actual recuperator exhaust gas temperature T_8 is

$$T_7 = T_6 - (T_6 - T_2)\varepsilon$$

The actual heat rejection per unit mass of working fluid is

$$q_{out} = c_p(T_7 - T_1) \quad \text{kJ/kg} \tag{5.48}$$

The net work output does not change if all parameters are identical with the nonregenerative cycle. Thus

$$u'_{net} = c_p[(T_5 - T_{6s})\eta_{it} - (T_{2s} - T_1)/\eta_{ic}] = \frac{c_p T_1}{\eta_{ic}}(r-1)\left(n\frac{\eta_{it}\eta_{ic}}{r} - 1\right) \quad \text{kJ/kg} \tag{5.49}$$

where $r = \beta^{(k-1)/k}$ and $n = T_5/T_1$.

The reduced heat addition to the combustor of the regenerative gas turbine causes an increase in the cycle thermal efficiency.

Rearranging Eq. (5.46) yields

$$q_{in.reg} = c_p(T_5 - T_3) = c_p[(T_5 - T_2) - \varepsilon(T_6 - T_2)] \tag{5.50}$$

With $r = \beta^{(k-1)/k}$, we obtain from Eqs. (5.14) and (5.17)

$$T_2/T_1 = (r - 1)/\eta_{ic} + 1 \tag{5.51}$$

and

$$T_6/T_5 = 1 - \eta_{it}(1 - 1/r) \tag{5.52}$$

Combining Eqs. (5.50)–(5.52) yields for the cycle with an ideal recuperator ($\varepsilon = 1$)

$$q_{in.reg} = c_p T_1 n \eta_{it}(1 - 1/r) \quad \text{kJ/kg} \tag{5.53}$$

The cycle thermal efficiency is given by

$$\eta_{th} = \frac{w_{net}}{q_{in}} = \left(n\frac{\eta_{it}}{r} - \frac{1}{\eta_{ic}}\right)\Big/\frac{n\eta_{it}}{r} = 1 - \frac{r}{n\eta_{it}\eta_{ic}} \tag{5.54}$$

If $\eta_{it} = \eta_{it} = 1$,

$$\eta_{th} = 1 - r/n = 1 - (T_1/T_5)\beta^{(k-1)/k} \tag{5.55}$$

Equation (5.55) shows that the thermal efficiency of a regenerative gas turbine cycle increases with increasing temperature ratio of TIT, T_5, and compressor inlet temperature, T_1, and with decreasing β. When isentropic efficiency of turbomachinery is taken into account, there is an optimum pressure ratio, β_{opt}, at which the efficiency has a maximum for a given T_5/T_1.

GAS TURBINE PLANTS WITH INTERCOOLING AND REHEAT

Gas Turbine Cycle with Intercooling

An additional method to improve the gas turbine cycle efficiency is to use multistage compression with air intercooling between compressor stages and multistage expansion with gas reheating between turbine stages (Cohen et al., 1987; Rolle, 1994; Eastop and McConkey, 1993; Rogers and Mayhew, 1992; Haywood, 1991; Wark, 1994). Air intercooling reduces the required compression work. Reheating increases the average temperature of heat addition in the cycle. Figure 5.11a shows the schematic of a two-stage compressor with intercooling as part of a gas turbine power plant. Figure 5.11a shows a T-s diagram of the gas turbine cycle with intercooling. The total compression work w_c in the two-stage compressor is the sum of the low-pressure (LP) stage work and the high-pressure (HP) stage work.

Assuming an ideal intercooling, i.e., the same air temperature at the inlet of each stage, yields the work required for the two-stage compression:

$$w_c = c_p T_1 \left(\frac{\beta_{LP}^{(k-1)/k} - 1}{\eta_{LP}} + \frac{\beta_{HP}^{(k-1)/k} - 1}{\eta_{HP}} \right) \tag{5.56}$$

where T_1 is the air temperature at the inlet of each stage, β_{LP} and β_{HP} are the pressure ratios in the LP and HP compressor stages, η_{LP} and η_{HP} are the isentropic efficiencies of the LP and HP compressor stages.

This work is a minimum when the pressure ratio is the same for each stage, i.e., $\beta_{LP} = \beta_{HP} = \beta_i$. Then the ideal pressure ratio β_i in one stage is

$$\beta_i = \frac{p_i}{p_1} = \frac{p_2}{p_i} = \sqrt{\beta} \tag{5.57}$$

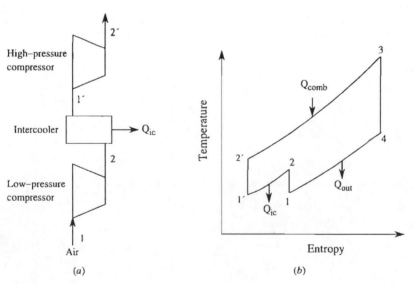

(a)   (b)

Figure 5.11. Gas turbine cycle with intercooling: (a) two-stage compressor with intercooling and (b) cycle on a T-s diagram.

where $\beta = p_2/p_1$ is the total pressure ratio. Then the ideal optimum intercooler pressure is

$$p_i = \sqrt{(p_1 p_2)} \tag{5.58}$$

For an ideal intercooler with the same air inlet temperature for each compressor stage having the same isentropic efficiency $\eta_{ic} = \eta_{LP} = \eta_{HP}$, the air discharge temperature for each compressor stage is

$$T_2 = T_1 \left(1 + \frac{\beta_i^{(k-1)/k} - 1}{\eta_{ic}} \right) \quad \text{K} \tag{5.59}$$

The power required for the two-stage compressor with the ideal optimum intercooler pressure is

$$P_c = m_a w_c = 2 m_a c_p T_1 \frac{\beta_i^{(k-1)/k} - 1}{\eta_{ic}} \quad \text{kW} \tag{5.60}$$

where m_a is the air mass flow rate and c_p is the specific heat of air.

The rate of heat removal from the intercooler is

$$Q_{ic} = m_a c_p (T_2 - T_1) = m_w c_{pw} \Delta t_w \quad \text{kW} \tag{5.61}$$

where m_w is the mass flow rate of cooling water, c_{pw} is the specific heat of cooling water, and Δt_w is the increase in the temperature of cooling water in the ideal intercooler.

Gas Turbine Reheat Cycle

Figure 5.12a shows the schematic of a reheat gas turbine, and Figure 5.12b shows the reheat cycle on a T-s diagram. The total heat addition in a reheat gas turbine cycle is the sum of heat supplied to the gas turbine combustor, Q_{comb}, and that supplied to the reheater, Q_{rh}:

$$Q_{in} = Q_{comb} + Q_{rh} = mc_p(T_3 - T_2) + mc_p(T_5 - T_4) \quad \text{kJ} \tag{5.62}$$

where T_2 is the air compressor discharge temperature, T_3 is the inlet temperature of the HP stage of the gas turbine, T_4 is the outlet temperature of the HP stage of the gas turbine, and T_5 is the reheat temperature.

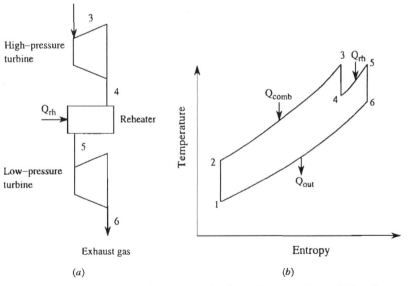

(a)   (b)

Figure 5.12. Gas turbine reheat cycle: (a) schematic of the reheat gas turbine and (b) cycle on a T-s diagram.

The specific work of the two-stage gas turbine is

$$w_t = c_p(T_3 - T_4) + c_p(T_5 - T_6) \quad \text{kJ/kg} \tag{5.63}$$

If the pressure ratio β_i for each stage of the gas turbine is the same and the reheat temperature T_5 is equal to the gas turbine inlet temperature T_3, then the outlet temperature T_6 of the gas turbine LP stage is equal to the outlet temperature T_4 of the gas turbine HP stage. Taking into account the irreversible expansion in both stages yields for the actual gas temperatures after the gas turbine HP and LP stages

$$T_4 = T_6 = T_3 \left[1 - \left(1 - \frac{1}{\beta_i^{(k-1)/k}} \right) \eta_{it} \right] \quad \text{K} \tag{5.64}$$

In this case, the gas turbine specific work output is

$$w_t = 2c_p T_3 \left(1 - \frac{1}{\beta_i^{(k-1)/k}} \right) \eta_{it} \quad \text{kJ/kg} \tag{5.65}$$

Gas Turbine with Intercooling and Reheat

Intercooling and reheat are usually employed simultaneously (see Figures 5.13a and 5.13b). Assuming equal pressure ratios for both the compressor stage and gas turbine stage yields for the net specific work output of the gas turbine plant:

$$w_{net} = w_t - w_c = 2c_p[(T_5 - T_6) - (T_2 - T_1)] \quad \text{kJ/kg} \tag{5.66}$$

Hence the plant net power output is

$$P_{net} = m w_{net} = 2mc_p T_3 \left[T_5 \left(1 - \frac{1}{\beta_i^{(k-1)/k}} \right) \eta_{it} - T_1 \frac{\beta_i^{(k-1)/k} - 1}{\eta_{ic}} \right] \quad \text{kW} \tag{5.67}$$

The total heat addition in the combustor and reheater is given by

$$Q_{in} = Q_{comb} + Q_{rh} = mc_p[(T_5 - T_4) + (T_7 - T_6)] \quad \text{kJ} \tag{5.68}$$

where m is the mass flow rate, T_5 and T_7 are the inlet temperatures of the HP and LP stages of gas turbine, T_4 is the compressor air discharge temperature, and T_6 is the exhaust temperature of the HP stage of the gas turbine.

The thermal efficiency is given by

$$\eta_{th} = \frac{P_{net}}{Q_{in}} \tag{5.69}$$

Practical aspects of intercooling, reheat, and regenerative gas turbine cycles will be discussed in the following sections.

COMPONENTS OF GAS TURBINE POWER PLANTS

Advanced Heavy-Duty Gas Turbines

As of May 1995, there were 5500 industrial heavy-duty gas turbines manufactured worldwide. The leading world manufacturers share the world's gas turbine market as follows:

- General Electric—3500 machines, or 64%
- Westinghouse—1300 machines, or 23%
- ABB—550 machines, or 10%
- Siemens—150 machines, or 3%

In recent years, significant advances have been made in the design of large-frame heavy-duty gas turbines. The world's leading gas turbine manufacturers, Siemens AG (Germany), ABB Power Generation Ltd. (Switzerland), General Electric Co. (United States), Westinghouse Co. (United States), and Mitsubishi Heavy Industries Ltd. (Japan), have introduced advanced gas turbines of

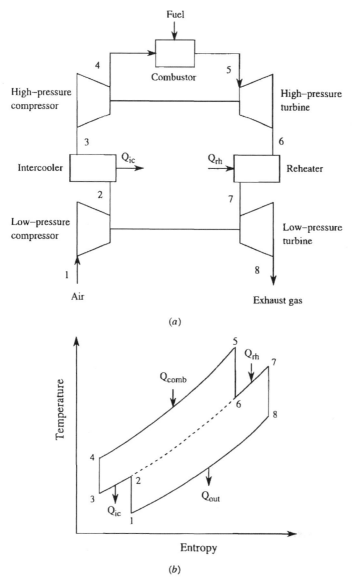

Figure 5.13. Gas turbine reheat cycle with intercooling: (a) schematic of the plant and (b) cycle on a T-s diagram.

the second and third generations. The gas turbine inlet temperature exceeds 1000°C for the former and 1100°C for the latter. The specifications of the most advanced gas turbines—ABB GT24 and GT26; Siemens 3A Series turbines V64.3A, V84.3A, and V94.3A; and General Electric turbines MS501G, MS7001G, and MS9001G—for both the 50-Hz and 60-Hz modes of operation—are presented in Tables 5.4–5.6 (ABB AG, 1991–1996; Becker and Finckh, 1995; Author, 1995). For comparison, Table 5.4 also contains data of two older ABB machines—GT8C and GT13E2. At the present time, the General Electric gas turbines of the G and H technologies are the world's largest and most efficient industrial gas turbines. Gas turbines MS7001H and MS9001H of

Table 5.4. Performance data of the ABB GT8C, GT13E2, GT24, and GT26 gas turbines

Performance data	50 Hz			60 Hz
	GT8C	GT13E2	GT26	GT24
Power output (gross), MW	56.8	165.1	254	173
Compressor pressure ratio	15.9	14.1	30	30
Mass flow rate, kg/s	179	502	562	390
Efficiency (LHV), %	34.1	35.7	38.3	38.1
Heat rate (LHV), kJ/kWh	10,400	10,084	9,400	9,450
Combustor/burner type	Silo/EV	Annular/EV	Annular/EV/SEV	Annular/EV/SEV
NO_x (at 15% O_2, dry), ppmv	25	<25	<25	<25

Fuel is natural gas; ISO standard conditions are 15°C, 1.013 bars; relative humidity 60%; sea level.
Source: ABB Power Generation Ltd., Baden, Switzerland (status: August 1996).
SEV, sequential EV.

Table 5.5. Simple-cycle performance of the Siemens gas turbines V64.3A, V84.3, and V94.3A

Performance data	V64.3A	V84.3A	V94.3A
Frequency, Hz	50/60	60	50
Power output, MW	70	170	240
Efficiency (LHV), %	36.8	38.0	38.0
Heat rate (LHV), kJ/kWh	9783	9474	9474
Exhaust gas flow rate, kg/s	194	454	640
Exhaust gas temperature, °C	565	562	562
NO_x (at 15% O_2, dry), ppmv	<25	<25	<25

Fuel is natural gas. ISO standard conditions are 15°C, 1.013 bars, relative humidity 60%, sea level.
Source: Siemens AG, Erlangen, Germany.

Table 5.6. Performance data of the General Electric advanced industrial gas turbines of the G technology

Performance data	MS501G	60 Hz MS7001G	50 Hz MS9001G
Net power output, MW	230	240	282
Firing temperature class, °C	—	1430	1430
Simple-cycle efficiency (LHV), %	38.5	39.5	39.5
Heat rate (LHV), kJ/kWh	9351	9115	9115
Heat rate (LHV), BTU/kWh	8860	8640	8640
Air flow rate, kg/s	545	558	685
Exhaust gas temperature, °C	593	593	—
NO_x emissions, ppm	25	25	25

Fuel is natural gas. ISO standard conditions are 15°C, 1.013 bars, relative humidity 60%, sea level.
Source: Power, 1995.

the H technology differ in terms of steam cooling of the gas turbine and lower NO_x emissions (9 ppm).

The performance data of simple-cycle power plants with second- and third-generation gas turbines are summarized in Tables 5.7 and 5.8 (ABB AG, 1991–1996; Becker and Finckh, 1995; Author, 1995). All the performance data are given at ISO standard conditions, i.e., at 15°C and 1013 mbar and air humidity of 60% for natural gas as fuel.

Based on the data given in the tables, the main performance features of advanced heavy-duty industrial gas turbines for power generation may be summarized as follows:

- high turbine power output (up to 230 MW at 60-Hz frequency and 282 MW at 50-Hz frequency)
- high specific power output (up to 443 kW per 1 kg/s of air flow)
- low heat rate (between 9770 and 9115 kJ/kWh)

Table 5.7. Comparsion of performance data of second-generation gas turbines

Performance data	General Electric Fiat, MHI		Westinghouse Nuovo Pignone	Siemens Ansaldo
	MS9001E[a]	MS7001EA	MW501D	V94.2[a]
Power output, MW	123.4	83.5	104.6	153.6
Simple-cycle efficiency, %	33.6		34.8	33.9
Compressor pressure ratio	12.3		14	10.9
Exhaust gas flow rate, t/h	1477		1649	1833
Turbine inlet temperature, °C	1124		1193	1125
Exhaust gas temperature, °C	539		509	546

[a] 50-Hz machine; all other gas turbines are 60-Hz machines.
Fuel is natural gas. ISO standard conditions are 15°C, 1.013 bars, relative humidity 60%, sea level.
Sources: ABB AG (1996), Becker and Finckh (1995), Dodero (1997), and Collins (1994).

- high pressure ratio (from 15 to 30)
- high gas turbine inlet temperature (1235°C or higher)
- high efficiency in the simple-cycle mode of operation (34.5–39.5%)
- low dry NO_x emissions (as low as 25–9 ppmv when burning natural gas, where ppmv denotes parts per million by volume)

Thus these gas turbines provide high efficiency, low NO_x emissions, and low cost of electricity. The advanced gas turbines can be efficiently used in power generation plants not only for peaking duty but also for intermediate and base-load duty. Figure 5.14 shows an advanced gas turbine design, the ABB GT26, with an annular combustor.

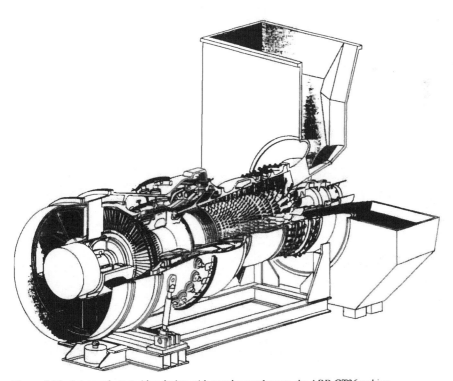

Figure 5.14. Advanced gas turbine design with annular combustor: the ABB GT26 turbine.

Table 5.8. Comparison of performance data of third-generation gas turbines

Performance data	Westinghouse Fiat, MHI			ABB				General Electric Nuovo Pignone			Siemens Ansaldo	
	TG50D5S6	FMW701F	MW501F[a]	GT13E2	GT11N2[a]	GT26	GT24[a]	MS9001FA	MS7001FA[a]	MS9001EC	V94.3A	V84.3A[a]
Power output, MW	143	237	153	164	109	254	173	226.5	159	219	240	170
Simple-cycle efficiency, %	34.8	37.2	35.3	35.7	34.2	38.3	38.0	35.7		34.9	38.0	38.0
Exhaust gas flow rate, kg/s	454	666		525	375	562	390	615		507	640	454
Turbine inlet temperature, °C	1250	1350		1100	1085			1235		1290	1204–1340	
Exhaust gas temperature, °C	528	550		525	524	608	610	589		558	562	
Compressor pressure ratio	14.1	15.9	16	15	15	30		15		14.2	16	

[a] 60 Hz: all other gas turbines are 50-Hz machines.
Fuel is natural gas. ISO standard conditions are 15°C, 1.013 bars, relative humidity 60%, sea level.
Sources: ABB AG (1996), Becker and Finckh (1995), Author (1995), and Collins (1994).

Table 5.9. Performance data of advanced
aeroderivative stationary gas turbines

Performance data	Value
Power output, MW	40–50
Turbine inlet temperature, °C	1280–1350
Compressor pressure ratio	30–60
Net specific work output, kJ/kg	350–370
Thermal efficiency, %	39.0–39.9
Air mass flow, kg/s	115–135
Gas turbine outlet temperature, °C	450–470

Advanced Aero-Derivative Stationary Gas Turbines

The aeroderivative gas turbines have much higher pressure ratios and higher gas turbine inlet temperatures than the heavy-duty gas turbines, and therefore they have higher thermal efficiency. Some innovative simple-cycle aeroderivative gas turbine configurations with higher efficiency have been developed recently. Steam injection, intercooling, and reheat are used to enhance their performance. They may be considered as alternatives to the combined-cycle plants. However, even the most advanced simple-cycle aeroderivative machines at very high turbine inlet temperatures (up to 1500°C) and very high compressor pressure ratios (up to 60) are less efficient than the best combined-cycle plants, which are discussed in Chapter 6. The net simple-cycle efficiency of aerodynamically optimized gas turbines with advanced hot gas path materials and cooling techniques can reach 50%. At the same time, combined-cycle plants based on high-performance, heavy-duty, large-frame industrial gas turbines with a simple-cycle efficiency of 38.5–39.5% (Becker and Finckh, 1995; Organowski, 1990) are capable of achieving overall efficiencies up to 58% (HHV) and 60% (LHV). Table 5.9 contains typical performance data of the state-of-the-art aeroderivative gas turbines.

Advanced aeroderivative gas turbine plants have additional components such as heat exchangers, reheat combustors, and air/water mixers or saturators. In order to reduce the compressor power consumption, multistage compression with intercooling is employed. Surface and evaporative intercoolers are used for that purpose. Another deployable modification is intake air conditioning, which includes air precooling or preheating as well as supercharging.

Advanced simple-cycle aeroderivative gas turbine configurations are as follows (Rice, 1995; Macchi et al., 1995):

- steam-injected gas turbine (STIG) cycle
- intercooled steam-injected gas turbine (ISTIG) cycle
- unmixed intercooled recuperated cycle (ICR)
- recuperated water-injected cycle (RWI)
- humid air turbine cycle (HAT)

Some of these configurations are discussed later in the chapter.

Advanced Materials and Technologies for Heavy-Duty Gas Turbines

Composite Materials

To achieve maximum efficiency and performance, gas turbines are now operated at firing temperatures of ~1400°C, which corresponds to 1150°C under ISO standard conditions. Turbine blades are subjected to the highest thermal and mechanical loads and chemical impact. Only composite materials are capable of withstanding these loads. The basic material provides the necessary mechanical properties, and the protective coating ensures the necessary corrosion and oxidation resistance. To achieve the anticipated life cycle of turbine blades of at least 50,000 hours, superalloys with high resistance to corrosion, mechanical fatigue, and thermal stress are required. Superalloys are nickel or cobalt based with the balance of such elements as chromium, molybdenum, tungsten, titanium, and aluminum. Owing to its outstanding resistance to corrosion and oxidation, chromium is an important element of superalloy compositions. Vacuum melting and precision casting allow the use of increased percentages of alloying elements, thus improving the blade's high-temperature strength. This technique also enables manufacturing of hollow blades with complex configurations for air cooling (ABB AG, 1991–1996; Becker and Finckh, 1995).

A 50 K increase in the gas turbine inlet temperature gives a 5 MW increase in power output of a 60-MW gas turbine. This, in turn, requires further increase in the strength and heat resistance of blade materials.

Advanced Manufacturing Methods

Gas turbine hot path components such as blades and vanes are subject to corrosion and thermal stress. Improved performance of gas turbines may be achieved through the utilization of advanced blade manufacturing methods. Vacuum precision casting and the process of directional solidification (DS) of blades allow the manufacture of hollow blades with complex configurations of internal channels for cooling air. The directionally solidified blades may be operated at temperatures about 25 K higher than conventionally cast blades. An additional temperature increase of about 25 K can be achieved by using blades made of materials containing no grain boundaries, i.e., single-crystal materials. Certain problems must be solved before this technology will be applied to manufacturing longer and heavier blades (up to 350 mm and 5 kg) for industrial combustion turbines. The required service life should be up to 100,000 hours.

Even higher material temperatures are possible using advanced oxide-dispersion-strengthened (ODS) and ceramic materials (Czech et al., 1994; Schulenberg, 1995; Bannister et al., 1995; Collins, 1994). The strength of ODS materials remains approximately constant up to 1000°C, but it is slightly lower than that of Ni-based precision cast alloys. Hence ODS materials are used in solid, slightly cooled stationary blades. These materials cannot be cast, and therefore a complicated manufacturing process is required to produce blades with cooling hollow spaces.

The use of ceramic materials for turbine blades may significantly improve the performance of gas turbines. These materials can enable the greatest increases in blade temperature. Thus they would offer a significant improvement in efficiency and performance. However, before they start to be applied, their reliability and mechanical properties must be substantially improved and the manufacturing costs reduced.

Protective Coatings

Special coatings on composite materials are used to protect the blade material against the chemical and thermal impact of hot gas. To prevent the high-temperature corrosion and/or oxidation of blades, anticorrosion coatings will be used. The corrosive impact depends on the contamination of hot gas and on the temperatures of the gas and the blade surface. Oxidation increases exponentially with temperature and is unavoidable due to the high O_2 content of the hot gas. At blade surface temperatures up to 800°C, chromium-based diffusion coatings provide effective protection for rear blade stages. At higher temperatures, more complex compositions such as MCrAlY (M is Co and/or Ni, Cr is chromium, Al is aluminum, and Y is yttrium) are used as protective coatings (Czech et al., 1994; Schulenberg, 1995; Bannister et al., 1995; Collins, 1994). Their protective effect is based on the formation of a dense aluminum oxide layer on the coating surface which should be thermally very stable. The remaining elements control the aluminum activity, hold the oxide layer in place, and adapt the coating to the properties of the base material.

The coatings are applied using the vacuum plasma-spray process, a method that produces a thick coating with excellent bonding properties and that prevents oxidation during application. For mechanical reasons, the thickness of the coating should not exceed approximately 0.4 mm.

Increasing thermal and mechanical stresses necessitate improvement of the first generation of MCrAlY materials, especially in terms of oxidation and thermal fatigue. All of the coatings described here are consumed during turbine operation. At present, their service life is still much shorter than that of the blade material; therefore the coatings must be refurbished after roughly 25,000 hours of operation.

Thermal Barrier Coatings

Thermal barrier coatings (TBCs) of turbine blades are used at high firing temperatures. Such a coating usually consists of an outer ceramic layer of zirconia (ZrO_2) stabilized with 7% yttria and a bonding metallic layer of an MCrAlY alloy (Czech et al., 1994). The ceramic layer has a low thermal conductivity and thus reduces the heat flow from the hot gas to the blade. The blade temperature is thus reduced.

TBCs have been used for several years in high-performance aircraft engines. The use of these coatings on blades of industrial gas turbines will enable an increase in the turbine inlet temperature of about 100 K with a further improvement of the efficiency.

Satisfactory service life for a TBC is achieved by limiting the thickness of the ceramic layer. A thickness of approximately 0.25 mm is sufficient to reduce the blade temperature by 100 K. For application of TBCs two principal techniques are used—thermal spraying and physical vapor deposition (PVD) (Czech et al., 1994; Schulenberg, 1995). PVD coatings currently yield better results but are more expensive. Therefore, in aircraft engines the rotor blades are provided with PVD coatings, while the stationary blade assembly has plasma-sprayed coatings. A similar use of the two methods is also anticipated for heavy-duty industrial gas turbines.

Advanced materials and manufacturing technologies of gas turbines described above increase the complexity and cost of gas turbines. It is anticipated that due to their application in industrial gas turbines along with rather involved cooling techniques, within the next few years, gas turbines with inlet temperatures greater than 1500°C will be developed (Collins, 1994). Combined-cycle plant efficiencies of about 60% are anticipated with such gas turbines.

Gas Turbine Cooling Technology

Hot path components of gas turbines, namely, hot gas casing, vanes, and blades should be cooled. The first purpose of gas turbine cooling is to achieve the highest possible gas turbine inlet temperature with permissible temperatures of hot path materials and least heat rejection. The second purpose is to achieve uniform distribution of gas turbine temperatures with an aim of NO_x emissions reduction. The highest gas turbine inlet temperatures enable a substantial increase in thermal efficiency.

The heat transfer rate to cooling medium is enhanced by using

- jet impingement of cooling medium
- extended surface area with pins and ribs
- cooling film on the surface to be cooled

This results in a higher cooling efficiency, which allows use of higher firing temperatures in the gas turbine combustor and therefore enhances the gas turbine plant efficiency.

However, gas turbine cooling causes a thermodynamic loss because of the heat transfer to the cooling medium. The cooling medium, i.e., air extracted from the compressor, is mixed with working fluid in the gas turbine. For cooling, large cooling air flow is required. Further discussion is given in the following sections.

Low-Emission Gas Turbine Combustors and Burners

Introduction

The major factors affecting NO_x production in the gas turbine combustor are (ABB AG, 1991–1996; Becker and Finckh, 1995; Lefebvre, 1995; Doebbeling et al., 1996) as follows:

- firing temperature
- oxygen availability
- duration of the combustion

NO_x is formed mainly when the temperatures are high, such as those found in the flame of a gas turbine combustor. The flame temperature depends on the excess air ratio λ. It is highest in the case of stoichiometric combustion ($\lambda = 1$). Below that level, the flame temperature is higher, but

DLN burner

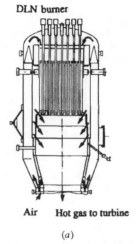

Air   Hot gas to turbine

(a)

Gas Turbine with
Silo-Type Combustion Chambers

(b)

Figure 5.15. (a) Schematic of a silo-type combustor and (b) photograph of a gas turbine with two silo-type combustion chambers.

there is less oxygen available to form NO_x, since most of it is used for combustion. Above that level, NO_x decreases because excess air within the flame lowers its temperature.

Very low excess air ratios are beneficial from the point of view of NO_x formation but are very detrimental to efficiency and cause the production of large amounts of CO and unburned hydrocarbons (UHC). Normally, gas turbine combustors operate with a high excess air ratio, ensuring complete and stable fuel combustion over the entire load range. Obviously, NO_x emissions will

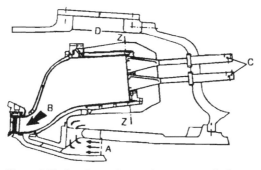

Figure 5.16. Annular combustor: A, compressor discharge; B, combustor exit; C, low-NO_x burner.

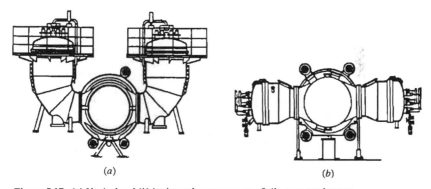

(a) (b)

Figure 5.17. (a) Vertical and (b) horizontal arrangements of silo-type combustors.

be very high unless special precautions are taken. Typically, NO_x levels in the exhaust gases after mixing with the cooling and secondary air are in the range 120–300 ppmv (ABB AG, 1991–1996; Lefebvre, 1995).

The main function of a gas turbine combustor is efficient and complete fuel combustion with minimum air pollutant emissions. There are two major types of gas turbine combustors: silo and annular (see Figures 5.15 and 5.16). Figure 5.17 shows two arrangements of large silo combustors—the old vertical and the novel compact horizontal arrangements.

Dry Low-NO_x (DLN) Combustor

Two types of burners are used: diffusion and premix. Present NO_x emission limits for gas turbines in Japan and the United States can hardly be satisfied with commonly used NO_x abatement techniques. Although the 75 ppm (at a 15% oxygen content) Environmental Protection Agency (EPA) benchmark value is the prevailing U.S. federal standard, new projects now specify an overall limit of under 25 ppm.

The formation of nitrogen oxides depends both on the firing temperature and the residence time of the fuel/air mixture in the combustion zone. To control NO_x emissions in conventional diffusion burners, wet NO_x abatement techniques are used, with water or steam injected into the combustor. Lowering the flame temperature by injecting water or steam into the flame reduces NO_x emissions up to 75–25 ppm (ABB AG, 1991–1996; Lefebvre, 1995). However, this method requires large amounts of treated water and causes a decrease in gas turbine efficiency and an increase in CO emission due to incomplete combustion. The steam injection affects the efficiency less and augments the power output. Another NO_x abatement technique, called selective catalytic reduction (SCR), increases operating costs and lowers efficiency.

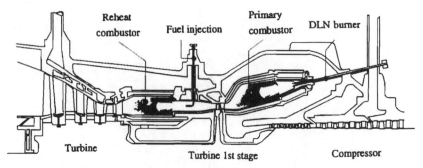

Figure 5.18. Dry, low-NO$_x$ combustor with sequential fuel combustion for reheat cycle.

The development of dry low-NO$_x$ (DLN) combustors and burner is one of the most important recent advances in gas turbine technology. DLN combustors enable a significant reduction of NO$_x$ emissions by burning a lean mixture of fuel and air. Complicated and expensive fuel and air staging schemes may thus be avoided. Fuel and air are premixed before they are introduced into the flame zone through multiple nozzles with precise control of both fuel and air. In DLN combustors, flame oscillations causing vibrations and affecting the life of the combustor will be avoided. DLN combustors are more complicated than diffusion combustors. The available DLN combustors and burners operate in the premix mode only in the range from around 50 to 100% of load. At lower power output levels, i.e., from start-up to around 50% of load, water/steam injection should be used to control NO$_x$ emissions.

Existing DLN combustors are well suited only for burning natural gas and light distillates. Hence DLN combustors and burners with enhanced fuel flexibility are needed. They should be capable of burning LHV fuel gases and low-quality heavy oils under the emissions limits over the full load range.

DLN combustors and premix burner designs were introduced by major world manufacturers of gas turbines. An environmentally beneficial annular combustor with a two-stage combustion of ABB (Switzerland) is shown in Figure 5.18 (ABB AG, 1991–1996; Doebbeling et al., 1996). A DLN EV burner of ABB is shown in Figure 5.19a. For comparison, Figure 5.19b shows the conventional diffusion combustor.

The EV burner has a simple design and can burn gaseous and/or liquid fuel. The NO$_x$ emissions can be reduced to under 10 ppm on natural gas without water/steam injection and to under 42 ppm on fuel oil with water injection. The EV burner also has very low CO and HC emissions, although the reduction of NO$_x$ emissions usually causes increased emissions of carbon monoxide CO and UHC.

The EV burner is a cone 400 mm long and 150 mm in diameter. Combustion air enters the burner tangentially through two inlet side slots, and the fuel gas is injected through a number of fine holes in their edges. A lean mixture of fuel with air (high excess air ratio) leaves the cone and enters the flame. At the exit of the burner the vortex breaks down, forming a recirculation zone that stabilizes the flame in free space. The combustion temperature is about 500 K lower than in a conventional diffusion burner, and thus low NO$_x$ emissions are achieved. Liquid fuel is sprayed through an atomizing nozzle and evaporated in the cone.

The EV burner is suited for both silo-type and annular combustors. Depending on gas turbine capacity, silo-type combustors have 19–54 burners. The NO$_x$ emissions decrease from 250 to 500 ppmv for diffusion burners in silo-type combustors to 10–25 ppmv for DLN burners. The annular combustor of the ABB heavy-duty gas turbine GT13E2 has 72 burners arranged in four rows around the annulus and achieves NO$_x$ values below 25 ppm on natural gas.

Large silo-type combustors ensure complete fuel combustion with minimum UHC and virtually no CO over the normal operating range. They can fire a wide variety of fuels, from LHV to HHV gaseous and/or liquid fuels, including treated heavy oils. Each combustor is normally equipped with multiple dual-fuel burners capable of firing natural gas and/or light distillate. The DLN burners can operate in the diffusion or premix combustion mode. In the premix mode the thermal

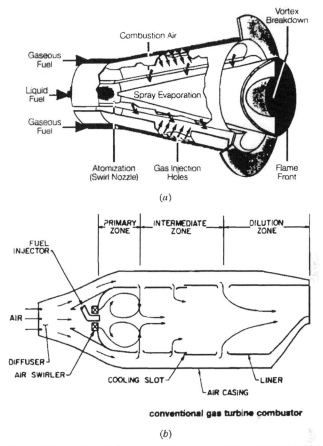

Figure 5.19. (*a*) Dual-fuel dry, low-NO$_x$ burner and (*b*) conventional diffusion combustor.

NO$_x$ formation is suppressed to very low values. However, the premix mode enables complete and stable fuel combustion with low NO$_x$ formation only in the load range from about 50% to 100%. At lower loads the diffusion combustion with steam or water injection must be employed.

A hybrid burner such as the one shown in Figure 5.20 enables the achievement of very low NO$_x$ emissions, below established limits without steam/water injection or a downstream SCR system. In the low power range, steam or water injection should be used to control NO$_x$ emissions.

Figure 5.21 shows the effect of firing temperature on NO$_x$ and CO emissions from gas turbines. In the temperature range 1400–1650°C, emissions of the two air pollutants are relatively low. Staged combustion reduces emissions (see Figure 5.22) because the firing temperature is kept within the optimum range. Figure 5.23 shows the effect of gas turbine load on emissions of NO$_x$, CO, smoke, and UHCs.

Catalytic Combustion in Gas Turbines

Catalytic combustion promises ultralow pollutant emissions (NO$_x$ below 3 ppm, CO and UHCs in the single-digit range). Similar to (DLN) technology, catalytic combustors burn ultralean fuel/air mixtures. However, ceramic-based embedded catalysts that are currently used cannot withstand the thermal cycling encountered by an operating gas turbine. To avoid the use of ceramic catalysts, the temperature in the catalyst zone must be reduced to a range suitable for catalysts on metallic

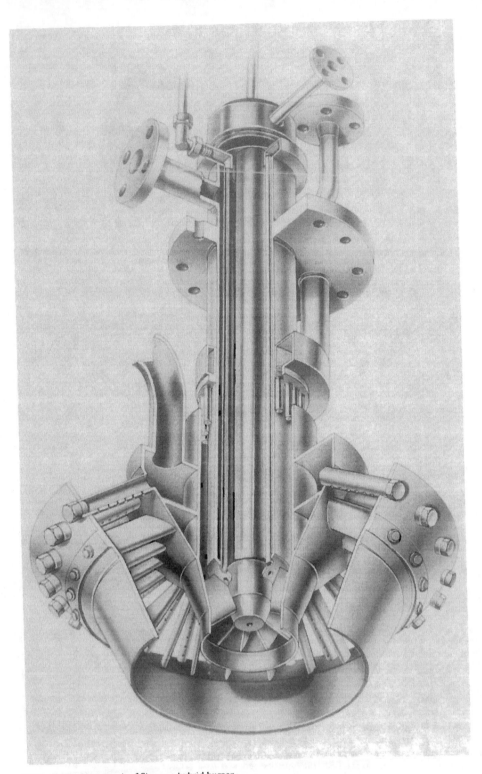

Figure 5.20. Photograph of Siemens hybrid burner.

116

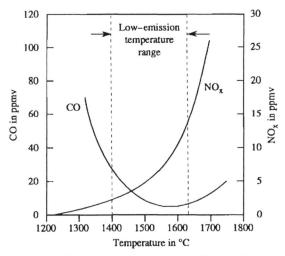

Figure 5.21. Effect of firing temperature on NO_x and CO emissions.

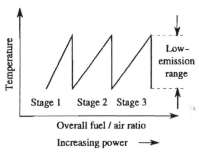

Figure 5.22. Staged combustion in the low-emission temperature range.

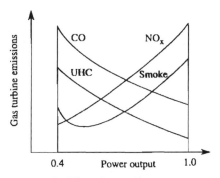

Figure 5.23. Effect of gas turbine power output on emissions of NO_x, CO, smoke, and unburned hydrocarbons (UHC).

substrates. This can be achieved by precise control of flow conditions through the catalytic combustor.

Catalytic combustors could avoid some performance and safety problems associated with DLN combustors such as accidental fuel ignition upstream of the burner or high-frequency pressure oscillations due to flame instability, which can cause machinery vibrations that may reduce the life of the gas turbine and the heat recovery steam generator.

If commercially proven, catalytic combustors could render DLN and SCR technologies obsolete because they would offer NO_x and CO emissions in the single-digit-ppm range at a substantially lower capital cost.

The catalytic combustor system Xonon developed by Catalytica Combustion Systems Inc. (Mountain View, California) and catalytic combustors of Precision Combustion Inc. (New Haven, Connecticut) were installed in 1996/1997 on some industrial gas turbines.

STEAM-INJECTED GAS TURBINES

As mentioned in the previous section, water or steam can be injected into the gas turbine combustor to control NO_x emissions. It must be stated that, although water injection reduces NO_x emissions, it exerts an adverse impact on gas turbine efficiency. However, steam provides both NO_x control and power output enhancement in a steam-injected gas turbine (STIG) plant (Rice, 1995; Macchi et al., 1995; Chiesa et al., 1995). Several configurations of gas turbine plants with steam or water injection have been developed in recent years. The injected steam produces more power for a given turbine inlet temperature (TIT) in all cases, but additonal heat input is required to heat the steam to full gas turbine inlet temperature. Therefore an increase in efficiency is not always attained. The injection of steam or water can occur at various points along the gas path in the gas turbine. However, due to the application of cooling, the firing temperature can be further increased, and thus the gas turbine efficiency would be enhanced.

A schematic diagram of the simple-cycle STIG plant is shown in Figure 5.24a. Steam may be injected around the fuel tips in the combustor and into the compressor discharge to mix with the discharge air. The mixture of steam and air can be used as a coolant for first-stage nozzle vanes. The steam flow entering the gas turbine produces work and thus enhances power output and helps to film cool the vanes. Thus increased mass flow rate through the gas turbine resulting from steam injection enhances the power output of the gas turbine by 10–40% at practically the same air flow. In the reheat gas turbine cycle the steam can be injected both in the main combustor and in the reheat stage (see Figure 5.24b).

It is clear that water or steam injection reduces NO_x emissions because of the lowered flame temperature. The effect of steam (water) injection on NO_x and CO emissions as well as on power output and heat rate of the gas turbine plant is illustrated in Figure 5.25. The NO_x emissions reduction depends on the mass of water or steam injected per unit mass of fuel. At a ratio of 1, the typical reduction factor is ~5 with water and ~3 with steam. Water is a more efficient flame coolant than steam owing to evaporation inside the flame.

From natural gas-fired turbine combustors, NO_x emissions as low as 40 ppmv or even 25 ppmv are attainable with water or steam injection. However, this method of emissions reduction has the following drawbacks: it requires large quantities of demineralized water, and it lowers the gas turbine efficiency, especially when water is injected. As was already mentioned, owing to an increase in mass flow through the gas turbine, its power output increases.

INTERCOOLING, INLET AIR CONDITIONING, AND SUPERCHARGING

Intercooling increases net specific work but reduces efficiency of the ideal gas turbine cycle. However, intercooling has a positive influence on the efficiency penalties of real cycles. Cooling air flow is required to cool the blades due to lower coolant temperature.

As it was discussed earlier, the ambient air condition greatly influences performance of gas turbine plants. Therefore inlet air conditioning and supercharging enable operating the machine at optimum inlet air temperature and pressure. The effect of air inlet temperature variations is significantly greater in advanced gas turbines. For maximum power output, the optimum inlet

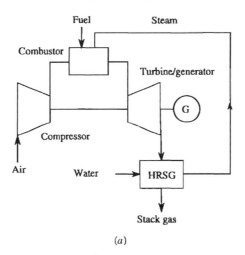

(a)

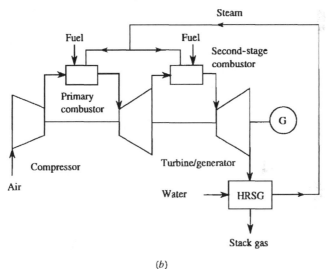

(b)

Figure 5.24. Schematic diagrams of (a) simple-cycle and (b) reheat cycle steam-injected gas turbine power plants.

temperature is 10°C. Hence it is beneficial to cool the inlet air at higher ambient temperatures and to heat it at lower temperatures. For instance, for the GE's 40-MW aeroderivative industrial gas turbine LM6000, a decrease in air inlet temperature from 38°C to 10°C can result in a 30% increase in power output and a 4.5% improvement in heat rate. Supercharging the inlet by a fan from 1.013 bar to 1.15 bars can increase output by nearly 20% (Kolp et al., 1995).

Figure 5.26 shows an advanced gas turbine cycle featuring the inlet air conditioning through supercharging and cooling with the goal of attaining improved performance of the gas turbine, particularly in the higher temperature range. Supercharging increases turbine output by raising compressor inlet pressure, but decreases net plant efficiency and thus increases plant net heat rate. A relatively large auxiliary power is required to drive the supercharger, e.g., about 2.4 MW for a 40-MW engine. At the same time, supercharging raises compressor inlet temperature and thus reduces the actual improvement in turbine power output. Therefore, after supercharging, air

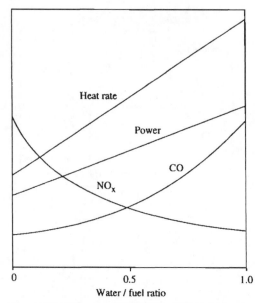

Figure 5.25. Effect of water/fuel ratio on NO_x and CO emissions, power output, and heat rate of the gas turbine power plant.

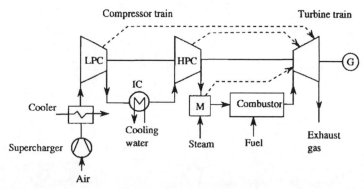

Figure 5.26. Supercharged intercooled aeroderivative turbine plant. LPC and HPC, low-pressure and high-pressure compressor; IC, intercooler; M, steam/air mixer.

should be cooled to remove the heat of compression. The combined effect of inlet air conditioning is improved performance

Evaporative cooling of the inlet air provides another cost-effective method for improving plant performance. Evaporative cooling does not increase turbine capacity but optimizes its operation. Unlike supercharging, adding evaporative cooling does not require increasing the size of gas turbine plant equipment. At ambient temperatures below 4°C, inlet heating should be employed. An evaporative cooler downstream of the supercharger removes the heat of compression introduced by the fan, which raises inlet pressure from, e.g., 101.3 to 116 kPa, and thus essentially improves gas turbine performance (Kolp et al., 1995).

To maintain the inlet at the optimum 10°C over the full operating range, an absorption chiller or a compression chiller should be supplemented by an inlet heater.

RECUPERATED WATER INJECTED AND HUMID AIR TURBINE CYCLES

The intercooled recuperated (ICR) cycle is a simple aeroderivative gas turbine with an added waste heat recuperator. It can significantly reduce specific fuel consumption. The ICR cycle supplemented by water injection makes up a recuperated water injected (RWI) cycle, which can remarkably improve gas turbine performance in terms of power output and efficiency (Macchi et al., 1995; Chiesa et al., 1995; Stecco et al., 1993; Cook et al., 1991).

Figure 5.27 shows two flow diagrams of the RWI plants. The RWI plants feature air intercooling in an evaporative or a surface heat exchanger and, after cooling, the air exiting the HP compressor in direct contact with water sprays in an air-water mixer. A waste heat economizer and a recuperator are used for heat recovery from the gas turbine exhaust gas. The economizer preheats the water to be injected into the evaporative intercooler and aftercooler. The amount of water to be injected to achieve maximum efficiency exceeds that required to saturate the air stream with water vapor. The air stream exiting the aftercooler contains very small water droplets that are evaporated in the recuperator, and thus a homogeneous air-water vapor mixture is formed and heated prior to entering the combustor. Fuel can also be preheated in the economizer and recuperator.

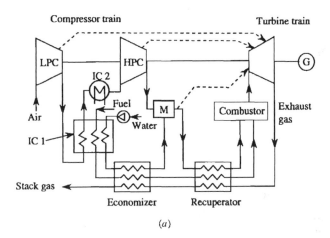

(a)

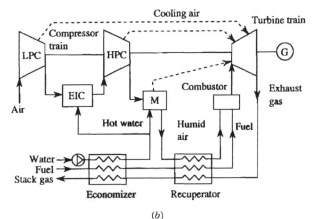

(b)

Figure 5.27. Recuperated water injected intercooled cycle plants with (a) two surface-type intercoolers, IC1 and IC2, and (b) an evaporative intercooler (EIC). LPC and HPC, low-pressure and high-pressure compressor; M, water/air mixer.

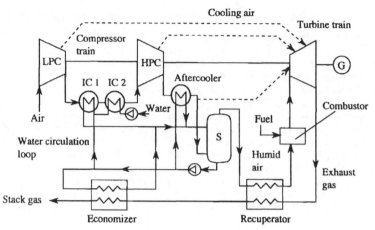

Figure 5.28. Layout of humid air turbine system. LPC and HPC, low-pressure and high-pressure compressor; IC1 and IC2, intercoolers; S, air saturator.

The surface intercooler cools air to about 25°C, while the evaporative intercooler should enable achieving a relative humidity of the air of 90%. For a given overall pressure ratio in a two-stage compressor, the flow rate of water injected into the aftercooler and the intercooling pressure must be optimized. For example, the RWI cycle with TIT of 1250°C and pressure ratio of 30 attains maximum efficiency with an intercooling pressure ratio of 3.3 (surface intercooler) and 5.7 (evaporative intercooler).

The humid air turbine (HAT) cycle is the gas turbine cycle with intercooling, recuperation, air/steam mixing in a saturator, and aftercooling (Macchi et al., 1995; Chiesa et al., 1995; Stecco et al., 1993; Cook et al., 1991). Thus the HAT plant consists of a number of intercoolers and aftercoolers, air saturator, economizer, recuperator, combustor, and gas turbine/generator. The general HAT system layout with saturator is depicted in Figure 5.28. Various possible HAT plant arrangements differ in the number and type of intercoolers and aftercoolers located at the exit of the LP and HP compressor stages. The air exiting the LP compressor may be cooled either by water recirculated from the saturator bottom in the intercooler IC1, or by make-up water in the heat exchanger IC2 with subsequent mixing with saturator water upstream of IC1. For cooling the HP compressor discharge air, several aftercoolers may be used. The preheated water sprayed at the top of the saturator humidifies the air stream ascending from the bottom of the saturator. Some predicted performance data of HAT cycles are presented in Table 5.10.

Various advanced aeroderivative gas turbine cycles are compared in Table 5.11. It includes the intercooled gas turbine cycle (ICC) with heat recovery, the intercooled gas turbine cycle with steam injection and heat recovery (ISTIG), the recuperated gas turbine cycle with intercooling, aftercooling, and water injection (RWI), and the humid air turbine (HAT) cycle.

Table 5.11 shows that intercooling significantly enhances efficiency and power output of simple-cycle plants based on aero-engines, and that the HAT cycle provides the best performance among

Table 5.10. Predicted performance data of HAT cycles based on advanced aeroderivative gas turbines

Performance data	Value
Pressure ratio	12.5–40
Turbine inlet temperature, °C	1160–1320
Specific work, kJ/kg	500–735
Efficiency (LHV), %	53.5–57.4

Table 5.11. Predicted performance data of various optimized mixed cycles

Performance data	Simple cycle	ICC	ISTIG	RWI	HAT
Gas turbine type	Heavy duty	Aeroderivative	Aeroderivative	Aeroderivative	Aeroderivative
Pressure ratio	30	46	45	33	48
Turbine inlet temperature, °C	1250	1500	1500	1500	1500
Water consumption, kg/kWh	0.892	0.74	0.74	0.61	0.72
Efficiency, %	39.7	55.5	53.2	55.1	57.0

ICC, intercooled combined cycle; ISTIG, intercooled steam injected cycle; RWI, recuperated water injected cycle; HAT, humid air turbine cycle.

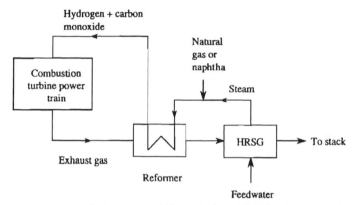

Figure 5.29. Chemically recuperated humid air turbine (CHAT) cycle power plant.

all the considered cycles. The HAT efficiency is at least as high as that of the best combined cycles based on large-frame heavy-duty gas turbines.

The HAT cycle has certain advantages as compared with the RWI cycle. The former has a more efficient water loop, which includes an intercooler, aftercooler, and economizer, and a saturator for air/water mixing. The hot water produced in various heat recovery processes is used to humidify the air in a counterflow direct-contact saturator. The heat released in the intercooler and by the low-temperature exhaust gases is more efficiently used in the HAT cycle.

The chemically recuperated humid air turbine (CHAT) cycle power plant consists of the following units:

- a gas turbine that generates electric power while burning a mixture of hydrogen and carbon monoxide
- a reformer to convert the natural gas or naphtha into a mixture of hydrogen and carbon monoxide
- a heat recovery steam generator (HRSG) to produce steam for natural gas reforming

Thus the heat of the gas turbine exhaust gas is recovered in the reformer and HRSG. This cycle has a potentially high thermal efficiency and low NO_x emissions. The schematic diagram of the CHAT plant is depicted in Figure 5.29.

CLOSURE

Modern large-frame heavy-duty industrial gas turbines as well as aeroderivative machines suitable for power generation in the intermediate-power output range have high efficiencies reaching that nearly 40%. The main contributors to this high efficiency are advanced materials and sophisticated cooling schemes that enable application of high compressor pressure ratios and, especially, of high gas turbine inlet temperatures, which currently are as high as 1250°C. A further increase in gas turbine inlet temperature up to 1500°C is anticipated.

Gas turbines are fired with clean fuels such as natural gas or light distillate, and therefore their major environmental impact is the NO_x emissions. Innovative combustors and burners along with water or steam injection enable achieving NO_x emission levels as low as 25 ppm or even 9 ppm.

For aeroderivative gas turbines, some special processes such as STIG, RWI, HAT, and CHAT have potential for achieving efficiencies over 40%. However, a real breakthrough is possible only in combined-cycle power plants, which are discussed in Chapter 6 and can achieve efficiencies of nearly 60%.

PROBLEMS

5.1. A simple-cycle gas turbine power plant operates with a compressor pressure ratio of 14, turbine inlet temperature of 1180°C, and compressor intake temperature of 10°C. The air mass flow rate is 500 kg/s. The fuel mass may be ignored. The compressor and turbine isentropic efficiencies are 0.87 and 0.92, respectively. Sketch the flow and T-s diagrams for this power plant. Calculate (a) the specific work of the compressor and the turbine, in kJ/kg, (b) the plant net specific work, in kJ/kg, (c) the plant power output, in MW, (d) the turbine exit temperature, in °C, and (e) the plant thermal efficiency. Assume an isentropic exponent of 1.4 and the specific heat of the air and gas of 1.05 kJ/(kg k).

5.2. A gas turbine with reheat has a turbine inlet temperature of 1230°C, a compressor pressure ratio of 16, and compressor and turbine efficiencies of 0.88 and 0.91, respectively. The pressure ratios of both the first and second turbine stages are the same. The reheat occurs up to a temperature of 1230°C. The compressor inlet conditions are 20 C and 1.01 bars. Assume that the working fluid heat capacity is 1.05 kJ/(kg K) and its isentropic exponent is $k = 1.4$. The fuel mass flow rate may be ignored. Determine (a) the pressures and temperatures at all real states of the cycle, (b) the turbine and compressor specific works and their ratio, and (c) the cycle thermal efficiency.

5.3. A simple-cycle gas turbine plant has compressor and turbine efficiencies of 0.88 and 0.91, respectively, and a compressor pressure ratio of 15. The air mass flow rate is 500 kg/s, and the fuel is natural gas with $LHV = 35$ MJ/m³ and a density of 0.76 kg/m³. Determine the specific work of the compressor and the turbine, the turbine exit temperature, the plant power output, fuel rate, thermal efficiency, and heat rate, for 15°C ambient and 1150°C turbine inlet temperature. Assume the compressor intake pressure of 1 bar, the isentropic exponent of 1.4 and the mean isobaric specific heat of working fluid of 1.05 kJ/(kg K).

5.4. A regenerative-cycle gas turbine plant has compressor and turbine efficiencies of 0.88 and 0.91, respectively, a recuperator effectiveness of 0.77, and a compressor pressure ratio of 16. Determine the specific work of the compressor and the turbine, the plant specific net work, the turbine and recuperator exit temperatures, and the cycle thermal efficiency for 15°C ambient and 1150°C turbine inlet temperature. Assume a compressor intake pressure of 1 bar, an isentropic exponent of 1.4, and a mean isobaric specific heat of working fluid of 1.05 kJ/(kg K).

5.5. A gas turbine plant with an intercooler and reheater has efficiencies of 0.87 and 0.9 for the LP and HP compressors and for the HP and LP turbines, respectively. The compressor pressure ratio is 6 both for LP and HP compressor stages. The HP and LP turbines have pressure ratios of 3 and 12, respectively. The air mass flow rate m_{air} is 360 kg/s and the fuel mass flow rate is 2.5% of that of air. Determine the compressor power input, the turbine and plant net power output, the plant thermal efficiency, and the plant heat rate for 15°C ambient and 1150°C turbine inlet temperatures, and the air mass flow rate of 360 kg/s, assuming that the fuel mass flow rate is 2.5% that of air. Assume a compressor intake pressure of 1 bar, an isentropic exponent of 1.4, and a mean isobaric specific heat of working fluid of 1.05 kJ/(kg K).

5.6. Consider a closed-cycle gas turbine plant with a coal-fired furnace used to heat the compressor discharge air in a high-temperature ceramic heat exchanger (heater). The compressor inlet, turbine inlet, and furnace exit temperatures are 35, 1200, and 1800°C, respectively. The efficiencies of the compressor, gas turbine, and furnace are 0.86, 0.9, and 0.95, respectively, and the compressor pressure ratio is 11. The heat of air leaving the turbine is partly used in the second heat exchanger (cooler) to preheat the combustion air for the furnace. If the coal has a heating value of 31.4 MJ/kg and the flue gas volume is 13.7 m³/kg coal, what is the coal consumption rate, in kg/s, for a 150-MW plant? Determine the plant net specific work per kg air, the mass flow of the fuel in kg/s, the thermal efficiency, and the heater gas exit temperature. Assume a compressor intake pressure of 1 bar, an isentropic exponent of 1.4, a mean isobaric specific heat of working fluid of 1.05 kJ/(kg K), and the mean isobaric specific heat of the flue gas of 1.35 kJ/(m³ K).

5.7. A 240-MW simple-cycle gas turbine plant operates with a 1250°C turbine inlet temperature, compressor pressure ratio of 13, and 1 bar and 10°C ambient conditions. The compressor and turbine efficiencies are 87% and 90%, respectively. The fuel is natural gas with $LHV = 48$ MJ/kg. Determine (a) the plant net specific work in kJ/kg air, (b) the air mass flow rate in kg/s, (c) the fuel mass flow rate in kg/s, and (d) the plant thermal efficiency. Assume a compressor intake pressure of 1 bar, an isentropic exponent of 1.4, and a mean isobaric specific heat of working fluid of 1.05 kJ/(kg K).

5.8. In a 200-MW simple-cycle gas turbine power plant that operates with a compressor pressure ratio of 16, turbine inlet temperature of 1450°C, and compressor intake temperature of 10°C, natural gas fuel with LHV = 45 MJ/kg is burned. The compressor and turbine isentropic efficiencies are 0.87 and 0.92, respectively. Calculate (a) the plant net specific work, in kJ/kg air, (b) the compressor air flow, in kg/s, (c) the fuel consumption rate, in kg/s, (d) the turbine exit temperature, in degrees centigrade, and (e) the plant thermal efficiency.

5.9. A 60-MW simple-cycle, aeroderivative gas turbine power plant uses a two-stage compressor with intercooling and a total pressure ratio of 36. Assume ideal intercooling to 15°C. The gas turbine inlet temperature is 1400°C, and the compressor and turbine efficiencies are 0.86 and 0.92, respectively. The ambient air temperature, pressure, and relative humidity are 15°C, 1.013 bars, and 60%, respectively. Calculate (a) the plant net specific work, in kJ/kg air, (b) the compressor air flow, in kg/s, and (c) the thermal efficiency.

REFERENCES

ABB AG. 1991–1996. Gas turbine specifications. Baden, Switzerland.

Bannister, R. L., Cheruvu, N. S., Little, D. A., and McQuiggan, G. 1995. Development requirements for an advanced gas turbine system. *Trans. ASME J. Eng. Gas Turbine Power* 117:724–735.

Becker, B., and Finckh, H. H. 1995. The 3A-series gas turbines. *Siemens Power J.* (August):13–17.

Chiesa, P., Lozza, G., Macchi, E., and Consonni, S. 1995. An assessment of the thermodynamic performance of mixed gas-steam cycles. Part B—Water-injected and HAT cycles. *Trans. ASME J. Eng. Gas Turbine Power* 117:499–508.

Cohen, H., Rogers, G. F. C., and Saravanamuttoo, H. I. H. 1987. *Gas turbine theory*, 3rd ed. New York: Longman.

Collins, S. 1994. Gas turbine power plants. *Power* 138(6):17–31.

Cook, D. T., McDaniel, J. E., and Rao, A. D. 1991. HAT cycle simplifies coal gasification power. *Mod. Power Syst.* 11(5):19, 21, 23, 25.

Czech, N., Esser, W., and Schmitz, F. 1994. Developments in materials for gas turbine blades. *Siemens Power J.* 4:22–28.

Dodero, 1997. *Italian power generation, transmission, and distribution handbook.* Italian Energy Handbook, ENEL.

Doebbeling, K., Knoepfel, H. P., Polifke, W. et al. 1996. Low NO_x premixed combustion of MBtu fuels using the ABB double cone burner (EV burner). *Trans. ASME J. Eng. Gas Turbine Power* 118(1):46–53.

Eastop, T. D., and McConkey, A. 1993. *Applied thermodynamics for engineering technologists*, 5th ed. New York: Longman, Harlow.

Haywood, R. W. 1991. *Analysis of engineering cycles.* New York: Pergamon.

Kano, K., Matsuzaki, H., Aoyama, K., et al. 1991. Development study of 1500°C class high temperature gas turbine. ASME paper 91-GT-297.

Kolp, D. A., Flye, W. M., and Guidotti, H. A. 1995. Advantages of air conditioning and supercharging an LM6000 gas turbine inlet. *Trans. ASME J. Eng. Gas Turbine Power* 117:513–520.

Lefebvre, A. H. 1995. The role of fuel preparation in low-emission combustion. *Trans. ASME J. Eng. Gas Turbine Power* 117:617–653.

Macchi, E., Consonni, S., Lozza, G., and Chiesa, P. 1995. An assessment of the thermodynamic performance of mixed gas-steam cycles. Part A—Intercooled and steam-injected cycles. *Trans. ASME J. Eng. Gas Turbine Power* 117:489–498.

Oganowski, G. 1990. GE LM6000 development of the first 40% thermal efficiency gas turbine. G.E. Marine and Industrial Engine Bulletin.

Rice, I. G, 1995. Steam-injected gas turbine analysis: Steam rates. *Trans. ASME J. Eng. Gas Turbine Power* 117(2):251–258.

Rogers, G. F., and Mayhew, Y. R. 1992. *Engineering thermodynamics: Work and heat transfer*, 4th ed. New York: Longman, Harlow.

Rolle, K. C. 1994. *Thermodynamics and heat power*, 4th ed. New York: Macmillan.

Schulenberg, T. 1995. Key technologies for improving gas turbine performance. *Siemens Power J.* 2:9–12.

Stecco, S. S., Desideri, U., and Bettagli, N. 1993. The humid air cycle: Some thermodynamic considerations. Humid air gas turbine cycle: A possible optimization. ASME papers 93-GT-77 and 93-GT-178.

Swanekamp, R. 1995. Enhancing gas turbine performance. *Power* 139(9):47–58.

Wark, K. 1994. *Advanced thermodynamics for engineers.* New York: McGraw-Hill.

Chapter Six

GAS TURBINE BASED COMBINED-CYCLE
POWER PLANTS

Combined-cycle power plants are the most efficient among all types of power plants. The basic configuration of a combined-cycle power plant consists of a gas turbine, a heat recovery steam generator (HRSG), and a steam turbine plant. This chapter discusses all major issues related to gas turbine and steam turbine combined-cycle power plants including thermodynamics, design, environmental impact, and economics. Thermodynamic analysis of a single-pressure combined-cycle power plant is given. To achieve the highest efficiency, ~58–60%, a combined cycle must employ the most efficient gas turbine and a rather sophisticated steam turbine plant including double- or triple-pressure HRSG with a reheat steam cycle. The newest trends in the development of even more efficient combined-cycle power plants are described. Numerous tables contain current specifications and performance data of advanced combined-cycle power plants.

TOPPING, BOTTOMING, AND COMBINED CYCLES

The principle of efficiency enhancement through increasing the mean temperature of the heat supply, T_{in}, and lowering the mean temperature of heat rejection, T_{out}, is consistently implemented in combined-cycle power plants. A combined-cycle power plant consists of a topping, high-temperature cycle and a bottoming, low-temperature cycle (see Figure 6.1). In the topping cycle, part of the fuel energy supplied to the cycle is converted into electricity, and the remaining heat is rejected from the topping cycle and further used to produce power in the bottoming cycle. In a combined cycle, fuel energy utilization is greater in both the topping and bottoming cycle. Hence the objective of a combined cycle is to achieve greater work output for a given heat (or fuel energy) input, or in other words, to achieve a higher energy conversion efficiency. In the search for higher conversion efficiencies, several innovative concepts of combined-cycle power plants have been developed in recent years. The majority of modern combined-cycle power plants are based on the utilization of gas turbines in the topping cycle and steam turbines in the bottoming cycle. Overall thermal efficiencies of combined-cycle power plants of 55–57% have been achieved thus far, and efficiencies of 58–60% have already been announced by some gas turbine manufacturers. The prerequisite for such high overall efficiencies is the high efficiency of modern advanced gas turbines. Today, simple-cycle gas turbine efficiency of 38.5–39.5% is achieved, and a further increase in efficiency is anticipated with even more advanced gas turbines (ABB AG, 1991–1996; Becker and Finckh, 1995; Balling et al., 1995; Bannister et al., 1995; Briesch et al., 1995).

Several other combined-cycle power plant concepts are now being further developed, e.g.,

- a magnetohydrodynamic (MHD)/steam turbine combined-cycle power plant in which the exhaust gas heat from an MHD generator, operating at high temperature, is used to raise steam for a steam cycle plant
- a fuel cell/steam turbine combined-cycle power plant (Maude, 1993)
- a combined-cycle power plant consisting of a gas turbine topping cycle and a Kalina bottoming cycle with a binary mixture of water and ammonia as a working fluid (Marston and Hyre, 1995)

127

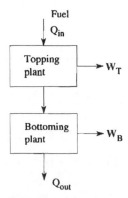

Figure 6.1. General concept of a combined-cycle power plant.

The latter two types of combined-cycle power plants are discussed in more detail in Chapter 8. Even more complex combined-cycle power plants are possible, e.g., a three-stage combined-cycle power plant consisting of a high-temperature topping (MHD) power plant, gas turbine cycle, and steam turbine process.

OVERALL EFFICIENCY OF COMBINED-CYCLE POWER PLANTS

Combined Cycles in General

Essential improvements in energy conversion efficiency are associated with combined-cycle power plants (Bolland, 1991; Haywood, 1991). In recent years, tremendous advances have been made in this area of power technology. To begin with, we shall derive a general expression for the overall efficiency of a combined-cycle power plant. Thus we consider an unspecified combined-cycle power plant shown in Figure 6.1 which consists of a topping cycle and a bottoming cycle. Assuming that heat is supplied only to the topping cycle, the overall efficiency of the combined cycle is derived as follows. The supplied heat Q_{in} is partially converted into net work output of the topping cycle W_T with a thermal efficiency η_T. The heat rejected from the topping cycle $Q_{T,out}$ is further converted into net work output of the bottoming cycle W_B with a thermal efficiency η_B. The remaining heat is then rejected from the bottoming cycle plant to the ambient.

The net work output of the topping cycle is

$$W_T = Q_{in}\eta_T \tag{6.1}$$

The heat rejected from the topping cycle is

$$Q_{T,out} = Q_{in}(1 - \eta_T) \tag{6.2}$$

The net work output of the bottoming cycle is

$$W_B = Q_{T,out}\eta_B = Q_{in}(1 - \eta_T)\eta_B \tag{6.3}$$

Then, the total net work output of the combined cycle is

$$W_{CC} = W_T + W_B = Q_{in}\eta_T + Q_{in}(1 - \eta_T)\eta_B \tag{6.4}$$

The thermal efficiency of the combined cycle is thus

$$\eta_{CC} = \frac{W_{CC}}{Q_{in}} \tag{6.5}$$

or

$$\eta_{CC} = \eta_T + \eta_B(1 - \eta_T) = \eta_T + \eta_B - \eta_T\eta_B \tag{6.6}$$

It is seen from Eq. (6.6) that the thermal efficiency of the combined cycle is higher than that of the topping cycle but smaller than the sum of the thermal efficiencies of the topping and bottoming cycles.

Gas Turbine Based Combined-Cycle Power Plants

The efficiency of the conversion of heat energy into work is enhanced when the mean temperature of heat supply increases and the mean temperature of heat rejection in the cycle decreases. For advanced heavy-duty gas turbines the exhaust gas temperature lies between 550 and 650°C (ABB AG, 1991–1996; Becker and Finckh, 1995; Balling et al., 1995; Bolland, 1991; Bannister et al., 1995; Briesch et al., 1995). If the exhaust gas heat is not used, a lot of energy is wasted.

The exhaust gas heat can be used in two different ways: (1) for preheating compressed air in a regenerative gas turbine cycle or (2) for raising steam in a combined cycle. In the former case, the addition of a recuperator for air preheating by the exhaust heat converts the simple-cycle gas turbine into the regenerative cycle with the resulting increased thermal efficiency.

However, fuel energy utilization efficiency is dramatically raised when the simple-cycle gas turbine is converted into a combined-cycle power plant. The gas turbine is used as the topping plant, and the steam turbine as the bottoming plant. The overall efficiency of the combined-cycle power plant is higher than the thermal efficiency of either the gas turbine or the steam power plant. It should be stressed that the most important prerequisite for higher efficiency of a combined-cycle power plant is the utilization of a highly efficient topping gas turbine. Conditions for achieving high energy conversion efficiencies in gas turbines are discussed in Chapter 5.

The gas turbine based combined-cycle power plant consisting of a topping gas turbine and a bottoming steam plant is usually designated as STAG. Heat rejected from the gas turbine plant is further utilized in the steam plant. The exhaust heat of the gas turbine plant is used in a heat recovery steam generator (HRSG) to raise steam for the steam plant. There are two basic types of gas turbine based combined-cycle power plants (STAG): (1) STAG without supplementary firing (see Figure 6.2) and (2) STAG with supplementary firing (see Figure 6.3). In a STAG without supplementary firing, fuel is burned only in the gas turbine combustor and heat is supplied to the steam cycle from the gas turbine exhaust gas.

As was described in Chapter 5, the simple-cycle gas turbine power plant consists of an air compressor, a combustion chamber, and a gas turbine with electric generator. The gas turbine drives both the compressor and generator. Heat is supplied to the cycle by fuel combustion and rejected to the ambient with gas turbine exhaust gas. A closed-cycle gas turbine plant that employs an external heater for heat addition to the working fluid and a cooler for heat rejection to the ambient is seldom considered for use in combined cycles.

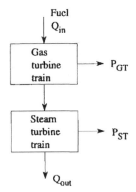

Figure 6.2. Gas and steam turbine combined-cycle power plant without supplementary firing.

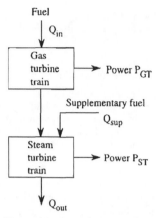

Figure 6.3. Gas and steam turbine
combined-cycle power plant with
supplementary firing.

It is known that the thermal efficiency of the Standard Joule cycle can be increased by increasing
the compression pressure ratio (Haywood, 1991). Real gas turbine plants differ from the Joule
cycle because of irreversibilities of processes in the compressor, combustor, and gas turbine. They
cause energy losses and thus decrease the gas turbine efficiency in comparison with the theoretical
cycle. All these aspects are considered in detail in Chapter 5.

The thermal efficiency η_{GT} of a simple-cycle gas turbine power plant depends on the pressure
ratio β, the maximum temperature ratio in the cycle T_3/T_1, as well as the isentropic efficiencies
of the compressor and gas turbine, η_{ic} and η_{it}, respectively (Haywood, 1991). Thus

$$\eta_{GT} = \frac{w_{net}}{q_{in}} = \frac{(r-1)(n\eta_{it}\,\eta_{ic}/r - 1)}{(n-1)\eta_{ic} - (r-1)} \tag{6.7}$$

where $r = \beta^{(k-1)/k}$, $n = T_3/T_1$, k is the isentropic exponent, T_3 is the gas turbine inlet temperature,
and T_1 is the air compressor intake temperature.

Raising the compression pressure ratio provides a substantial improvement in thermal efficiency of gas turbine power plants operating on the Joule cycle. However, the efficiency of real
simple-cycle gas turbine plants is enhanced predominantly by increasing the gas turbine inlet
temperature. Further improvements in the gas turbine efficiency are attained through the use of
additional components such as intercoolers, reheaters, and recuperators for air preheating. All
these modifications to the basic gas turbine power plant cycle are discussed in Chapter 5.

Let us derive an expression for the overall efficiency, η_{CC}, of a gas turbine based combined-cycle
power plant without supplementary firing shown in Figure 6.4. Figure 6.5 shows a T-s diagram of

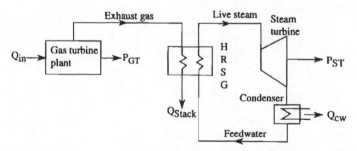

Figure 6.4. Flow diagram of a combined-cycle plant without supplementary firing.

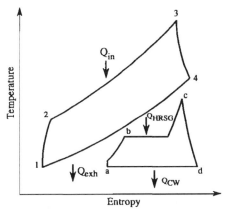

Figure 6.5. T-s diagram of a combined-cycle power
plant (irreversible processes).

this combined-cycle power plant. Substituting the thermal efficiency of gas turbine η_{GT} and that
of steam turbine η_{ST} for the efficiencies of topping and bottoming cycle plants, η_T and η_B, in Eq.
(6.6) yields

$$\eta_{CC} = \eta_{GT} + \eta_{ST}(1 - \eta_{GT}) = \eta_{GT} + \eta_{ST} - \eta_{GT}\eta_{ST} \tag{6.8}$$

It is obvious that

$$\eta_{GT} = P_{GT}/Q_{in} \tag{6.9}$$

where P_{GT} is the gas turbine power output and Q_{in} is the heat input to the gas turbine combustor.
The gas turbine exhaust gas heat is

$$Q_{exh} = (1 - \eta_{GT})Q_{in} \tag{6.10}$$

At the same time, it is the heat addition to the steam cycle. Therefore the thermal efficiency of the
steam cycle is

$$\eta_{ST} = P_{ST}/Q_{exh} \tag{6.11}$$

where P_{ST} is the power output of the bottoming steam turbine plant.

To prove the sensitivity of the overall thermal efficiency to changes in the thermal efficiency of
the topping cycle, Eq. (6.8) will be differentiated as follows:

$$d\eta_{CC}/d\eta_{GT} = 1 + d\eta_{ST}/d\eta_{GT}(1 - \eta_{GT}) - \eta_{ST} \tag{6.12}$$

The overall efficiency increases only if

$$d\eta_{CC}/d\eta_{GT} > 0 \tag{6.13}$$

It follows for the relative decrease in the bottoming cycle:

$$-d\eta_{ST}/d\eta_{GT} < (1 - \eta_{ST})/(1 - \eta_{GT}) \tag{6.14}$$

From Eq. (6.14), it follows that increasing the thermal efficiency of the topping cycle improves
the overall thermal efficiency of the combined cycle, provided that the thermal efficiency of the
bottoming cycle thereby does not decrease too much.

Similar to single-cycle gas turbine power plants, the overall thermal efficiency of the combined
cycle η_{CC} depends on the compressor pressure ratio β and the gas turbine inlet temperature (TIT)
(Haywood, 1991).

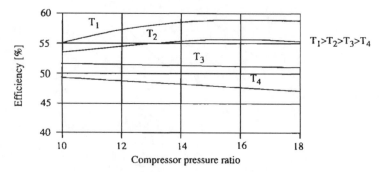

Figure 6.6. Combined-cycle efficiency as a function of compressor pressure ratio and turbine inlet temperature.

For simple-cycle as well as for combined-cycle power plants, the maximum gas turbine inlet temperature is limited by the applied materials and the cooling efficiency of the gas turbine blades. The gas turbine inlet temperature lies between 1100 and 1200°C for second-generation gas turbines. For third-generation gas turbines it exceeds 1200°C, and in some advanced gas turbines it attains values above 1350°C (Author, 1995; Bannister et al., 1995). As a result of further advances in turbine blade materials and their improved cooling, gas turbine firing temperatures up to 1500°C are anticipated in the next couple of years (Maude, 1993; Author, 1995). As soon as the gas turbine inlet temperature is fixed at a certain value, the thermal efficiency of the simple-cycle power plant is mainly determined through the pressure ratio and the ratio of the gas turbine inlet temperature and the compressor intake temperature. The maximum simple-cycle efficiency is reached when the exhaust gas temperatures are quite low, as low exhaust temperature means a high pressure ratio.

The overall efficiency of the combined-cycle power plant is shown in Figure 6.6, which gives the relationship between the efficiency, compressor pressure ratio, and TIT.

On the basis of economical considerations, gas turbines are usually optimized with respect to maximum specific power output per a unit mass flow of working fluid in the gas turbine. The optimum coincides fairly accurately with the optimum overall efficiency of the combined-cycle power plant. However, these conditions differ from those for the maximum efficiency of the simple cycle.

Example 6.1 illustrates the effect of changes in thermal efficiencies η_{GT} and η_{ST} on the overall efficiency and power output of the combined cycle. The calculations are based on Eqs. (6.8) and (6.14).

Example 6.1

Estimate the effect of changes in the thermal efficiency of the simple-cycle gas turbine plant, η_{GT}, and in the thermal efficiency of the bottoming steam turbine plant, η_{ST}, on the overall efficiency of the combined-cycle power plant.

Assuming initially $\eta_{GT} = 0.3$ and $\eta_{ST} = 0.28$, what is the maximum allowable decrease in η_{ST} when η_{GT} is increased to 0.38?

Solution

1. Substituting the initial values of η_{GT} and η_{ST} in Eq. (6.8) yields

$$\eta_{CC} = \eta_{GT} + \eta_{ST}(1 - \eta_{GT}) = 0.3 + 0.28(1 - 0.3) = 0.496$$

2. Increasing η_{GT} to 0.38 gives the maximum allowable decrease in η_{ST} related to the change in η_{GT}:

$$-d\eta_{ST}/d\eta_{GT} < (1 - 0.28)/(1 - 0.38) < 1.16$$

Then the maximum allowable relative decrease in η_{ST} is $1.16 \times (0.38 - 0.3) = 0.09$

Table 6.1. Maximum allowable relative
decrease in the thermal efficiency of the
steam process as a function of gas
turbine efficiency

η_{GT}	$-d\eta_{ST}/d\eta_{GT}$
0.3	1
0.35	1.08
0.4	1.17

Steam process efficiency is 0.3.

3. Now, assuming $\eta_{GT} = 0.38$ and $\eta_{ST} = 0.25$,

$$\eta_{CC} = \eta_{GT} + \eta_{ST}(1 - \eta_{GT}) = 0.38 + 0.25(1 - 0.38) = 0.535$$

This means an increase by $(0.535 - 0.496) = 0.039$ in the overall thermal efficiency of the combined-cycle power plant.

Table 6.1 shows the maximum allowable relative decrease in the thermal efficiency of the steam process $-d\eta_{ST}/d\eta_{GT}$ as a function of gas turbine efficiency. Thus at higher thermal efficiency of the gas turbine, the maximum allowable decrease in the thermal efficiency of the steam process is larger. Example 6.2 shows how the increase in η_{GT} affects the overall efficiency and the power output of the combined cycle.

Example 6.2

Consider a gas turbine based combined cycle without supplementary firing. Calculate the power output of the simple-cycle gas turbine and that of the combined cycle if the rate of heat supply to the gas turbine combustor is $Q_{GT} = 600$ MW. Compare the increases in the power output of the simple cycle and combined cycle power plants if the initial values of the thermal efficiencies of the gas turbine cycle $\eta_{GT} = 0.31$ and that of the bottoming steam process $\eta_{ST} = 0.29$ are changed to $\eta'_{GT} = 0.38$ and $\eta'_{ST} = 0.26$.

Solution

1. The power output of the simple-cycle gas turbine power plant with the initial and changed thermal efficiency of the gas turbine, respectively, is

   (i) $P_{GT} = \eta_{GT} Q_{GT} = 0.31 \times 600 = 186$ MW

   (ii) $P'_{GT} = \eta'_{GT} Q_{GT} = 0.38 \times 600 = 228$ MW

2. The overall thermal efficiency of the combined cycle with the initial and changed thermal efficiencies of the gas turbine and of steam turbine, respectively, is

   (i) $\eta_{CC} = \eta_{GT} + \eta_{ST}(1 - \eta_{GT}) = 0.31 + 0.29(1 - 0.31) = 0.51$

   (ii) $\eta'_{CC} = \eta'_{GT} + \eta'_{ST}(1 - \eta'_{GT}) = 0.38 + 0.26(1 - 0.38) = 0.54$

3. Correspondingly, the overall power output of the combined-cycle power plant is

   (i) $P_{CC} = \eta_{CC} Q_{GT} = 0.51 \times 600$ MW $= 306$ MW

   (ii) $P'_{CC} = \eta'_{CC} Q_{GT} = 0.54 \times 600$ MW $= 324$ MW

4. An increase in the overall power output of the combined-cycle power plant is

   $$\Delta P_{CC} = P'_{CC} - P_{CC} = 324 - 306 = 18 \text{ MW}$$

   or $100\% \times \Delta P_{CC}/P_{CC} = 5.55\%$.

5. The power output of the bottoming steam process is

   (i) $P_{ST} = P_{CC} - P_{GT} = 306 - 186 = 120$ MW

   (ii) $P'_{ST} = P'_{CC} - P'_{GT} = 324 - 228 = 96$ MW

Hence there is an increase in the proportion of the overall power output being provided by the gas turbine, thus increasing the efficiency of the bottoming steam process.

Combined-Cycle Power Plants with Supplementary Firing

Let us now consider a combined-cycle power plant with supplementary firing depicted in Figure 6.3. Here fuel is burned not only in the combustor of the gas turbine, but also in a supplementary furnace of the HRSG, generating steam for the steam turbine. If more heat is required for the steam bottoming cycle than the gas turbine exhaust gas can provide, the overall combined-cycle efficiency would be increased if the additional work output is large enough and supplementary heat is small enough. The gas turbine exhaust gas usually has a high residual oxygen content, around 15–17%, because of the high air-fuel ratio required to limit the turbine inlet temperature. This hot, oxygen-rich exhaust gas can be used instead of air to burn additional fuel in a steam generator. This is particularly advantageous for repowering old steam power plants by converting them to combined-cycle power plants, since, owing to supplementary firing, the existing power train can be further utilized.

In general, the overall thermal efficiency of a gas turbine based combined-cycle power plant is defined as the ratio of the total power output of the combined-cycle power plant to the total rate of heat input to the plant:

$$\eta_{CC} = (P_{GT} + P_{ST})/(Q_{GT} + Q_{SF}) \tag{6.15}$$

where P_{GT} is the gas turbine power output, P_{ST} is the steam turbine power output, Q_{GT} is the rate of heat addition in the gas turbine combustor, and Q_{SF} is the rate of heat supply to the supplementary firing.

The thermal efficiencies of simple-cycle gas turbine plant and steam power plant are, respectively,

$$\eta_{GT} = P_{GT}/Q_{GT} \qquad \eta_{ST} = P_{ST}/Q_{ST} \tag{6.16}$$

Therefore the gas turbine exhaust heat is

$$Q_{exh} = Q_{GT}(1 - \eta_{GT}) \tag{6.17}$$

The total rate of heat input to the steam turbine process is then

$$Q_{ST} = Q_{exh} + Q_{SF} = Q_{GT}(1 - \eta_{GT}) + Q_{SF} \tag{6.18}$$

The thermal efficiency of the steam process in a combined cycle with supplementary firing is

$$\eta_{ST} = P_{ST}/Q_{ST} = P_{ST}/[Q_{GT}(1 - \eta_{GT}) + Q_{SF}] \tag{6.19}$$

The power output of the gas turbine is

$$P_{GT} = \eta_{GT}Q_{GT} \tag{6.20}$$

Now, the power output of the steam turbine is

$$P_{ST} = \eta_{ST}Q_{ST} = \eta_{ST}Q_{GT}[(1 - \eta_{GT}) + f_{SF}] \tag{6.21}$$

where f_{SF}/Q_{GT} is the ratio of heat inputs to the supplementary firing and to the gas turbine combustor.

Thus the total power output of the combined-cycle power plant with supplementary firing is

$$P_{CC} = P_{GT} + P_{ST} = Q_{GT}\{\eta_{GT} + \eta_{ST}[(1 - \eta_{GT}) + f_{SF}]\} \tag{6.22}$$

Therefore the overall efficiency of the combined cycle with supplementary firing is

$$\eta_{CC} = \{\eta_{GT} + \eta_{ST}[(1 - \eta_{GT}) + f_{SF}]\}/(1 + f_{SF}) \tag{6.23}$$

Examples 6.3 and 6.4 show how the performance of the combined cycle changes when supplementary firing is used.

Example 6.3

For the combined-cycle power plant consisting of a gas turbine with a thermal efficiency $\eta_{GT} = 0.39$ and a steam turbine with a thermal efficiency $\eta_{ST} = 0.31$, calculate the overall efficiency and the plant net heat rate.

Solution

The overall efficiency of the combined-cycle power plant is

$$\eta_{CC} = \eta_{GT} + \eta_{ST} - \eta_{GT}\eta_{ST} = 0.39 + 0.31 - 0.39 \times 0.31 = 0.579$$

The plant net heat rate HR is

$$HR = 3600/\eta_{CC} = 3600/0.579 = 6216.5 \text{ kJ/kWh}$$

Example 6.4

Calculate the overall efficiency, the heat rate, and the total fuel rate of a combined-cycle power plant (1) if no supplementary firing is used and (2) if 30% of the gas turbine fuel demand is additionally burned in the supplementary firing of the HRSG. The gas and steam turbine efficiencies are 37.5% and 26% [both based on lower heating value (LHV)], respectively. Assume that the power plant has an electrical power output of 380 MW and that the fuel HFO No. 2 has the lower heating value of 43 MJ/kg.

Solution

1. Combined-cycle power plant without supplementary firing
   Plant overall efficiency

   $$\eta_{CC} = \eta_{GT} + \eta_{ST}(1 - \eta_{GT}) = 0.375 + 0.26(1 - 0.375) = 0.5375$$

   Heat rate

   $$HR = 3600/\eta_{CC} = 3600/0.5375 = 6697.7 \text{ kJ/kWh}$$

   Fuel rate

   $$m_f = P_{CC}/(\eta_{CC}\text{LHV}) = 380 \text{ MW}/(0.5375 \times 43 \text{ MJ/kg}) = 16.44 \text{ kg/s}$$

2. Combined-cycle power plant with supplementary firing
   Plant overall efficiency

   $$\eta_{CC} = (\eta_{GT} + \eta_{ST} - \eta_{GT}\eta_{ST} + f_{SF}\eta_{ST})/(1 + f_{SF})$$
   $$= (0.375 + 0.26 - 0.375 \times 0.26 + 0.3 \times 0.26)/(1 + 0.3) = 0.473$$

   Heat rate

   $$HR = 3600/\eta_{CC} = 3600/0.473 = 7611 \text{ kJ/kWh}$$

   Fuel rate

   $$m_f = P_{el}/(\eta_{CC}\text{LHV}) = 380 \text{ MW}/(0.473 \times 43 \text{ MJ/kg}) = 18.66 \text{ kg/s}$$

Thus, in comparison with the combined-cycle power plant without supplementary firing, the combined-cycle power plant with supplementary firing has a lower efficiency but higher heat and fuel rates.

COMBINED-CYCLE POWER PLANT CONFIGURATIONS

Single-Pressure Combined-Cycle Power Plants

Depending on the complexity of the HRSG and steam turbine, there are three basic configurations of combined-cycle power plants (ABB AG, 1991–1996; Becker and Finckh, 1995; Balling et al., 1995; Bolland, 1991; Taft, 1991; Swanekamp, 1995; Croonenbrock et al., 1996):

- single-pressure combined-cycle power plant
- dual-pressure combined-cycle power plant
- triple-pressure combined-cycle power plant

The single-pressure combined cycle power plant is shown in Figure 6.7. Figure 6.8 shows the temperature profiles of exhaust gas and water/steam working fluid in the single-pressure HRSG. Exhaust gas cools as it passes through the HRSG, where feedwater is heated and evaporated and steam is superheated.

Applying the First Law of Thermodynamics to the superheater and evaporator together, the energy balance for this section of the HRSG is given by

$$m_g c_{pg}(t_1 - t_2) = m_s(h_6 - h_5) \quad \text{kJ/s} \tag{6.24}$$

where m_g and m_s are the mass flow rates for the exhaust gas and steam in kg/s, respectively, h_5 is the enthalpy of saturated liquid at the HRSG pressure in kJ/kg, and h_6 is the enthalpy of superheated steam in kJ/kg.

The flue gas temperature at state 2 is

$$t_2 = t_5 + \text{PP} \quad °\text{C} \tag{6.25}$$

where t_5 is the saturation temperature at the HRSG pressure in °C and PP is the pinch point temperature difference in K.

The PP temperature difference is the minimum temperature difference between the two fluids—exhaust gas and water/steam—in the HRSG, i.e., $t_2 - t_5$. It should exceed some minimum design value (about 10 K) for all operating conditions of the system to make effective use of all of the HRSG heat transfer surface. Smaller PP values would substantially increase the heat transfer surface area needed, while significantly larger values would necessitate the reduction of the saturation temperature and, subsequently, pressure, which would adversely affect the combined-cycle thermal efficiency (Bolland, 1991; Horlock, 1995).

From Eq. (6.24) the rate of steam production in a single-pressure HRSG is

$$m_s = m_g c_{pg}(t_1 - t_2)/(h_6 - h_5) \quad \text{kg/s} \tag{6.26}$$

Similarly, the energy balance for the economizer is

$$m_g c_{pg}(t_2 - t_3) = m_s(h_5 - h_4) \quad \text{kJ/s} \tag{6.27}$$

where t_3 is the flue gas temperature at the HRSG exit in °C and h_4 is the enthalpy of feedwater at the HRSG inlet in kJ/kg.

By adding Eqs. (6.24) and (6.27), we obtain the energy balance for the entire HRSG. Thus the heat transfer rate for the HRSG is

$$Q_{\text{HRSG}} = m_g c_{pg}(t_1 - t_3) = m_s(h_6 - h_4) \quad \text{kJ/s} \tag{6.28}$$

From Eq. (6.27) or (6.28), the temperature of the flue gas leaving the HRSG is

$$t_3 = t_2 - m_s(h_5 - h_4)/m_g c_{pg} = t_1 - m_s(h_6 - h_4)/m_g c_{pg} \quad °\text{C} \tag{6.29}$$

This temperature is limited because there is danger of corrosion from water vapor condensation in the presence of sulfur oxides. The temperature must be higher than the dewpoint temperature of the flue gas, which is determined by the partial pressure of water vapor and is affected by the sulfur content of the fuel. On the other hand, the feedwater temperature at the inlet of the HRSG should be as low as possible to allow optimum utilization of gas turbine exhaust heat. Since the heat transfer rate on the gas side in the HRSG is poor in comparison with that on the water/steam side, the surface temperature at the cold end of the HRSG is only slightly higher than the feedwater temperature.

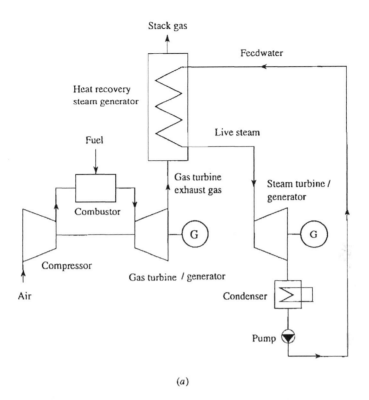

(a)

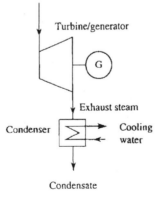

(b)

Figure 6.7. Single-pressure combined-cycle power plant: (a) flow diagram and (b) schematic of steam turbine.

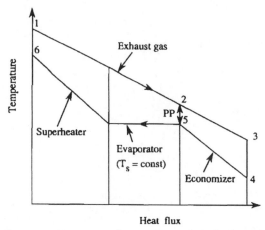

Figure 6.8. Temperature–heat flux diagram of single-pressure
HRSG.

Example 6.5

Calculate the rate of steam production, the stack gas temperature, and the stack gas heat rejection
rate for a single-pressure HRSG of a combined-cycle power plant (see Figures 6.7 and 6.8). The
conditions are as listed below:

Gas turbine exhaust gas mass flow rate $m_g = 540$ kg/s
Exhaust gas temperature $t_1 = 610°C$
Superheated steam pressure $p_6 = 60$ bars and temperature $t_6 = 500°C$
Feedwater temperature $t_4 = 110°C$

Assume that the HRSG PP is 10 K and that the exhaust gas specific heat c_{pg} is 1.05 kJ/(kg K).

Solution

1. Superheated steam enthalpy $h_6 = 3425$ kJ/kg (from h-s diagram)
   Feedwater enthalpy $h_4 = c_{pw}t_4 = 4.187 \times 110 = 460.6$ kJ/kg
   Saturation temperature at $p_5 = 60$ bars is $t_5 = 275.55°C$, and saturated liquid enthalpy
   $h_5 = 1213.7$ kJ/kg (from water/steam table)
2. Designating the states as shown in Figure 6.8, we first calculate the rate of steam production
   per unit mass of the exhaust gas. The flue gas temperature at state 2 is

   $$t_2 = t_5 + PP = 275.55 + 10 = 285.55°C$$

   From the energy balance for the superheater and evaporator of the HRSG, the steam production
   rate is

   $$m_s = m_g c_{pg}(t_1 - t_2)/(h_6 - h_5)$$
   $$= 540 \times 1.05 \times (610 - 285.55)/(3425 - 1213.7) = 83.18 \text{ kg/s}$$

3. The rate of heat recovery in the HRSG is

   $$Q_{HRSG} = m_s(h_6 - h_4) = 83.18 \times (3425 - 460.6) = 246{,}574 \quad kW$$

4. The stack gas temperature is

   $$t_3 = t_1 - Q_{HRSG}/(m_g c_{pg}) = 610 - 246{,}574/(540 \times 1.05) = 175.1°C$$

5. The stack gas heat rejection rate is

   $$Q_{stack} = m_g c_{pg} t_3 = 540 \times 1.05 \times 175.1 = 99{,}296 \text{ kW}$$

Example 6.6

For the single-pressure combined-cycle power plant of Example 6.5, the following conditions are given (states are designated as shown in Figure 6.8):

Live steam parameters: $t_1 = 500°C$, $p_1 = 60$ bars, enthalpy $h_1 = 3425$ kJ/kg, mass flow rate $m_s = 83.18$ kg/s
Steam turbine isentropic efficiency $\eta_{it} = 0.9$
Condenser pressure $p_2 = 0.04$ bar
Gas turbine exhaust temperature $t_{exh} = 610°C$ and mass flow rate $m_g = 540$ kg/s

Calculate the actual steam turbine specific work, steam turbine power output, and thermal efficiency of the steam cycle.

Solution

1. For the isentropic expansion, the exhaust steam enthalpy is taken from the h-s diagram:

    $h_{2s} = 2075$ kJ/kg

2. The actual specific work of the steam turbine is

    $w_{t,a} = (h_1 - h_{2s})\eta_{it} = (3425 - 2075)0.9 = 1215$ kJ/kg

3. The steam turbine power output is

    $P_{t,a} = m_s w_{t,a} = 83.18 \times 1215 = 101,064$ kW

4. The steam cycle heat input rate is equal to the heat of the gas turbine exhaust gas, i.e.,

    $Q_{in} = m_g c_{pg} t_{exh} = 540 \times 1.05 \times 610 = 345,870$ kW

5. The thermal efficiency of the steam cycle is

    $\eta_{th} = P_{t,a}/Q_{in} = 101,064/345,870 = 0.292$

Dual- and Triple-Pressure Combined-Cycle Power Plants

To enhance the utilization of the gas turbine exhaust heat, dual- and triple-pressure combined-cycle power plants are employed. Consider the dual-pressure combined-cycle power plant shown in Figure 6.9. Figure 6.10 shows a temperature–heat flux diagram for the dual-pressure combined-cycle power plant. The energy balance for the high-pressure (HP) evaporator and superheater in the dual-pressure HRSG is given by

$$m_g c_{pg}(t_1 - t_2) = m_{HP}(h_{10} - h_9) \quad \text{kJ/s} \tag{6.30}$$

where m_{HP} is the steam mass flow in the HP section of the HRSG in kg/s, h_{10} is the enthalpy of superheated steam leaving the HRSG in kJ/kg, and h_9 is the enthalpy of saturated liquid at high pressure in kJ/kg.
    Then the rate of HP steam production in the HRSG is

$$m_{HP} = m_g c_{pg}(t_1 - t_2)/(h_{10} - h_9) \quad \text{kg/s} \tag{6.31}$$

From the energy balance for the HP economizer, the gas temperature at the HRSG high-pressure section exit is

$$t_3 = t_2 - m_{HP}(h_9 - h_8)/(m_g c_{pg}) \quad °C \tag{6.32}$$

where h_8 is the enthalpy of HP feedwater in kJ/kg. Similarly, the rate of low-pressure (LP) steam production in the HRSG is

$$m_{LP} = m_g c_{pg}(t_3 - t_4)/h_{fg,LP} \quad \text{kg/s} \tag{6.33}$$

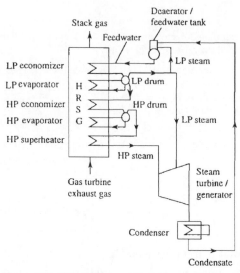

Figure 6.9. Steam/water portion of a dual-pressure combined-cycle power plant.

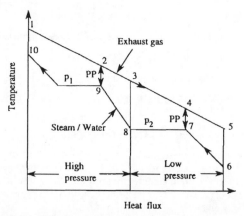

Figure 6.10. Temperature–heat flux diagram of a dual-pressure HRSG.

where $h_{\text{fg,LP}}$ is the vaporization enthalpy at LP in the HRSG in kJ/kg. Hence the total rate of steam production of the HRSG is

$$m_s = m_{\text{HP}} + m_{\text{LP}} \quad \text{kg/s} \tag{6.34}$$

The heat output rate of the HRSG is

$$Q_{\text{HRSG}} = m_g c_{pg}(t_1 - t_5) = m_{\text{HP}}(h_{10} - h_8) + m_{\text{LP}}(h_8 - h_6) \quad \text{kW} \tag{6.35}$$

From Eq. (6.35), the stack gas temperature is

$$t_5 = t_1 - Q_{\text{HRSG}}/(m_g c_{pg}) \quad {}^\circ\text{C} \tag{6.36}$$

Alternatively, it is also

$$t_5 = t_3 - m_{\text{LP}}(h_8 - h_6)/(m_g c_{pg}) \quad {}^\circ\text{C} \tag{6.37}$$

Example 6.7

For a combined-cycle power plant with dual-pressure HRSG, the following conditions are given (see Figures 6.9 and 6.10):

High pressure $p_{HP} = 120$ bars and low pressure $p_{LP} = 9$ bars
Gas turbine exhaust gas flow rate $m_g = 540$ kg/s and temperature $t_1 = 610°C$
Feedwater temperature $t_6 = 110°C$
HP superheated steam pressure $p_{10} = 120$ bars and temperature $t_{10} = 545°C$

Assuming a PP $= 10$ K and an exhaust gas specific heat $c_{pg} = 1.05$ kJ/(kg K), calculate the general steam production rate of the HRSG, the gas turbine exhaust heat rate, and the stack gas temperature.

Solution

1. The steam and water parameters at various locations in the HRSG, taken from the h-s diagram and steam table, are given below.
   The low-pressure parameters at $p_{LP} = 9$ bars are saturation temperature $t_7 = t_8 = 175.36°C$, vaporization enthalpy $h_{fg,LP} = 2029.5$ kJ/kg, saturated liquid enthalpy $h_7 = 742.64$ kJ/kg, saturated steam enthalpy $h_8 = 2772.1$ kJ/kg, and feedwater enthalpy $h_6 = c_{pw}t_6 = 4.187 \times 110 = 460.6$ kJ/kg.
   Live steam enthalpy at 120 bars and 545°C is $h_{10} = 3465$ kJ/kg. Feedwater temperature is $t_{fw,HP} = 175.36°C$ and its enthalpy is $h_{fw,HP} = c_{pw}t_{fw,HP} = 4187 \times 175.36 = 734.2$ kJ/kg. Saturated liquid parameters at $p_{HP} = 60$ bars are temperature $t_9 = 324.65°C$ and enthalpy $h_9 = 1491.8$ kJ/kg.

2. The flue gas temperature at states 2 and 4 are

$$t_2 = t_9 + PP = 324.65°C + 10 \text{ K} = 334.65°C$$

$$t_4 = t_7 + PP = 175.36°C + 10 \text{ K} = 185.36°C$$

3. From the energy balance for the high-pressure superheater and evaporator, the high-pressure steam production rate of the HRSG is

$$m_{HP} = m_g c_{pg}(t_1 - t_2)/(h_{10} - h_9)$$
$$= 540 \times 1.05 \times (610 - 334.65)/(3465 - 1491.8) = 79.12 \text{ kg/s}$$

4. From the energy balance for the high-pressure economizer, the gas temperature at the HRSG high-pressure section exit is

$$t_3 = t_2 - m_{HP}(h_9 - h_{fw,HP})/(m_g c_{pg})$$
$$= 334.65 - 79.12(1491.8 - 734.2)/(540 \times 1.05) = 228.9°C$$

5. The low-pressure steam production rate of the HRSG is

$$m_{LP} = m_g c_{pg}(t_3 - t_4)/h_{fg,LP}$$
$$= 540 \times 1.05 \times (228.9 - 185.36)/2029.5 = 12.16 \text{ kg/s}$$

6. The general steam production rate of the HRSG is

$$m_s = m_{HP} + m_{LP} = 79.12 + 12.16 = 91.28 \text{ kg/s}$$

7. The heat output rate of the HRSG is

$$Q_{HRSG} = m_{HP}(h_{10} - h_{fw,HP}) + m_{LP}(h_8 - h_6)$$
$$= 79.12(3465 - 734.2) + 12.16(2772.1 - 460.6)$$
$$= 216,061 + 28,108 = 244,169 \text{ kW}$$

8. The stack gas temperature is

$$t_5 = t_1 - Q_{HRSG}/(m_g c_{pg}) = 610 - 244,169/(540 \times 1.05) = 179.3°C$$

or

$$t_5 = t_3 - m_{LP}(h_8 - h_6)/(m_g c_{pg})$$

$$= 228.9 - 12.16(2772.1 - 460.6)/(540 \times 1.05) = 179.3°C$$

Example 6.8

For a combined-cycle power plant with dual-pressure HRSG, the following conditions are given (see Figure 6.10):

High pressure $p_{HP} = 60$ bars and low pressure $p_{LP} = 5$ bars
Gas turbine exhaust gas flow rate $m_g = 540$ kg/s and temperature $t_1 = 610°C$
Live steam pressure $p_{10} = 60$ bars and temperature $t_{10} = 500°C$
Feedwater temperature $t_6 = 110°C$

Assuming PP $= 10$ K and $c_{pg} = 1.05$ kJ/(kg K), calculate the general steam production rate of the HRSG, the gas turbine exhaust heat rate, and the stack gas temperature.

Solution

1. The steam and water parameters at various locations in the HRSG, taken from the h-s diagram and steam table, are given below.
   The low-pressure parameters at $p_{LP} = 5$ bars are saturation temperature $t_7 = t_8 = 151.84°C$, vaporization enthalpy $h_{fg,LP} = 2107.4$ kJ/kg, saturated liquid enthalpy $h_7 = 742.64$ kJ/kg, saturated steam enthalpy $h_8 = 2747.5$ kJ/kg and feedwater enthalpy $h_6 = c_{pw}t_6 = 4.187 \times 110$ $= 460.6$ kJ/kg.
   The live steam enthalpy at 60 bars and 545°C is $h_{10} = 3425$ kJ/kg. The feedwater temperature is $t_{fw,HP} = 152°C$, and its enthalpy is $h_{fw,HP} = c_{pw}t_{fw,HP} = 4,187 \times 152 = 636.4$ kJ/kg. Saturated liquid parameters at $p_{HP} = 60$ bars are temperature $t_9 = 275.55°C$ and enthalpy $h_9 = 1213.7$ kJ/kg.
2. The gas temperature at states 2 and 4 are

$$t_2 = t_9 + PP = 275.55 + 10 = 285.55°C$$

$$t_4 = t_7 + PP = 151.84 + 10 = 161.84°C$$

3. From the energy balance for the high-pressure superheater and evaporator, the high-pressure steam production rate of the HRSG is

$$m_{HP} = m_g c_{pg}(t_1 - t_2)/(h_{10} - h_9)$$

$$= 540 \times 1.05 \times (610 - 285.55)/(3425 - 1213.7) = 83.2 \text{ kg/s}$$

4. From the energy balance for the high-pressure economizer, the gas temperature at the HRSG high-pressure section exit is

$$t_3 = t_2 - m_{HP}(h_9 - h_{fw,HP})/(m_g c_{pg})$$

$$= 285.55 - 83.2(1213.7 - 636.4)/(540 \times 1.05) = 200.8°C$$

5. The low-pressure steam production rate of the HRSG is

$$m_{LP} = m_g c_{pg}(t_3 - t_4)/h_{fg,LP}$$

$$= 540 \times 1.05(200.8 - 161.84)/2107.4 = 10.48 \text{ kg/s}$$

6. The general steam production rate of the HRSG is

$$m_s = m_{HP} + m_{LP} = 83.2 + 10.48 = 93.68 \text{ kg/s}$$

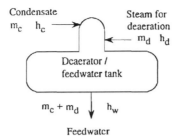

Condensate m_c h_c → Steam for deaeration ← m_d h_d

Deaerator / feedwater tank

$m_c + m_d$ ↓ h_w

Feedwater

Figure 6.11. Deaerator/feedwater tank of a combined-cycle power plant.

7. The heat output rate of the HRSG is

$$Q_{HRSG} = m_{HP}(h_{10} - h_{fw,HP}) + m_{LP}(h_8 - h_6)$$

$$= 83.2(3425 - 636.4) + 10.48(2747.5 - 460.6) = 280{,}651 \text{ kW}$$

8. The stack gas temperature is

$$t_5 = t_1 - Q_{HRSG}/(m_g c_{pg}) = 610 - 280{,}651/(540 \times 1.05) = 158°C$$

or

$$t_5 = t_3 - m_{LP}(h_8 - h_6)/(m_g c_{pg})$$

$$= 200.8 - 10.48(2747.5 - 460.6)/(540 \times 1.05) = 158°C$$

Example 6.9

The condensate from a steam turbine condenser is directly supplied to a deaerator (see Figure 6.11). Assuming the condensate mass flow m_c is 93.68 kg/s, the deaerator pressure p_d is 1.5 bars, and the condenser pressure p_c is 0.04 bar, estimate the deaerator steam requirements.

Solution

1. The enthalpies of saturated steam for deaeration (h_d), condensate (h_c), and feedwater (h_w) are taken from the steam/water table:

$$h_d = 2693.4 \text{ kJ/kg} \qquad h_c = 121.41 \text{ kJ/kg} \qquad h_w = 467.13 \text{ kJ/kg}$$

2. From the energy balance for the deaerator, the steam consumption is given by

$$m_d = m_c(h_w - h_c)/(h_d - h_c)$$

$$= 102.6(467.13 - 121.41)/(2693.4 - 121.41)$$

$$= 12.59 \text{ kg/s}$$

Example 6.10

Consider the dual-pressure steam turbine of the combined-cycle power plant of Examples 6.3 and 6.4. The conditions (see Figure 6.12) are as given below:

Gas turbine exhaust gas mass flow $m_g = 540$ kg/s
Exhaust gas temperature $t_{exh} = 610°C$
Live (HP) steam condition: $p_1 = 60$ bars, $t_1 = 500°C$, and enthalpy $h_1 = 3425$ kJ/kg
LP steam condition: $p_2 = 5$ bars, $h_{2s} = 2775$ kJ/kg (from h-s diagram)
Condenser pressure $p_3 = 0.04$ bar

The rate of steam generation in the HP and LP sections of the HRSG are $m_{HP} = 83.2$ kg/s and $m_{LP} = 10.48$ kg/s, respectively. After throttling, the following portion of LP steam is used for the deaeration: $m_d = 12.59$ kg/s.

Steam from a dual-pressure
HRSG

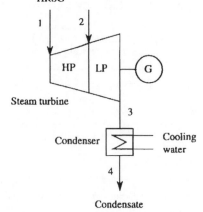

Figure 6.12. Dual-pressure steam turbine of a combined-cycle power plant.

Assuming a steam turbine isentropic efficiency $\eta_{it} = 0.9$ and an isobaric specific heat of exhaust gas $c_{pg} = 1.05$ kJ/(kg K), calculate the power output and efficiency of the steam turbine.

Solution

1. The enthalpy of steam, after isentropic expansion, in the HP and LP turbine sections is found from the h-s diagram to be

   $$h_{2s} = 2775 \text{ kJ/kg} \qquad h_{3s} = 2120 \text{ kJ/kg}$$

2. The actual steam enthalpies at states 2 and 3 are

   $$h_2 = h_1 - (h_1 - h_{2s})\eta_{it} = 3425 - (3425 - 2775)0.9 = 2840 \text{ kJ/kg}$$

   $$h_3 = h_2 - (h_2 - h_{3s})\eta_{it} = 2840 - (2840 - 2120)0.9 = 2192 \text{ kJ/kg}$$

3. The actual specific works of the HP and LP turbine sections are

   $$w_{HP} = h_1 - h_2 = 3425 - 2840 = 585 \text{ kJ/kg}$$

   $$w_{LP} = h_2 - h_3 = 2840 - 2192 = 648 \text{ kJ/kg}$$

4. The steam mass flow rates in the HP and LP turbine sections are

   $$m_{HP} = 83.2 \text{ kg/s}$$

   $$m'_{LP} = m_{HP} + m_{LP} - m_d = 83.2 + 10.48 - 12.59 = 87.61 \text{ kg/s}$$

5. The steam turbine power output is

   $$P_{ST} = m_{HP}w_{HP} + m'_{LP}w_{LP} = 83.2 \times 585 + 87.61 \times 648 = 105{,}443 \text{ kW}$$

6. The steam turbine efficiency defined as the ratio of power output to heat input is:

   $$\eta_{ST} = P_{ST}/Q_{in} = P_{ST}/m_g c_{pg} t_{exh} = 105{,}443/540 \times 1.15 \times 610 = 0.305$$

Example 6.11

For a triple-pressure combined cycle with the temperature–heat flux diagram shown in Figure 6.13, the following conditions are given:

Gas turbine exhaust gas flow rate $m_g = 540$ kg/s
Temperature $t_1 = 610°C$

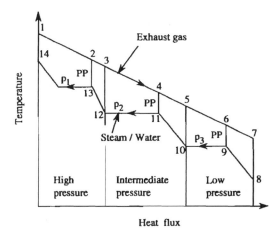

Figure 6.13. Temperature–heat flux diagram of a triple-pressure HRSG.

Specific heat $c_{pg} = 1.05$ kJ/(kg K)
Steam cycle parameters:

HP live steam condition: $p_1 = 120$ bars, $t_{14} = 545°C$
IP steam pressure: $p_2 = 20$ bars
LP steam pressure: $p_3 = 4$ bars
Condenser pressure $p_c = 0.04$ bar
Steam turbine isentropic efficiency $\eta_{it} = 0.9$
Deaerator pressure $p_d = 1.5$ bars
Feedwater temperature 110°C
Pinch point PP $= 10$ K
Low-pressure steam production rate $m_{LP} = 10.48$ kg/s
Steam consumption for the deaerator $m_d = 12.59$ kg/s

Calculate (1) steam production rate, in kg/s, (2) steam turbine power output, in kW, and (3) steam plant efficiency.

Solution

Steam and water conditions (from water/steam table and h-s diagram):

State	Condition	Pressure, bars	Temperature, °C	Specific enthalpy, kJ/kg
8	Feedwater	4	110	460.6
9	Sat. liquid	4	143.62	604.67
10	Sat. steam	4	143.62	2737.6
11	Sat. liquid	20	212.37	908.59
12	Sat. steam	20	212.37	2797.2
13	Sat. liquid	120	324.65	1491.8
14	Live steam	120	545	3465

Sat., saturated

Flue gas temperatures are $t_1 = 610°C$, $t_2 = t_{13} + PP = 324.65 + 10 = 334.65°C$, $t_4 = t_{11} + PP = 212.37 + 10 = 222.37°C$, and $t_6 = t_9 + PP = 143.62 + 10 = 153.62°C$.

HP steam production rate is

$$m_{HP} = m_g c_{pg}(t_1 - t_2)/(h_{14} - h_{13})$$

$$= 540 \times 1.05 \times (610 - 334.65)/(3465 - 1491.8) = 79.12 \text{ kg/s}$$

Actual steam exhaust enthalpy is

$$h_3 = h_2 - (h_2 - h_{3s})\eta_{it} = 2840 - (2840 - 2120)0.9 = 2192 \text{ kJ/kg}$$

Actual specific work for the HP and the LP steam turbine are

$$w_{HP} = h_1 - h_2 = 3465 - 2840 = 585 \text{ kJ/kg}$$

$$w_{LP} = h_2 - h_3 = 2840 - 2192 = 648 \text{ kJ/kg}$$

Steam mass flow rate for the HP and the LP steam turbine are

$$m_{HP} = 79.12 \text{ kg/s}$$

$$m'_{LP} = m_{HP} + m_{LP} - m_d = 79.12 + 10.48 - 12.59 = 77.01 \text{ kg/s}$$

Steam turbine power output is

$$P_{ST} = m_{HP}w_{HP} + m_{LP}w_{LP} = 79.12 \times 585 + 77.01 \times 648 = 96,188.8 \text{ kW}$$

Steam plant efficiency is

$$\eta_{ST} = P_{ST}/Q_{in} = P_{ST}/m_g c_{pg} t_1 = 96,188.8/540 \times 1.05 \times 610 = 0.278$$

Tables 6.2–6.7 contain the design performance data of combined-cycle power plant configurations based on gas turbines of leading manufacturers. The corresponding data of the simple-cycle

Table 6.2. Performance data (gross) of combined-cycle power plants KA13E2 with one to three gas turbines GT13E2 and single-, dual-, or triple-pressure reheat or nonreheat steam cycle

	Power output (gross), MW			
Combined-cycle configuration	Gas turbine	Steam turbine	Combined cycle	Efficiency (LHV), %
KA13E2 with one gas turbine GT13E2				
Single-pressure steam cycle	160.2	67.1	227.3	51.3
Dual-pressure steam cycle	160.2	82.9	243.1	53.3
Triple-pressure steam cycle	159.4	92.1	251.5	55.1
KA13E2-2 with two gas turbines GT13E2				
Single-pressure cycle	318.6	148.1	466.7	51.6
Dual-pressure cycle	318.6	172.0	490.6	54.2
Triple-pressure cycle	318.6	179.2	497.8	55.0
KA13E2-3 with three gas turbines GT13E2				
Triple-pressure reheat steam cycle	3×159.4	281.2	759.4	55.5

Frequency is 50 Hz. Fuel is natural gas. Condenser pressure is 0.028 bar. ISO standard conditions are 15°C, 1.013 bars, relative humidity 60%, sea level.
Source: ABB Power Generation Ltd., Baden, Switzerland (status: August 1996).

Table 6.3. Performance comparison of combined-cycle power plants KA8C-2 and KA26 with two gas turbines GT8C and GT26 and single-, dual-, or triple-pressure reheat or nonreheat steam cycle

	Power output (gross), MW			
Configuration	Gas turbine	Steam turbine	Combined cycle	Efficiency (LHV), %
KA8C-2, Frequency 50/60 Hz				
Single-pressure cycle	101.5	48.7	150.2	49.4
Dual-pressure cycle	101.5	54.8	156.3	51.4
Triple-pressure cycle	101.5	58.0	159.5	52.4
KA26, Frequency/50 Hz				
Triple-pressure reheat cycle	240	125	365	58.5

Fuel is natural gas. Condenser pressure is 0.028 bar. ISO standard conditions are 15°C, 1.013 bars, relative humidity 60%, sea level.
Source: ABB Power Generation Ltd., Baden, Switzerland (status: August 1996).

Table 6.4. Performance data of combined-cycle power plants based on the Siemens 3A Series gas turbines

	Gas turbine model		
	V64.3A	V84.3A	V94.3A
Power output, MW	103	254	359
Efficiency (LHV),%	54.8	57.9	58.1
Heat rate, kJ/kWh	6566	6208	6187
Exhaust gas flow, kg/s	194	454	640
Specific power, kW/(kg/s)	361	374	375
Exhaust gas temperature, °C	565	562	562

Dual-pressure reheat steam cycle, condenser pressure 0.04 bar. Fuel is natural gas. Condenser pressure is 0.028 bar. ISO standard conditions are 15°C, 1.013 bars, relative humidity 60%, sea level.
Source: Siemens AG, Erlangen, Germany.

Table 6.5. Net performance data of combined cycle (STAG) power plants based on the General Electric G and H gas turbine technologies

	Gas turbine model			
	MS7001G[a]	MS9001G[b]	MS7001H[a]	MS9001H[b]
Combined-cycle power plant	S107G[a]	S109G[b]	S107H[a]	S109H[b]
Power output, MW	350	420	400	480
Heat rate (LHV), BTU/kWh	5883	5883	5687	5687
Heat rate (LHV), kJ/kWh	6207	6207	6000	6000
Overall efficiency (LHV), %	58	58	60	60
Gas turbine cooling	air	air	steam	steam
NO_x emissions, ppmv	25	25	9	9

Fuel is natural gas. ISO standard conditions are 15°C, 1.013 bars, relative humidity 60%, sea level.
Source: Power (1995).
[a] 50-Hz machine. [b] 60-Hz machine.

gas turbine power plants are given in Tables 5.4–5.8 in Chapter 5. To date, the most efficient industrial gas turbines are large-frame General Electric G and H series machines. Major design performance data of combined-cycle plants based on these machines are given in Table 6.5. They can achieve an overall efficiency of 60% and a heat rate of 6000 kJ/kWh (both are based on LHV of natural gas fuel).

HEAT RECOVERY STEAM GENERATOR DESIGN

In conventional steam power plants, multistage feedwater preheating up to the highest temperature level is used to enhance the cycle efficiency, and the rate of steam production depends on the feedwater temperature at the economizer inlet. The feedwater temperature in the HRSG of a combined-cycle power plant must be above the dew point temperature to prevent low-temperature corrosion. The rate of steam production is independent of the inlet feedwater temperature and is higher at higher live steam pressures.

An HRSG recovers the heat energy from the gas turbine exhaust. It consists of a number of heat exchangers that heat water, evaporate it, and superheat steam. Cold feedwater is fed from the feedwater tank to the economizer. The exhaust gas temperature of advanced gas turbines is 600°C and above (ABB AG, 1991–1996; Author, 1995; Bannister et al., 1995). With higher exhaust gas temperatures, the HRSG becomes even more complex, and thus dual-pressure and triple-pressure HRSGs are required.

HRSGs for combined-cycle applications are natural-and forced-circulation, unfired and supplementary fired steam boilers that cover the turbine power range up to 280 MW and have main steam pressures up to 140 bars (2030 psia) and temperatures up to 545°C (1010°F) (Author, 1995; Swanekamp, 1995; Croonenbrock et al., 1996).

Table 6.6. Performance of combined-cycle triple-pressure power plants with third-generation gas turbine combined-cycle power plants

	Manufacturer and gas turbine model						
	Westinghouse Fiat, MHI		ABB		GE, Nuovo Pignone		Siemens Ansaldo
	TG50D5S6	FMW701F MW501F[a]	GT13E2 GT11N[a]	GT26 GT24[a]	MS9001FA MS7001FA[a]	MS9001EC	V94.3 V84.3[a]
Combined-cycle triple-pressure, one-gas-turbine power plant							
Steam cycle	Reheat	Reheat	Nonreheat	Nonreheat	Reheat	Reheat	Reheat
Net output, MW	212.3	349.4	244.2	361.5	348.5	259.6	328
Net heat rate (LHV), kJ/kWh	6977	6569	6680	6218	6560	6720	6569
Net efficiency (LHV), %	51.6	54.8	53	56.9	54.8	53.5	54.8
Combined-cycle triple-pressure, two-gas-turbine power plant							
Steam cycle	Reheat	Reheat	Nonreheat	Nonreheat	Reheat	Reheat	Reheat
Net output, MW	427.2	702.4	490.8	723.6	700.8	521.9	657
Net heat rate (LHV), kJ/kWh	6936	6533	6640	6220	6530	6690	6557
Net efficiency (LHV), %	51.9	55.1	53.3	56.9	55.1	53.8	54.9

Fuel is natural gas. ISO standard conditions are 15°C, 1.013 bars, relative humidity 60%, sea level.
Sources: ABB AG (1991–1996), Becker and Finckh (1995), Author (1995).
All gas turbines are 50-Hz machines except where rated.
[a]60 Hz.

Table 6.7. Performance data of Siemens combined-cycle single and multishaft blocks based on V84.3A and V94.3A gas turbines

Gas turbine model	V84.3A[a]	V94.3A
Combined-cycle single-shaft unit with one gas turbine		
Block net output, MW	250	354
Net efficiency, %	57.0	57.2
Exhaust gas flow, kg/s	454	640
Specific power, kW/(kg/s)	374	375
Exhaust-gas temperature, °C	562	562
Combined-cycle multishaft unit with two gas turbines		
Gas turbine output, MW	2 × 165	2 × 233
Steam turbine output, MW	176	249
Block net output, MW	499	705
Net efficiency, %	56.9	57.0

Triple-pressure reheat steam cycle, condenser pressure 0.04 bar. Fuel is natural gas. Condenser pressure is 0.028 bar. ISO standard conditions are 15°C, 1.013 bars, relative humidity 60%, sea level.
Source: Siemens AG, Erlangen, Germany.
[a] 60-Hz machine.

In the simplest case of a single-pressure HRSG, the natural-circulation boiler consists of an economizer, evaporator, and superheater. The feedwater from a deaerator/feedwater tank is pumped through the economizer into the boiler drum, then is evaporated in the natural-circulation loop, and the steam is separated from water in the drum and superheated to a specified temperature in the superheater. Horizontally configured tube bundles with vertical gas flow need forced circulation to assist the water flow (Dodero, 1997).

Single-pressure combined cycle has a low efficiency of the steam cycle because of poor utilization of the gas turbine exhaust heat. For efficient use of exhaust gas heat of advanced gas turbines with their higher exhaust temperatures, dual-pressure and triple-pressure HRSGs are employed. Steam pressures and temperatures in the HRSG and steam turbine of a combined-cycle power plant are modest compared to those of conventional steam power plants. The highest pressure in HRSGs is about 120 bars (1740 psia), and the highest temperature is about 540°C (1004°F). In order to increase the efficiency of the steam plant, more complex steam turbine schemes such as those with one or two reheats are used. Reheat HRSGs have been deployed in the latest combined-cycle systems with 540°C and higher gas turbine exhaust. Thus advanced HRSG designs employ all methods of efficiency enhancement that are already proven in modern steam generators at utility power stations except for regenerative feedwater heating.

Two HRSG configurations are used: horizontal and vertical (Figure 6.14). The standard configuration in North America is a natural-circulation HRSG consisting of vertically hung heat transfer tube bundles with exhaust gas flowing horizontally. In Europe, natural-circulation or forced-circulation HRSGs with horizontally configured tube bundles and vertical gas flow are employed (Jones, 1996).

Further differentiation occurs on the lines of application of reheat. Reheat HRSGs are now employed in combined-cycle systems with 600°C and higher gas turbine exhaust. They are about 2% more efficient than the nonreheat combined-cycle systems (Balling et al., 1995; Bolland, 1991; Bolland and Stadaas, 1995; Noymer and Wilson, 1993; Klara, 1995; Klara and Ward, 1992; Cook et al., 1991; Dörr, 1996). In HRSGs the feedwater is usually preheated only in a deaerator, since it must have the lowest possible temperature in order to attain high utilization of exhaust gas heat. In single-pressure HRSGs the stack temperature is normally about 200 C (392°F).

The pinch point is the minimum temperature difference between the exhaust gas and water/steam in the HRSG. Reducing the pinch point yields a higher rate of steam production but requires greater heat transfer surface areas. The optimum pinch point values in HRSGs are about 10–15 K (18–27 °F) (Bolland, 1991).

The approach is the temperature difference between saturation temperature and water temperature at the economizer outlet. The pinch point and approach determine the size of the heat transfer

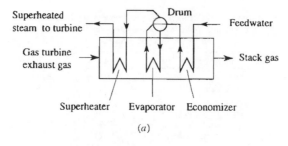

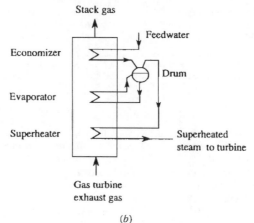

Figure 6.14. (a) Horizontal and (b) vertical HRSG configurations.

surface area of the economizer and evaporator of the HRSG. The choice of feedwater temperature is based on the consideration that no condensation of water vapor should take place at the cold end of the economizer. Thus low-temperature corrosion of the heat transfer surface caused by formation of sulfuric and sulfurous acids will be prevented. In more advanced designs of the heat transfer surface of the economizer, exhaust gases will be allowed to cool below the dew point due to application of corrosion-resistant materials.

In advanced HRSGs, finned tubes are employed to enhance the heat transfer rate on the flue gas side. Steaming in the economizer during low loads and start-up is prevented by maintaining feedwater pressure high enough to keep the water below the saturation temperature in the whole range of loads. The feedwater supply control valve may be located at the outlet of the economizer, the intermediate-pressure (IP) and high pressure (HP) economizers may be configured in parallel with the HP feedwater control valve located at the outlet of the low pressure (LP) economizer, and the IP feedwater control valve placed after the IP economizer.

Although the unfired HRSG combined cycle exhibits the best efficiency, many combined cycles include supplementary fired burners for operating flexibility, especially to generate power and steam (in cogeneration applications) if the gas turbine is down. Supplementary firing may also be advantageous in increasing steam temperature and pressure for a more efficient steam cycle.

Although combined-cycle power plants without supplementary firing have higher efficiencies, supplementary firing may enhance operating flexibility, especially in cogeneration and repowering applications. Most HRSGs now operate at 110-bar steam pressure. Supplementary firing could increase the main steam pressure up to 140–165 bars (2000 and 2400 psig). This pressure level is optimum for repowering of existing steam power plants. The main concern with units for 165-bar pressure is maintaining proper natural circulation.

The application of once-through HRSGs has been demonstrated in a number of smaller combined cycle power plants. A once-through HRSG contains vertically configured tube bundles with an inside flow of water/steam and horizontal gas flow across the tubes. With supplementary firing, steam temperatures up to 600°C are attainable (Balling et al., 1995; Swanekamp, 1995). Advanced corrosion-resistant alloys are used as the tubing material. Using deionized water eliminates the elaborate water treatment systems normally used in carbon-steel natural-circulation HRSGs.

In cogeneration plants, an auxiliary boiler fired with an alternative fuel, e.g., coal, is also an important component.

In order to prevent cold-end corrosion, the HRSG exhaust gas temperature must be kept above the dew point temperature of the flue gas, which depends on the sulfur content in the fuel. Using condensing heat exchangers made of corrosion-resistive materials enables the recovery of heat of condensation. Lowering the HRSG temperature exhaust from 100°C (212°F) to 40°C (104°F) improves the overall efficiency by 3–5%.

An alternative bottoming cycle, the Kalina cycle, using an ammonia/water mixture at a concentration of 70% by mass as a working fluid may improve the overall efficiency of the combined cycle by about 2% (Maude, 1993).

Selective catalytic reduction (SCR) elements for NO_x control can be placed between the tube bundles at an appropriate location.

COMBINED-CYCLE SYSTEM DESIGN CONSIDERATIONS

Advanced combined-cycle power plants can have a single-shaft or multishaft configuration (Dörr, 1996; Swanekamp, 1995; Anon, 1995). Multishaft combined-cycle blocks allow the construction of the combined cycle to be accomplished in two phases: first, gas turbines and, then, HRSG and steam turbines. The construction of the relatively expensive steam cycle will then be financed from the revenues of the gas turbine operation. Bypass stacks are required for the initial operation of gas turbines and also improve the combined-cycle power plant's operating flexibility, as the gas turbines can then be started up rapidly without any restriction on maximum permissible loading gradient (13 MW/min for Model V94.3A machines) (Becker and Finckh, 1995; Dörr, 1996).

Multishaft combined-cycle blocks are preferred over single-shaft combined-cycle blocks by utilities that use cheap natural gas. In this case, the gas and steam turbines are connected with one electrical generator to form a single-train block, which ensures optimum operation during fully automatic start-up, loading, and shut-down of the block.

For utilities that are confronted with particularly high prices of natural gas or distillate, large-capacity single-shaft combined-cycle plants with the highest possible efficiency are needed in the continuous base-load operating mode. Combined-cycle blocks equipped with Siemens 3A series advanced gas turbines (such as the first 50/60-Hz gas turbine V84.3A of the new 3A series) are capable of achieving a net efficiency of 57% (LHV), which is roughly equivalent to 58% gross efficiency. The single-shaft combined-cycle concept provides higher efficiency, e.g., the single-shaft GUD 1S.84.3A block is 5% more efficient than the multishaft GUD 2.84.2 block owing to

- higher turbine inlet temperature and higher pressure ratio (efficiency gain of 3.1%)
- triple-pressure reheat steam cycle instead of dual-pressure nonreheat cycle (efficiency gain of 1.2%)
- preheating the natural gas with water from the HRSG (gain of 0.4%)
- reduction in steam-exit losses by employing a single-flow rather than a double-flow low-pressure turbine (additional gain of 0.4%)

The first GUD 1S.94.3 block of this design for the 340-MW King's Lynn station in England is scheduled to go on line in 1996. Three GUD 1S.94.3A blocks are due to start operation in 1998 at the 1000-MW Tapada do Outeiro plant in Portugal.

In the single-shaft configuration the generator is placed between the gas and steam turbines, each of which has its own axial-thrust bearing to maintain the rotor blades in the correct position relative to the stationary vanes and thus to minimize losses due to leakage between the moving blades and the casing (Dörr, 1996). Beneficial features of the single-shaft configuration are elimination of HRSG bypass stacks and augmentation of efficiency (Swanekamp, 1996; Feenstra, 1995; Anon, 1995).

A synchronous clutch between generator and steam turbine enables initial gas turbine operation in the simple-cycle mode while warming up the HRSG pipework and steam turbine valves. When the steam turbine shaft revolves at the same rate as the generator shaft, the clutch engages, and the steam turbine will be loaded. Thus, owing to this clutch, a simple-cycle gas turbine operation is also possible.

The two-casing reheat steam turbine configuration consists of a single-flow barrel-type HP section and an opposed-flow IP/LP section with axial diffuser directly connected to an axial-flow steam condenser, which minimizes the steam-exit losses. The selection of a two-casing or single-casing unit should be based on the consideration of economical and energetical issues such as capital costs and performance of nonreheat plants. For reheat combined-cycle plants the two-casing steam turbine design is more efficient as throttle pressure increases.

Advanced materials technology enables deployment of single-casing combined-cycle plants for 100-MW to 250-MW capacities with main steam temperatures of 600°C (Dodero, 1997).

COMBINED-CYCLE VERSUS CONVENTIONAL POWER STATIONS

Comparison Criteria

The selection of power plant type is based on thermal efficiency, cost-effectiveness, and environmental impact. The performance parameters, i.e., thermal efficiency and heat rate, are the most important factors in evaluation and comparison of various types of power plants. High efficiency is the primary prerequisite for making an economical choice of power plant.

The thermodynamic superiority of combined-cycle power plants is their outstanding feature. The efficiencies of steam power plants and single-cycle gas turbine power plants are surpassed. The best gas-fired steam power plants can attain efficiencies of about 45%. The simple-cycle advanced gas turbine efficiency at a turbine inlet temperature of over 1100°C is around 37–38.5%, whereas advanced combined-cycle power plants attain efficiencies of 55% or even higher (up to 58–60%) (ABB AG, 1991–1996; Becker and Finckh, 1995; Balling et al., 1995).

Let us consider the following types of power plants: steam turbine, gas turbine, and diesel engine. The efficiency of advanced diesel engines is comparable to that of gas turbines of equal power capacity, and therefore diesel engine power plants may appear to be the optimum option for smaller to medium power outputs, e.g., up to 30 MW, and in the most favorable cases, even up to 50 MW. At higher capacities, diesel engine power plants have higher capital and maintenance costs than gas turbine power plants. However, diesel engines have greater environmental impact, especially with the emissions of NO_x and unburned hydrocarbons.

The comparison below is restricted to steam turbine, simple-cycle gas turbine, and gas turbine based combined-cycle power plants.

The power plants under consideration include those having a total capacity of up to 1000 MW. The specific capital costs per kW of power output of combined-cycle power plants increase as the plant power rating decreases. Therefore combined-cycle power plants of smaller power outputs are better suited for industrial or district heating cogeneration plants discussed in Chapter 7. However, the minimum economical size of the combined-cycle cogeneration plants based on the utilization of gas turbines is ~10MW.

The second important criterion for comparing types of power plants is the economic factor. Steam power plants are significantly more expensive than combined-cycle power plants. A coal-fired power plant, for example, costs 2–3 times as much as a combined-cycle power plant with the same power output. Advanced combined-cycle power plants are therefore simpler and less expensive than steam power units. Their construction period is shorter than that of steam power plants. A possibility of progressive staged construction of combined-cycle power plants is yet another advantage. At the first stage, the gas turbine plant is installed and commissioned. During the second stage, the steam plant train is installed. Construction costs of the steam plant will be financed from the revenues of electric power produced by the gas turbine plant.

Comparing both the performance and economic criteria shows that combined-cycle power plants have an evident advantage over simple-cycle plants such as steam or gas turbine power plants. Therefore combined-cycle power plants represent the optimum energy system type that is suitable both for the construction of new power plants and for upgrading and repowering of existing steam power plants.

Operating Costs

Because of the high reliability of advanced gas turbines, simple-cycle gas turbine plants have the lowest operating and maintenance (O and M) costs, although they require more spare parts than steam turbines. A steam power plant requires more staff, and its maintenance costs are higher. The O and M costs of combined-cycle power plants depend on the complexity of the steam portion and are between those of simple-cycle gas turbine plants and steam power plants.

Availability and Reliability

The major factors determining power plant availability are

* design of the major components
* mode of operation (base, medium, or peak load)
* type of fuel

All the power plants under consideration have similar availabilities when used under the same operating conditions. Typical figures for the availability of base-load power plants are as follows (ABB AG, 1991–1996; Becker and Finckh, 1995; Dodero, 1997):

* gas-fired gas turbine power plants 88–95%
* oil- or gas-fired steam turbine power plants 85–90%
* coal-fired steam turbine power plants 80–85%
* gas-fired combined-cycle power plant 85–90%

The availability of peak and medium load machines is lower because of frequent start-ups and shutdowns that reduce life expectancy of the machine and thus increase the scheduled maintenance and forced outage rates.

Unlike steam power plants that can be fired with any fuel, gas turbine power plants employ only natural gas or light distillate (fuel oil No. 2) as fuel. These fuels are more expensive than coal for steam power plants. The fuel price is an essential constituent of electricity costs. Therefore, as a rule, coal-fired reheat steam power plants produce electricity cheaper but at a higher environmental impact than power plants based on gas-fired gas turbines. Among all oil- or gas-fired power-stations, the combined-cycle power plant is the most economical technology for electricity generation. For short utilization periods of peak-load plants with annual service of up to 2000 h/yr for gas-fired large-capacity power plants and up to 1500 h/yr for oil-fired smaller units, the gas turbine is the most economically viable choice. Coal-fired steam power plants are suitable for use as base-load plants if the price difference between the coal and the gas turbine fuel is sufficiently large (around US\$ 3–6/GJ).

Combined-cycle power plants with supplementary firing represent a viable option for repowering as well as for application in cases when the gas or oil supply is scarce and the more easily available fuel is coal for use in supplementary firing.

Fuel Selection

The selection of the fuel and the corresponding type of power plant is determined not only by short-term economic considerations but also by long-term developments in the prices for the various possible fuels. In this regard, the following aspects can become important in selecting the type of power plant to be built: long-term availability of the fuel at a favorable price and environmental considerations. It should be stressed that the fuel selected may appear to be the best at the time of plant construction. Therefore the greater the fuel flexibility of the power plant chosen, the less risk from possible increases in fuel prices.

The fuel flexibility of combined-cycle power plants is less than that of steam power plants. Normally, gas turbines burn natural gas or light oil; some industrial gas turbines are designed to burn heavy oil or crude. It is easier to burn these fuels in turbines with large single-burner combustors than in annular combustors with multiple small burners (ABB AG, 1991–1996; Becker and Finckh, 1995). An appropriate pretreatment is required for heavy oil or crude to remove or inhibit harmful constituents such as vanadium and sodium, which cause high-temperature corrosion.

The gas turbine can be incorporated into combined-cycle and cogeneration plants with the triple-pressure steam train comprising an HRSG and the HP, IP, and LP steam turbine cylinders. The last guide blade rows of the LP sections with an exhaust annulus area of 8 m^2 are fitted with longitudinally curved blading to enhance machine efficiency. Useful heat can be extracted from the IP turbine (one bleed point) and from the LP turbine with two bleed points.

Net efficiency of a simple-cycle gas turbine power plant of 38.5% and that of a combined cycle of 58% based on HHV of fuel or 60% based on HHV of fuel can be attained with the gas turbine 501G (General Electric).

INTERCOOLED REHEAT COMBINED-CYCLE POWER PLANTS

Figures 6.15a–6.15c show temperature-entropy diagrams for combined-cycle power plants with a reheat gas turbine. They are drawn for single-pressure (Figur 6.15a), dual-pressure (Figure 6.15b), and triple-pressure (Figure 6.15c) combined-cycle power plants.

The following thermodynamic analysis is applicable to all three configurations. The energy balance for the reheat gas turbine is given by

$$Q_{in} = Q_{comb} + Q_{reheat} = P_{GT} + Q_{HRSG} + Q_{exh} \quad kW \tag{6.38}$$

where Q_{comb} and Q_{reheat} are the rate of heat addition to the gas turbine combustor and reheater, respectively, P_{GT} is net gas turbine power output, Q_{HRSG} is the rate of heat transfer in the HRSG, and Q_{exh} is the rate of heat loss with the exhaust gas to the stack.

The energy balance for the steam plant is given by

$$Q_{HRSG} = P_{ST} + Q_{cw} \quad kW \tag{6.39}$$

where P_{ST} is net steam turbine power output and Q_{cw} is rate of heat rejection in the condenser to the cooling water.

Equations (6.39) and (6.40) yield the energy balance for the whole combined-cycle power plant:

$$Q_{in} = Q_{comb} + Q_{reheat} = P_{GT} + P_{ST} + Q_{exh} + Q_{cw} = P_{CC} + Q_{loss} \quad kW \tag{6.40}$$

where P_{CC} is net power output of the combined-cycle power plant and Q_{loss} is total rate of heat loss of the power plant including exhaust gas and cooling water heat losses.

Figure 6.16 depicts a T-s diagram for a triple-pressure combined-cycle power plant with a reheat gas turbine and reheat steam turbine.

The energy balance equations, Eqs. (6.38) to (6.40) remain the same. The only difference is that the term QHRSG contains not only the rates of heat transfer in the economizer, evaporator, and superheater of the HRSG, but also the rate of heat addition to the steam reheater. The efficiency of the combined-cycle power plant is given by Eq.(6.5). The reheat combined-cycle power plants are more efficient than the nonreheat combined-cycle power plants.

All the methods of enhancement of performance of gas turbines discussed in Chapter 5 are also beneficial for combined-cycle power plants. Advanced aeroderivative gas turbines may employ intercooling, inlet supercharging and cooling, humification, and steam or water injection. Intercooled combined cycle (ICC) power plants are depicted in Figures 6.17a and 6.17b. The plant shown in Figure 6.17a consists of a two-stage compressor train with a surface-type or evaporative intercooler, whereas the plant in Figure 6.17b employs intercooling and aftercooling.

The effect of intercooling on the performance of combined-cycle power plants is similar to that for a simple-cycle gas turbine described in Chapter 5. The benefit of intercooling for combined

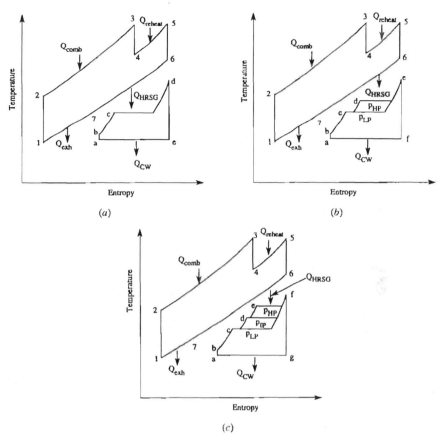

Figure 6.15. *T-s* diagram of a combined-cycle power plant with reheat gas turbine: (*a*) single pressure, (*b*) dual pressure, and (*c*) triple pressure.

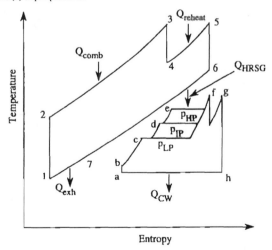

Figure 6.16. *T-s* diagram of a triple-pressure combined-cycle power plant with reheat gas turbine and reheat steam turbine.

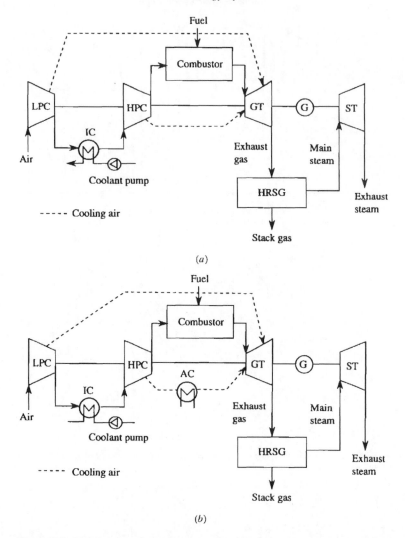

Figure 6.17. Intercooled combined-cycle power plants (*a*) with intercooling (IC) and (*b*) with intercooling (IC) and aftercooling (AC).

cycles based on heavy-duty gas turbines is the augmentation of power output. For example, for a combined cycle with a rated power of 350 MW at ISO standard conditions and an inlet air flow of 600 kg/s, an increase in electric power to about 480 MW through the application of intercooling is predicted when the TIT reaches 1500°C (Macchi et al., 1995; Chiesa et al., 1995).

The rate of heat removal from the air stream in the surface-type intercooler placed between the HP and LP compressor stages is given by

$$Q_{ic} = m_a c_{pa}(T_i - T_c) = m_w c_{pw} \Delta t_w \quad \text{kW} \tag{6.41}$$

where m_a and m_w are mass flow rates of air and cooling water, respectively, c_{pa} and c_{pw} are constant pressure specific heat of air and cooling water, respectively, T_i and T_c are intercooler inlet and exit air temperature, respectively, and Δt_w is cooling water temperature increase in the intercooler.

The maximum (theoretical) reduction in the specific work of compression may be evaluated as follows:

$$\Delta w_c = w_c - w_{c,ic} = c_p(T_2 - T_1) - 2c_p(T_2' - T_1) \quad \text{kJ/kg} \tag{6.42}$$

where w_c and $w_{c,ic}$ are specific work of compression in the one- and two-stage compressor with intercooling, respectively, T_1 is LP compressor inlet temperature, T_2 and T_2' are discharge air temperature in the one- and two-stage compressor with intercooling, respectively.

The above equations are valid under the assumption that the pressure ratios of LP and IP compressors are the same and that the intercooling is ideal, i.e., the inlet temperatures of both compressor stages and of one stage are the same. The practically achievable work reduction is smaller and depends on the pressure ratios of the two compressor stages.

Similar to simple-cycle gas turbine power plants, steam injection is also used in combined-cycle power plants for NO_x emissions control and for augmentation of the power output (Bolland and Stadaas, 1995; Noymer and Wilson, 1993; Rice, 1995).

The performance of combined-cycle power plants is affected by the ambient condition. At higher ambient temperatures the density of the air is lower, and thus the air mass flow drawn into the air is lower, which reduces the net power output of the gas turbine. Similarly, at higher elevations the air pressure is lower and therefore the air mass flow is lower. This results in a decrease in the power output. To optimize gas turbine performance, additional components for inlet air conditioning such as air coolers and superchargers may be used.

HIGH-PERFORMANCE POWER SYSTEM

The high-performance power system (HIPPS) is the indirectly fired combined-cycle power plant for clean and effective power generation from coal (Klara, 1995; Klara and Ward, 1992). It employs a closed-cycle gas turbine (see Figure 6.18). The plant consists of a natural gas fired supplementary furnace for achieving the optimum gas turbine inlet temperature of 1260°C, a high-temperature air heater, and an HRSG for raising steam to drive the steam turbine.

HIPPS plants built with state-of-the-art materials and technology can achieve competitive plant efficiencies. The performance of state-of-the-art configurations meets or exceeds that reported for pulverized coal-fired steam, integrated gasification combined-cycle (IGCC), and advanced pressurized fluidized bed combustion plants (discussed in Chapter 9). The combined-cycle power plant should achieve thermal efficiency of 47% and reduce electricity costs by 10% and CO_2 emissions by 25–30%. The efficiency of HIPPS is higher than that of fluidized bed combustion or coal gasification plants, and the exhaust gas emissions are much lower. Commercialization within 3–5 years is anticipated.

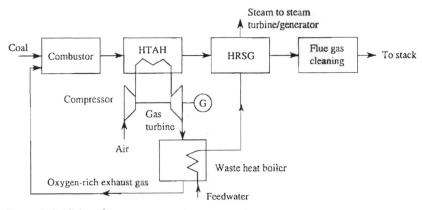

Figure 6.18. High-performance power system.

EXTERNALLY FIRED COMBINED CYCLE

The use of coal in gas turbine based combined-cycle configuration would greatly enhance the economics of electricity generation. Three technologies for coal utilization in combustion turbine combined cycles are being developed at the present time (Cook et al., 1991; Taft, 1991; Maude, 1993):

- direct firing of pulverized coal in the gas turbine combustion chamber
- integrated coal gasification and subsequent firing of the fuel-gas in the gas turbine combustion chamber of a combined-cycle power plant (IGCC)
- burning coal in a combustion chamber external to the gas turbine cycle and utilizing the hot flue gases to heat the gas turbine working fluid in an externally fired combined cycle (EFCC)

All three technologies are attractive options for power generation owing to their advantages of low fuel cost and high thermal efficiency. Both direct pulverized coal fired and IGCC systems utilize standard gas turbines. However, the former is seriously handicapped by the solid-particle erosion in the gas turbine, and the latter by a complex and expensive gasification plant. EFCCs need a high-temperature heat exchanger, e.g., HTAH, for indirect heating of the turbine working fluid.

EFCCs with a gas turbine inlet temperature of 1250°C at a pressure ratio of 10 are capable of achieving plant heat rates of 7600–7900 kJ/kWh at relatively low capital costs (Solomon et al., 1996). Thus they can be competitive with other power generation technologies. The steam cycle portion of the EFCC is essentially the same as for standard oil/gas-fired combined cycle.

In the basic configuration of the EFCC plant the gas turbine working fluid (air) is compressed from ambient conditions up to a pressure of about 10 bars and a temperature of approximately 350°C. The compressed air is then heated in the HTAH. If the required gas turbine inlet temperature (~1250°C) cannot be achieved in the HTAH, it will be increased by burning clean fuel such as natural gas in the air stream. After expansion through the turbine, the air exits at about 620°C. The air is then passed through an HRSG, which produces steam for the steam cycle. A portion of the air is used as hot combustion air in the slagging combustor, which produces flue gas with temperatures of ~1900°C. The flue gas is passed through the HTAH, the separate air HRSG, and the flue gas cleanup system to reduce the concentrations of NO_x, SO_x, and particulate matter to acceptable levels. Finally, the warm air from the air HRSG is mixed with the relatively cool flue gas from the gas cleanup system and discharged via the plant stack.

Only the HTAH and the combustor are nonstandard components in the EFCC system. In the EFCC design of the U.S. Department of Energy (DOE) Morgantown Energy Technology Center, an advanced heat exchanger HTAH exposed to 1930°C gases is used to heat air to 1250°C that can be used to drive a gas turbine generator (Solomon et al., 1996). The program of the U.S. DOE aimed at the development of a direct coal-fired regenerative HTAH was terminated without success in 1984.

Advanced ceramic composite materials promise that the recuperative HTAH would be more feasible than the regenerative one. At high temperatures in the HTAH, radiation is the predominant mode of heat transfer. An advanced composite tubular component made of silicon carbide particulate reinforced alumina has been exposed to combustion gases at a temperature of about 1900 °C. It attained a temperature of 1425°C and survived with minimal damage.

A tube wall concept of radiant recuperative HTAHs has been proposed (Solomon et al., 1996).

REPOWERING AND UPRATING

Repowering is used as a means for performance enhancement of older steam power plants fired by fossil fuel. By adding a topping cycle, typically a gas turbine with an HRSG, a conventional steam power plant is converted to a combined-cycle power plant. Repowering extends the service life of a steam plant and significantly increases power output and efficiency. In addition, it allows the utility to reduce environmental impact of the power plant through the application of advanced combustion technologies and improved utilization of fuel energy. Boilers fired by coal or heavy oil in existing steam plants will be replaced with gas turbines and HRSGs fired by natural gas or light distillate that are used in a topping cycle. Figure 6.19 shows a repowering flow diagram.

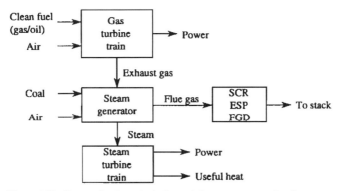

Figure 6.19. Repowering (uprating) of an existing steam power plant by means of a topping gas turbine.

Several approaches can be used for repowering steam turbines with gas turbines (Stambler, 1993; Makansi, 1994). The efficiency of vintage steam power plants can be significantly improved in the so-called parallel-powered combined-cycle power plants. These consist of a natural gas fired gas turbine generating about 1/5 of the total power output of the combined-cycle plant, an HRSG providing superheated steam for the intermediate-pressure steam turbine with an inlet condition of 70 bars/550°C, and a coal-fired steam generator producing main steam with a condition of 300 bars/580°C for the high-pressure steam turbine. The steam power plant train generates about 4/5 of the total power output of the combined-cycle plant. Overall net efficiencies of 50% or more can be achieved in parallel-powered combined-cycle power plants (Balling et al., 1995).

Another repowering method is the use of a topping gas turbine with either single-pressure reheat HRSG or even triple-pressure reheat HRSG. Quite a different approach is the wind box repowering that involves injecting gas turbine exhaust directly into the boiler furnace, thus replacing the forced-draft system. At an incremental cost of approximately $800/kW, a 42% increase in power output and an 8% reduction of heat rate are claimed to be achieved (Makansi, 1994). One more repowering method involves feedwater heating by the gas turbine exhaust heat. Thereby, all or part of the feedwater heaters normally heated with steam bled from the turbine will be replaced by exhaust gas–water heat exchangers. In this case a substantial power output enhancement and a 6% reduction in the heat rate can be achieved at an incremental cost of $700/kW.

Uprating is used for the power output enhancement of existing power plants, including simple-cycle gas turbine and combined-cycle power plants, through the implementation of more advanced technologies and configurations without changing the type of plant. In most cases the efficiency will thus also be improved, and pollutant emissions reduced.

Large-frame gas turbines with efficiencies up to 39.5% (LHV) now offer gross efficiencies up to 58% (based on HHV) and 60% (based on LHV) at the ISO standard conditions (15°C and 101.3 kPa). As a result of this high efficiency potential, and because natural gas prices remain relatively low, the combined-cycle option is projected to account for more than 60% of all NUG capacity additions and 30% of all electric utility capacity additions this decade. The following case histories detail some of the benefits and features of recently installed combined cycles. Combined-cycle plants achieve high overall availabilities up to 98%.

Another essential advantage of combined-cycle power plants is phased construction: first, gas turbines will be installed, then HRSGs and steam plants will be installed while the gas turbines generate electricity and finance construction from the revenues. The first phase duration is at the present time about 1 year, while steam plant erection can take 2–3 years.

CLOSURE

This chapter discusses combined-cycle power plant technology in full detail. Equation (6.6) gives the overall efficiency applicable to combined-cycle power plants of any kind, such as gas turbine based, fuel cell based, or MHD generator based combined-cycle power plants. Advanced gas turbine based combined-cycle power plants have already achieved efficiencies of over 55%, and in the near-term perspective they will attain 60% efficiency. The basis for such high efficiency values is the application of advanced gas turbines described in chapter 4, and of sophisticated double-pressure and triple-pressure reheat steam power plant cycles. When fuel costs are high, the most efficient combined-cycle power plants should be employed, although they have high initial costs due to complicated steam plant schemes. The main advantages of combined-cycle power plants are their high efficiency and favorable economic and environmental characteristics. An additional benefit is the possibility of staged construction of the combined-cycle power plant with the gas turbine installed first and the steam power plant construction financed by the revenues from gas turbine power generation. A further development tied to coal-fired power plants such as HIPPSs or EFCCs is HTAHs, which are described in this chapter. Repowering and uprating of available steam power plants by converting them to combined-cycle power plants will enable significant improvements in efficiency, economics, and environmental impact.

PROBLEMS

6.1. A combined-cycle (CC) power plant consists of a topping gas turbine (GT) plant with a thermal efficiency of 0.385, a heat recovery steam generator (HRSG), and a bottoming steam turbine (ST) plant with a thermal efficiency of 0.31. Calculate (a) the overall efficiency of the CC power plant, (b) the plant net heat rate in kJ/kWh, (c) the heat input rate in MW, and (d) the GT and ST power outputs in MW and their ratio, if the CC plant power output is 400 MW.

6.2. Consider a single-pressure combined-cycle power plant consisting of a gas turbine (GT), heat recovery steam generator (HRSG), and steam turbine (ST) plant. The GT exhaust gas flow rate is 600 kg/s, temperature is 600°C, live steam condition is 70 bars/500°C, condenser pressure is 0.04 bar, and feedwater temperature is 140°C. Assuming a pinch point of 12 K for both high and low pressure, an exhaust gas specific heat of 1.05 kJ/(kg K) and an isentropic efficiency of the ST of 0.9, calculate (a) the rate of steam production in kg/s, (b) the stack gas temperature in °C, (c) the steam turbine power output in MW, and (d) the heat transfer rate in the HRSG in kJ/s.

6.3. Consider a dual-pressure combined-cycle power plant consisting of a gas turbine (GT) heat recovery steam generator (HRSG) and steam turbine (ST) plant. The GT exhaust gas flow rate is 590 kg/s, temperature is 608°C, high-pressure superheated steam condition is 140 bars and 550°C temperature, LP saturated steam is at 8 bars, condenser pressure is 0.05 bar, and feedwater temperature is 120°C. Assuming a pinch point of 11 K for both high and low pressure, an exhaust gas specific heat of 1.05 kJ/(kg K), and an isentropic efficiency of the ST of 0.9, calculate (a) the total steam production rate of the HRSG in kg/s, (b) the HRSG heat transfer rate in kJ/s, (c) the stack gas temperature in °C, and (d) the steam turbine power output in kW.

6.4. Consider a single-pressure combined-cycle power plant consisting of a gas turbine (GT), heat recovery steam generator (HRSG), supplementary firing (SF), and steam turbine (ST) plant. The compressor inlet temperature is 15°C, its pressure ratio is 15, and isentropic efficiency is 0.88. The 200-MW GT burns natural gas with a 380% content of theoretical air, and the HHV of the gas is 49.5 MJ/kg. The GT has an inlet temperature of 1270°C and isentropic efficiency of 0.92. Supplementary firing uses a bituminous coal with an HHV of 34 MJ/kg to raise the gas temperature to 900°C before entering the HRSG. Steam is generated at 9 MPa and 540°C from feedwater at 110°C. The pinch point in the HRSG is 13 K. The flue gas leaves the HRSG to the stack at 250°C. Assume an isentropic exponent of 1.4 and a specific heat of 1.05 for the working fluid of the GT and of 4.187 kJ/kg for water. Calculate (a) the heat addition in supplementary firing per kg of the GT exhaust gas, (b) the net special work of the GT plant w_{GT}, (c) the air mass flow rate m_{air}, (d) the thermal efficiency of the GT plant, (e) the rate of steam generation m_s in the HRSG, and (f) the ratio m_s/m_{air}.

REFERENCES

ABB AG. 1991–1996. Gas turbine specifications. Baden, Switzerland.

Anon. 1995. A 1000-MW single-shaft combined-cycle(GUD) power plant. Siemens.

Balling, L., Joyce, J. S., and Rukes, B. 1995. The new generation of advanced GUD combined-cycle power plants. *Siemens Power J.* 2:18–22.

Bannister, R.L., Cheruvu, N.S., Little, D.A., and McQuiggan, G. 1995. Development requirements for an advanced gas turbine system. *Trans. ASME J. Eng. Gas Turbine Power* 117:724–732.

Becker, B., and Finckh, H. H. 1995. The 3A series gas turbines. *Siemens Power J.* (August):13–17.

Bolland, O. 1991. A comparative evaluation of advanced combined cycle alternatives. *Trans. ASME J. Eng. Gas Turbine Power* 113:190–197.

Bolland, O., and Stadaas, J. F. 1995. Comparative evaluation of combined cycles and gas turbine systems with water injection, steam injection, and recuperation. *Trans. ASME J. Eng. Gas Turbine Power* 117(1):138–145.

Briesch, M.S., Bannister, R.L., Diakunchak, I.S., and Huber, D.J. 1995. A combined cycle designed to achieve greater than 60 percent efficiency. *Trans. ASME J. Eng. Gas Turbine Power* 117:733–740.

Chiesa, P., Lozza, G., Macchi, E., and Consonni, S. 1995. An assessment of the thermodynamic performance of mixed gas-steam cycles. Part B: Water-injected and HAT cycles. *Trans. ASME J. Eng. Gas Turbine Power* 117:499–508.

Cook, D. T., McDaniel, J. E., and Rao, A. D. 1991. HAT cycle simplifies coal gasification power. *Mod. Power Syst.* 11(5):19, 21, 23, 25.

Croonenbrock, R., et al. 1996. Abhitzedampferzeuger fuer Gasturbinen moderner Kraftwerksprozesse. *VGB Kraftwerkstechnik* 76(2):97–101.

Dodero, 1997. *Italian energy handbook.*

Dörr, H. 1996. Die neue Generation der leistungsstarken Gasturbinen für den Einsatz in GUD/Kombi-Kraftwerken in Einwelleanordnung. *BWK* 48(1/2):47–53.

Feenstra, J. 1995. The single-shaft units of the EEMS 95/96 project. Paper presented at PowerGen Europe.

Haywood, R. W. 1991. *Analysis of engineering cycles.* New York: Pergamon.

Horlock, J. 1995. Combined cycle power plants—Past, present and future. *Trans. ASME J. Eng. Gas Turbine Power* 17:608–615.

Jones, C. 1996. Competitive realities change focus of boiler HRSG design. *Power* 140(2):33–37.

Klara, J. M. 1995. HIPPS can compete with conventional PC systems. Part II: Power engineering. *Barrington* 98(1):33–36.

Klara, J. M., and Ward, J. H. 1992. High-performance power systems: State-of-the-art configurations. ASME paper. 92 JPGC FACT-19, pp. 1–8.

Macchi, E., Consonni, S., Lozza, G., and Chiesa, P. 1995. An assessment of the thermodynamic performance of mixed gas-steam cycles. Part A: Intercooled and steam-injected cycles. *J. Eng. Gas Turbine Power* 117:489–498.

Makansi, J. 1994. Repowering. *Power.* 138(6):33–40.

Marston, C.H., and Hyre, M. 1995. Gas turbine bottoming cycle: Triple-pressure steam versus Kalina. *Trans. ASME J. Eng. Gas Turbine Power* 117(1):10–15.

Maude, C. 1993. *Advanced power generation.* London: IEA Coal Research.

Noymer, P.-D., and Wilson, D.-G. 1993. Thermodynamic design considerations for steam-injected gas turbines. ASME paper 93-GT-432, pp. 1–7.

Rice, I.G. 1995. Steam-injected gas turbine analysis: Steam rates. *Trans. ASME J. Eng. Gas Turbine Power* 117:251–258.

Solomon, P.R., et al. 1996. A coal-fired heat exchanger for an externally fired gas turbine. *Trans. ASME J. Gas Turbine Power* 118:22–32.

Stambler, I. 1993. Syngas repowering breathes new life into old steam plants. *Gas Turbine World* 23(4):12–17.

Swanekamp, R. 1995. Gas turbine/combined cycle power systems. *Power* 139(4):15–26.

Swanekamp, R. 1996. Single-shaft combined cycle packs power in at low cost. *Power* 140(1):24–28.

Taft, M. 1991. A comprehensive classification of combined cycle and cogeneration plants. Part 2: Introducing the integrated steam cycle in practice. *Proc. Inst. Mech. Eng. Part A. J. Power Eng.* 205(A3):145–159.

Chapter Seven

COGENERATION

To deliver electricity and useful heat simultaneously, cogeneration plants can be profitably used. This chapter deals with advanced types of cogeneration plants based on the utilization of back-pressure or extraction steam turbines, and gas turbines or gas/diesel engines with waste-heat boilers (WHBs) or heat recovery steam generators (HRSGs). Unlike plants for power generation only, the efficiency of cogeneration plants is characterized by such criteria as energy utilization factor (EUF), heat-to-power ratio, and fuel energy savings ratio (FESR), in addition to the electrical efficiency that characterizes the heat-to-power conversion. Some cases of existing and planned cogeneration plants as well as some new projects are analyzed.

GENERAL PRINCIPLE OF COGENERATION

Cogeneration is the simultaneous production of work and useful heat in a single cogeneration plant or combined heat and power plant (CHP) (Horlock, 1987; Wills et al., 1989; Haywood, 1991). Figure 7.1 illustrates the general principle of cogeneration. The fuel energy input is efficiently converted into electrical energy and useful heat. Under the most favorable conditions, the fraction of the useful energy output can reach up to 85–90% of the fuel energy input.

Cogeneration plants may supply useful heat in the form of hot water for district heating and/or in the form of steam for industrial processes. In a cogeneration plant the heat input Q_{in} to the furnace or combustor is used for both production of work W_{net} and also of useful heat Q_u for district heating or industrial heat supply. Although synergies of energy generation and production of useful chemicals in one plant are also possible, e.g., coproduction of electricity and methanol (Brown et al., 1989), they will not be considered in this chapter. Cogeneration is beneficial because it enables a better utilization of the fuel energy and therefore allows achievement of considerable fuel energy savings when compared with separate production of electricity and useful heat in a conventional power plant and in a boiler, respectively. In comparison with the two reference plants—a conventional steam power plant for electric energy generation and a heating plant or boiler for production of useful heat—the cogeneration plant provides a much higher efficiency of fuel energy utilization.

There are cogeneration plants with a topping or bottoming configuration. In the topping configuration the fuel energy is used to produce working fluid (steam, gas, or products of combustion) with high enthalpy suitable for electricity generation in a heat engine (steam turbine, gas turbine, and diesel or gas engine). The exhaust heat of the working fluid leaving the heat engine is further used in a waste-heat recuperator (WHR) or waste-heat boiler (WHB) to produce useful heat. This is the most common configuration of cogeneration plants. In an industrial process at a high temperature the bottoming configuration can be used to generate electric power by means of the process waste heat. In this case, an organic Rankine cycle (ORC) with an alternative working fluid such as propane will be employed to generate electricity. High cost-effectiveness can be achieved practically only in cogeneration plants with the topping configuration.

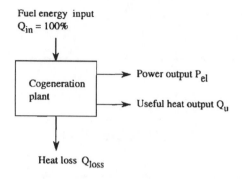

$$EUF = (P_{el} + Q_u) / Q_{in} = \text{up to } 90\%$$

Figure 7.1. Principle of a cogeneration plant.

STEAM TURBINE BASED COGENERATION PLANTS

Cogeneration plants can be classified depending on the type of prime mover that is used in the plant. There are cogeneration plants based on the utilization of steam turbines, gas turbines, and diesel or gas engines. The cogeneration plants based on steam and gas turbines have the following major configurations (Weston, 1992; Taft, 1991; Schnitz and Koch, 1995; Bohn, 1991; Vulkan-Verlag, 1991; Adlhoch and Bolt, 1994):

- back-pressure steam turbine based cogeneration plant
- extraction-condensing steam turbine based cogeneration plant
- cogeneration plant based on a gas turbine with a WHB that is used to supply process heat
- cogeneration plant based on a gas turbine with a WHR that is used to produce hot water
- combined-cycle (steam and gas turbine) cogeneration plant with the steam turbine of either the back-pressure type or extraction-condensing type

As mentioned above, there are also cogeneration plants based on diesel or gas engines. In this case, mechanical power produced in the engine is converted into electric power in a generator, and the exhaust gas heat along with jacket cooling water heat, lubricating oil heat, and charge loading heat are used to produce hot water for decentralized heat supplies. Most process applications require low-grade (temperature) steam produced in a topping cycle. Depending on process requirements, process steam can either be (1) extracted from the steam turbine at an intermediate stage, similar to steam extraction for feedwater heating, or (2) taken at the turbine exhaust, in which case it is called a back-pressure turbine. Process steam pressure requirements vary widely, say, between 0.5 and 40 bars.

There are two basic configurations of cogeneration plants with steam turbines (Peraiah, 1993). Figure 7.2 shows a cogeneration plant based on the utilization of a back-pressure steam turbine. This type of steam turbine differs from steam turbines of the condensing type by the level of exhaust steam pressure. Unlike the condensing turbines from which the steam is exhausted into a condenser with a pressure of around 0.04 bar, the exit pressure in the back-pressure steam turbine is higher than 1 bar. This enables using the exhaust steam directly in some industrial processes or indirectly via a heat exchanger that will be employed to heat water for district heating. The cycle of a steam power plant with a back-pressure steam turbine is shown in Figure 7.3 as compared to the cycle of a plant with a condensing steam turbine. Thereby, cycle 12345 refers to the back-pressure turbine plant, whereas cycle 12'3'4'5 refers to the condensing turbine plant. It is seen that due to higher back-pressure, the cycle net heat (and the equivalent net work correspondingly) in the case of the back-pressure steam turbine plant is less than that of the condensing turbine plant by the area of 22'3'4'4. The area under the line 2'3' is the available heat, part of which can be used for industrial or district heating purposes.

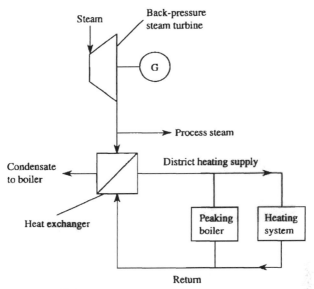

Figure 7.2. Simplified diagram of a back-pressure steam turbine based cogeneration plant for power generation, process heat production, and district heating.

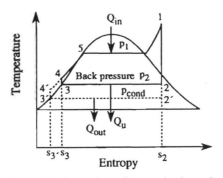

Figure 7.3. Comparison of the condensing and back-pressure steam turbine cycles on a T-s diagram.

Thus the back-pressure level directly affects the enthalpy drop in the steam turbine. The higher the back-pressure, the less electricity is produced. If this pressure is very high, the use of a steam turbine becomes questionable, since its pressure differential is then too small. In such a case, the steam process reduces to a waste-heat boiler.

A more detailed block diagram of the cogeneration plant configuration with an extraction-condensing steam turbine, heat exchanger, and district heating system is presented in Figure 7.4. Cogeneration plants with back-pressure steam turbines are most suitable when the electric power demand is low in comparison with the heat demand. This is the configuration that is widely used for industrial cogeneration. The cogeneration plant with extraction-condensing turbine is applicable in a wide range of useful heat-to-power ratios. Figure 7.5 shows a schematic diagram of a cogeneration plant based on an extraction-condensing steam turbine with blocks of high-pressure and low-pressure feedwater heaters.

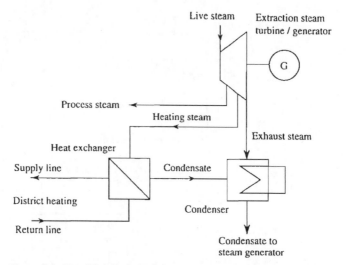

Figure 7.4. Simplified diagram of a cogeneration plant based on an extraction steam turbine for power generation, process heat production, and district heating.

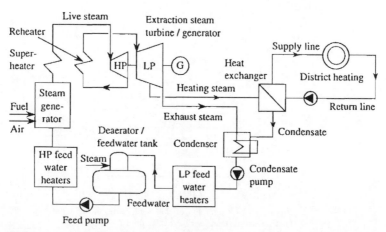

Figure 7.5. Schematic diagram of a cogeneration plant for power generation and district heating based on an extraction-condensing steam turbine with high-pressure (HP) and low-pressure (LP) feedwater heaters.

PERFORMANCE CRITERIA OF COGENERATION PLANTS

Performance Criteria of Conventional Power Plants

A higher degree of overall utilization of fuel energy is achieved in cogeneration plants in comparison with conventional power plants and heating plants. The performance of cogeneration plants is characterized by means of certain criteria of performance such as those presented below (Horlock, 1987; Haywood, 1991).

As described in Chapter 4, the most important performance criterion of conventional power plants is overall efficiency, that is, the product of the cycle thermal efficiency η_{th} and the combustor (boiler) efficiency η_{comb}:

$$\eta_o = \eta_{th}\eta_{comb} = W_{net}/Q_f \tag{7.1}$$

where W_{net} is the plant net work and Q_f is the fuel energy input to the combustor (boiler). The fuel energy input is given by

$$Q_f = m_f LHV \tag{7.2}$$

where m_f is the mass of the fuel burned and LHV is the lower heating value of the fuel.

Another criterion of performance of a conventional power plant is the plant net heat rate, which is defined as

$$HR = \text{Heat input/Net power output} \tag{7.3}$$

It is then

$$HR = 3600/\eta_o \quad kJ/kWh \tag{7.4}$$

or

$$HR = 34,120/\eta_o \quad BTU/kWh \tag{7.5}$$

The heat input to the cycle is less than Q_f. It is given by

$$Q_{in} = Q_f \eta_{comb} \tag{7.6}$$

Energy Conversion Efficiency

The conversion of the fuel energy to power in a cogeneration (CG) or CHP plant is characterized by the energy conversion efficiency:

$$\eta_{CG} = W_{net}/Q_{in} \tag{7.7}$$

Energy Utilization Factor

In order to characterize the overall efficiency of the fuel energy utilization for both the power generation and useful heat production, the EUF is used (Heywood, 1988). It is a proper criterion of performance of a cogeneration plant as a whole. EUF is defined as the ratio of the total useful energy, i.e., the sum of the net work output W_{net} and the useful heat Q_u to the fuel energy input to the cogeneration plant, $Q_{f,CG}$. Thus

$$EUF = (W_{net} + Q_u)/Q_{f,CG} \tag{7.8}$$

or

$$EUF = (P_{el} + Q_u)/Q_{f,CG} \tag{7.9}$$

In Eq. (7.9), electrical power P_{el} and useful heat rate Q_u are related to the rate of fuel energy input $Q_{f,CG}$ of the cogeneration plant.

For open-cycle power plants, such as gas turbine or diesel engine power plants, the fuel heat released in a furnace or combustor Q_f is a proper input energy quantity. In closed-cycle plants such as steam power plants, the heat supplied from combustion gases to the working fluid of the cycle Q_{in} is more adequate as a basis value. Thus for steam turbine based cogeneration plants,

$$EUF = (W_{net} + Q_u)/Q_{in,CG} \tag{7.10}$$

However, it should be stressed that the two products of a cogeneration plant—work W_{net} and useful heat Q_u—have different values. To produce 1 kWh of electrical power in a conventional power plant, 2.5–3 kWh of fuel energy are usually required. Thus work and electrical power are higher grade energy forms. They have a higher quality and therefore a higher price, whereas the heat is a lower grade energy with a lower price. The EUF does not take into account this difference in the quality of energies W_{net} and Q_u. Therefore EUF may not be the only criterion of performance of cogeneration plants, and additional criteria are needed.

Heat-to-Power Ratio

An additional important performance criterion of cogeneration plants is the cogeneration ratio or the ratio of the useful heat to electric power generated, i.e., the heat-to-power ratio

$$HPR = Q_u/P_{el} \tag{7.11}$$

where P_{el} is the plant power output and Q_u is the useful heat rate of the cogeneration plant.

The fuel energy supplied to a boiler, gas turbine combustor, or diesel engine of a cogeneration plant, $Q_{f,CG}$, is converted to heat energy and transferred to the working fluid with an efficiency of the combustion device η_{comb}. Then the energy balance of the cogeneration plant may be written as

$$Q_{f,CG}\eta_{comb} = Q_{in} = P_{el} + Q_u + Q_{out} \tag{7.12}$$

where Q_{out} is the rejected (nonuseful) heat of the cogeneration plant.

Incremental Heat Rate

The total energy input Q_{in} may be arbitrarily broken down into one portion that is used in the boiler to produce the useful heat Q_u with an efficiency η_{comb} and the second portion that is used for generation of electrical power P_{el} at an incremental heat rate (IHR) (Horlock, 1987; Li Kam and Priddy, 1985). Thus

$$Q_{in} = IHR \times P_{el}/3600 + Q_u/\eta_{comb} \tag{7.13}$$

Therefore

$$IHR = 3600[Q_{in}/P_{el} - Q_u/((P_{el}\eta_{comb})] \quad kJ/kWh \tag{7.14}$$

Using the HPR for the cogeneration plant as a whole and the overall efficiency, $\eta_{CG} = P_{el}/Q_{in}$, for the generation of electrical power in the cogeneration plant, we obtain

$$IHR = 3600[1/\eta_{CG} - HPR/\eta_{comb}] \tag{7.15}$$

Rational Criterion of Performance

A so-called rational criterion of performance (RC) is defined as the ratio of the sum of net work and useful heat outputs of the cogeneration plant to that of a reversible plant (Horlock, 1987). Thus

$$RC = (P_{el} + Q_u)_{CG}/(P_{el} + Q_u)_{rev} \tag{7.16}$$

For a given heat load Q_u and a given electrical power demand P_{el}, RC is the ratio of fuel energy demand in the reversible plant to that in the actual cogeneration plant:

$$RC = Q_{f,rev}/Q_{f,CG} \tag{7.17}$$

The value of RC can be related to other performance criteria of the cogeneration plant, such as the EUF, overall efficiency η_{CG}, HPR, and the ratio of the ambient temperature T_a to the temperature of useful heat supply T_u, as follows:

$$RC = EUF - HPR \times \eta_{CG}(T_a/T_u) \tag{7.18}$$

The following example gives a performance evaluation of a cogeneration plant with a back-pressure steam turbine.

Example 7.1

A back-pressure turbine cogeneration plant has the following parameters:

Main steam condition at turbine throttle $p_1 = 120$ bars and $t_1 = 550°C$
Back-pressure $p_2 = 4$ bars
Steam mass flow $m = 150$ kg/s
Turbine isentropic efficiency $\eta_{it} = 0.91$
Generator efficiency $\eta_g = 0.99$

Calculate

- power output of the plant
- electrical efficiency
- heat-to-power ratio
- energy utilization factor of the cogeneration plant if the heat loss η_{dh} in the district heating system is 20%

Solution

1. From the Mollier h-s diagram:
   Main steam enthalpy at $p_1 = 120$ bars and $t_1 = 550°C$ is $h_1 = 3480$ kJ/kg.
   Exhaust steam enthalpy at isentropic expansion from p_1 to p_2 is $h_{2s} = 2635$ kJ/kg.
2. Actual exhaust steam enthalpy at irreversible expansion

$$h_2 = h_1 - \eta_{it}(h_1 - h_{2s}) = 3480 - 0.91(3480 - 2635) = 2711 \text{ kJ/kg}$$

3. Feedwater enthalpy at p_2 (look up in the water/steam table):

$$h_{fw} = 604.67 \text{ kJ/kg}$$

4. Actual specific turbine work, i.e., the enthalpy drop in the turbine

$$w_t = h_1 - h_2 = 3480 - 2711 = 769 \text{ kJ/kg}$$

5. Heat addition per kg steam

$$q_{in} = h_1 - h_{fw} = 3480 - 604.67 = 2875.33 \text{ kJ/kg}$$

6. Actual thermal efficiency

$$\eta_{th} = w_t/q_{in} = 769/2875.33 = 0.267$$

7. Electrical efficiency

$$\eta_{el} = \eta_{th}\eta_g = 0.267 \times 0.99 = 0.264$$

8. Electrical power

$$P_{el} = m w_t \eta_g = 150 \text{ kg/s} \times 769 \text{ kJ/kg} \times 0.99 = 114{,}196.5 \text{ kW}$$

9. Useful heat output

$$Q_u = m(h_2 - h_{fw})(1 - \eta_{dh}) = 150 \text{ kg/s}(2711 - 604.67)(1 - 0.2) = 252{,}759.6 \text{ kW}$$

10. Heat-to-power ratio

$$Q_u/P_{el} = 252{,}759.6/114{,}196.5 = 2.2$$

11. Energy utilization factor of the cogeneration plant

$$EUF = (P_{el} + Q_u)/(m q_{in}) = (114{,}196.5 + 252{,}759.6)/(150 \times 2875.33) = 0.851$$

Fuel Energy Savings Ratio

FESR is another performance criterion that is used to evaluate the performance of a CHP in terms of the fuel demand as compared to a conventional power station and a heating plant. This comparison involves estimation of the fuel quantity required to meet the given power and heat loads in the cogeneration plant as well as the calculation of the total fuel demand necessary to produce equal quantities of electrical energy and useful heat in two separate conventional plants. Let us consider a conventional power station with an overall efficiency η_{el} and a heating plant or boiler with an efficiency η_b. Then the total fuel energy requirements $Q_{f,c}$ of these two conventional plants are

$$Q_{f,c} = P_{el}/\eta_{el} + Q_u/\eta_b \tag{7.19}$$

Similarly, the fuel energy required in the cogeneration plant with an energy utilization factor (EUF) is

$$Q_{f,CG} = (P_{el} + Q_u)/EUF \tag{7.20}$$

For the fuel energy saved in the cogeneration plant as compared to a conventional plant, we have the following equation:

$$\Delta Q_f = Q_{f,c} - Q_{f,CG} \tag{7.21}$$

Now the FESR is defined as the ratio of the fuel energy savings of the cogeneration plant to the fuel energy required in two conventional plants. Hence

$$FESR = \Delta Q_f/Q_{f,c} \tag{7.22}$$

COGENERATION PLANTS BASED ON GAS TURBINE WITH WASTE-HEAT BOILER OR RECUPERATOR

Plant Configurations

A gas turbine power plant may be equipped with a WHB or a WHR. The power plant is thus converted into a gas turbine based cogeneration plant that generates electrical power and produces useful heat. After expansion in the gas turbine, the hot exhaust gas of the gas turbine will be used in a WHR or WHB either for district heating or for heat and steam supply of industrial processes. Gas turbine based cogeneration plants are characterized by high-energy utilization factors (up to 90%) and low capital costs.

In order to further increase the production of useful heat in the form of hot water and/or steam, the heat recovery unit may be enhanced by means of a supplementary firing such as that shown in Figure 7.6. Additional improvement in the cogeneration plant performance will be brought about if a peaking boiler and thermal storage are integrated into the system as shown in Figure 7.7.

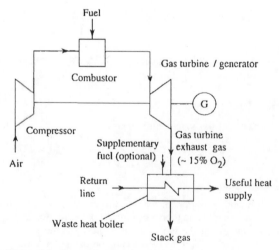

Figure 7.6. Gas turbine based cogeneration plant with supplementary firing of the waste-heat boiler.

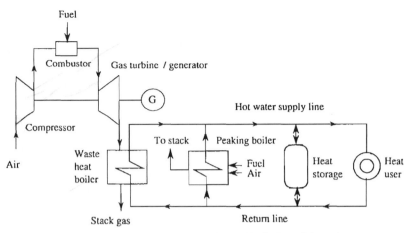

Figure 7.7. Gas turbine based cogeneration plant with peaking boiler and thermal energy storage.

Performance of Gas Turbine Based Cogeneration Plants

Consider a cogeneration unit consisting of a gas turbine plant and a WHR. The irreversibilities in the compressor and gas turbine are accounted for by their isentropic (more correctly, polytropic) efficiencies, η_{ic} and η_{it}. The power outputs of reversible and irreversible gas turbine plants are

$$P_{rev} = P_t - P_c = m_g \Delta h_t - m_a \Delta h_c \tag{7.23}$$

and

$$P = m_g \Delta h_t \eta_{it} - m_a \Delta h_c / \eta_{ic} \tag{7.24}$$

The mass flow rate in the gas turbine is the sum of the mass flow rates of the air and the fuel:

$$m_g = m_a + m_f \tag{7.25}$$

The mass flow rate of the fuel burned in the gas turbine combustor is given by

$$m_f = Q_f / LHV \tag{7.26}$$

The heat addition in the gas turbine combustor to the actual cycle is smaller than that in the reversible one. Thus

$$Q_f = Q_{f,rev} - P_c(1/\eta_{ic} - 1) \tag{7.27}$$

The enthalpy of the gas turbine exhaust gas in the actual cycle is higher than that in the reversible cycle, and the useful heat is therefore greater:

$$Q_u = Q_{u,rev} + P_t(1 - \eta_{it}) \tag{7.28}$$

The thermal efficiency of the gas turbine plant is reduced from

$$\eta_{th,rev} = P_{rev}/Q_{f,rev} \tag{7.29}$$

to

$$\eta_{th} = P/Q_f \tag{7.30}$$

The following example illustrates the procedure for the performance evaluation of cogeneration plants based on a gas turbine and a WHR.

Example 7.2

Consider a cogeneration plant consisting of a gas turbine for electric power generation and WHR producing useful heat for district heating. The compressor intake temperature T_1 is 290 K, the gas turbine inlet temperature T_3 is 1450 K, the compressor pressure ratio β is 11.78, $\eta_{it} = 0.92$, $\eta_{ic} = 0.89$, and the stack gas temperature is 130°C. Assuming the mass flow rate of air of 540 kg/s and the heating value of natural gas fuel of 47 MJ/kg, calculate (1) the conversion efficiency, (2) the energy utilization factor, (3) the power output, (4) the useful heat output, and (5) the heat-to-power ratio of the cogeneration plant.

Assume the specific heat of air [1.01 kJ/(kg K)] and of gas [1.05 kJ/(kg K)]. The heat loss of the WHR may be ignored.

Solution

1. Compressor discharge air temperatures for isentropic and actual compression are respectively

$$T_{2s} = T_1\beta^{(k-1)/k} = 290 \times 11.78^{(1.4-1)/1.4} = 586.7 \text{ K}$$

$$T_2 = T_1 + (T_{2s} - T_1)/\eta_{ic} = 290 + (586.7 - 290)/0.89 = 623.4 \text{ K}$$

2. Specific compressor work input

$$w_c = c_p(T_2 - T_1) = 1.05(623.4 - 290) = 350.08 \text{ kJ/kg}$$

3. Gas turbine exhaust gas temperature for isentropic and actual expansion are respectively

$$T_{4s} = T_3/\beta^{(k-1)/k} = 1450/11.78^{(1.4-1)/1.4} = 716.7 \text{ K}$$

$$T_4 = T_3 - (T_3 - T_{4s})\eta_{it} = 1450 - (1450 - 716.7)0.92 = 775.4 \text{ K} = 502.25°\text{C}$$

4. Specific gas turbine work output

$$w_t = c_p(T_3 - T_4) = 1.05(1450 - 775.4) = 708.33 \text{ kJ/kg}$$

5. Plant net specific work output for

$$w_{net} = w_t - w_c = 708.33 - 350.08 = 358.25 \text{ kJ/kg}$$

6. Specific heat addition

$$q_{in} = c_p(T_3 - T_2) = 1.05(1450 - 623.4) = 867.93 \text{ kJ/kg}$$

7. Plant conversion efficiency

$$\eta_{th} = w_{net}/q_{in} = 358.25/867.93 = 0.4128$$

8. Specific useful heat of the cogeneration plant

$$q_u = c_p(t_{exh} - t_{stack}) = 1.05(502.25 - 130) = 390.86 \text{ kJ/kg}$$

9. Energy utilization factor of the cogeneration plant with irreversible cycle

$$\text{EUF} = (w_{net} + q_u)/q_{in} = (358.25 + 390.86)/867.93 = 0.863$$

10. Heat-to-power ratio

$$\text{HPR} = q_u/w_{net} = 390.86/358.25 = 1.09$$

The gas turbine based cogeneration plants can be utilized both as district energy supply plants and as industrial cogeneration plants. The former produce power and domestic heat; the latter provide power and process heat or steam.

COMBINED-CYCLE COGENERATION PLANTS

As indicated in Chapter 6, combined cycles for power generation are thermodynamically superior in comparison with both gas turbine and steam turbine simple cycles. Owing to the higher average temperature of heat addition and lower average temperature of heat rejection in combined-cycle plants, their electrical efficiency is higher than that of simple-cycle power plants. Their superiority becomes even more pronounced when combined cycles are used in cogeneration plants to produce power and heat. In combined-cycle cogeneration plants the total energy utilization factor usually approaches 90% (Schnitz and Koch, 1995; Bohn, 1992; Steimle, 1995; Adlhoch and Bolt, 1994).

Cogeneration plants based on gas turbine simple and combined cycles are suitable options for relatively large electric and heating demands and therefore can be employed for energy heat supply of a district or a town (Roy, 1993; Anon, 1995; Kehlhofer, 1991).

Combined-cycle cogeneration plants usually have the following configuration (Plohberger et al., 1995; Kehlhofer, 1991):

- a gas turbine
- an HRSG
- an extraction-condensing or back-pressure steam turbine

Figure 7.8 shows a schematic diagram of a combined-cycle cogeneration plant with topping gas turbine, HRSG, and bottoming extraction-condensing steam turbine.

Simple configurations of combined-cycle plants usually will be adopted. Thus an HRSG design without supplementary firing will normally be chosen. The district heating water will be heated to the required supply temperature by steam bled from the extraction-condensing steam turbine.

Both the pressure level of the process steam and the heat-to-power (HPR) are significant design parameters because the pressure of the process steam directly affects the enthalpy drop in the steam turbine. The higher the pressure, the less electricity is produced. If the process steam pressure is very high, the use of a steam turbine becomes questionable, since its enthalpy differential is then too small. In such a case, the steam process reduces to a WHB.

In cogeneration plants for district heating, independent control of electricity and heat production is generally not required, since such plants are usually integrated into large utility grids, which

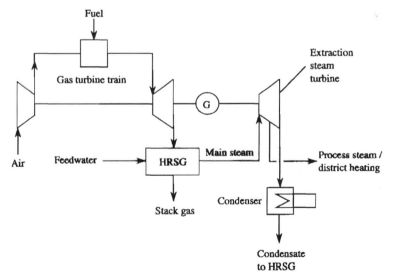

Figure 7.8. Combined-cycle cogeneration plant with topping gas turbine, HRSG, and bottoming steam turbine.

compensate the possible deficiency of power generation in comparison with the actual power demand. The heat output of a cogeneration plant for district heating is usually controlled as a function of the ambient air temperature. In combined-cycle cogeneration plants for district heating the strong impact of the ambient air temperature on the gas turbine power output may have an overall positive effect, since the maximum heat demand corresponds to the lowest ambient air temperature.

As far as the supply water temperature for district heating is concerned, it should be selected as low as possible. However, the choice of the design water temperature is a compromise between high power output associated with a low water supply temperature and low costs of district heat transport at high water supply temperature.

Combined-cycle cogeneration plants are most favorably employed in cases of prevailing electric loads, which corresponds to low HPRs. An extraction-condensing steam turbine enables increasing the power output and offers greater design and operating flexibility than the back-pressure turbine.

In combined-cycle cogeneration plants with supplementary firing, live steam parameters are similar to those used in conventional steam power plants. To attain a more complete utilization of the gas turbine exhaust heat in the WHB or HRSG, the feedwater temperature must be as low as possible. However, it must be higher than the dew point of the flue gas (see Chapter 2) in order to prevent corrosion.

HRSG designs for combined-cycle cogeneration plants without supplementary firing are similar to those discussed in Chapter 6 for single-pressure and double-pressure combined-cycle power plants. However, the live steam pressure in the single-pressure configurations should be high enough to assure a reasonably high enthalpy differential by the steam expansion between the live steam and process steam pressure levels. If a relatively high process steam pressure level is required, then the EUF is rather low. Applying a double-pressure combined-cycle cogeneration plant improves the EUF value. The low-pressure portion of the HRSG will then be used only for generating the process steam.

As an example of a gas turbine-based cogeneration plant with HRSG, the Bewag cogeneration district heating plant that began operation in 1996 in Berlin, Germany, will be considered (Anon, 1995). The plant consists of two 165-MW ABB GT13E2 gas turbine units, each with an unfired HRSG. A gas turbine unit consists of a 21-stage compressor and a 5-stage gas turbine. The compressor pressure ratio is 15, and the gas turbine inlet temperature is 1100°C. Fuel is burned in the annular combustor with 72 dry, low-NO$_x$ (DLN) dual-fuel EV burners, which are described in Chapter 5. The main fuel is natural gas, and the backup fuel is light distillate (fuel oil No. 2). Owing to the large number of individually controlled burners, the gas temperature in front of the first row of gas turbine blades is uniform. The DLN EV burners operate on natural gas with low NO$_x$ emissions levels (below 25 ppmv), whereas water injection is used to reduce NO$_x$ emissions when the gas turbine is to be operated on light distillate (fuel oil No. 2). Each of the two dual-pressure unfired vertical HRSGs is a natural-circulation boiler with pump circulation enhancement. The furnace water walls are made of spiral finned tubes. An HRSG consists of two parallel units and has a length of 22 m, width of 4 m, and height of 3 m. During start-up, the gas turbine exhaust gas is led via bypass to a chimney. The main technical data of the combined-cycle cogeneration plant for power generation and district heating in Berlin are presented in Table 7.1.

In the parallel-powered combined-cycle block the steam will be supplied to the steam turbine in parallel from a fully fired steam generator and from a HRSG associated with the gas turbine, which will also preheat part of the condensate and feedwater. The utility optimized the power plant configuration in cooperation with the suppliers of the turbine generators and steam generators to meet the station's particular needs. A greater cost-effectiveness and operational flexibility is achieved by means of a possible fully independent parallel operation of steam and gas turbine trains. This also enables better performance in regard to part-load power and heat production. The extraction of steam for district heating is possible in both parallel-powered mode and during independent operation of the steam or gas turbine plants. An example of steam power plant repowering with a combined-cycle gas turbine plant is the Deutsch 400-MW Altbach-Deizisau cogeneration plant in Germany. This parallel-powered combined-cycle cogeneration plant burns coal in the boiler furnace and natural gas in the gas turbine combustor. Examples 7.3 and 7.4 present the calculation of the performance characteristics of combined-cycle cogeneration plants.

Table 7.1. Design data of the Bewag combined-cycle cogeneration plant in Berlin, Germany

Parameter	Value
Gas turbine number and type	2 × ABB GT 13E2
Plant power output, MW	380
District heating thermal output, nominal/maximum, MJ/s	340/380
Energy conversion efficiency, %	47.4
Heat rate, kJ/kWh	7595
Energy utilization factor, %	89.2
HRSG	
HP steam outlet pressure and temperature, bars/°C	76.9/525
HP steam mass flow rate, t/h	210
Feedwater temperature, °C	110
LP steam outlet pressure and temperature, bars/°C	5.3/203
LP steam mass flow rate, t/h	48
District heating supply/return water temperature, °C	105/55

Example 7.3

A combined-cycle cogeneration plant consists of two gas turbines with electrical power P_{GT} of 140 MWe each, two HRSGs, and a 136-MWe extraction steam turbine. The plant is used for power generation and useful heat production for district heating. The maximum useful heat output is 280 MW. The gas turbine efficiency is 35%.

Calculate the electrical efficiency of the combined cycle, the efficiency of the steam turbine, the EUF, and the fuel demand if the fuel is natural gas with LHV = 50 MJ/kg.

Solution

Fuel energy input to the gas turbine combustor

$$Q_f = P_{GT}/\eta_{GT} = 2 \times 140/0.35 = 800 \text{ MJ/s}$$

Electrical efficiency of the combined-cycle power plant

$$\eta_{CC} = P_{el}/Q_f = (2 \times 140 + 136)/800 = 0.52$$

Efficiency of the steam turbine power plant

$$\eta_{ST} = (\eta_{CC} - \eta_{GT})/(1 - \eta_{GT}) = (0.52 - 0.35)/(1 - 0.35) = 0.2615$$

EUF of the cogeneration plant

$$\text{EUF} = (P_{el} + Q_u)/Q_f = (2 \times 140 + 136 + 280)/800 = 0.8575$$

Fuel demand for the combined-cycle cogeneration plant

$$m_f = Q_f/\text{LHV} = 800/50 = 16 \text{ kg/s}$$

Example 7.4

A combined-cycle cogeneration plant burns natural gas with LHV = 50 MJ/kg and delivers electrical power and useful heat for district heating. It consists of three units, each with a 240-MWe gas turbine with a thermal efficiency of 39%, an HRSG, and a 110-MWe steam turbine. The power plant heat rate is 6330 kJ per kWh of electric energy. The EUF is 0.88. The plant burns natural gas with LHV = 48.7 MJ/kg and delivers electrical power and useful heat for district heating.

Assuming the stoichiometric air-fuel ratio of 17.5 kg per kg fuel, the percentage excess air of 240%, and the flue gas specific heat c_{pg} of 1.05 kJ/(kg K), calculate

- the electrical efficiency of the combined-cycle power plant
- the thermal efficiency of the steam turbine plant

- the HPR of the cogeneration plant
- the useful heat production rate
- the HRSG exit gas temperature

Solution

1. The electrical efficiency of the combined cycle power plant is

   $\eta_{CC} = 3600/HR = 3600/6330 = 0.569$

   where HR is the heat rate.
2. The thermal efficiency of the steam turbine plant is

   $\eta_{ST} = (\eta_{CC} - \eta_{GT})/(1 - \eta_{GT}) = (0.569 - 0.39)/(1 - 0.39) = 0.293$

3. The total fuel heat input to all the three gas turbine combustors is

   $Q_f = P_{GT}/\eta_{GT} = 3 \times 240/0.39 = 1846.15$ MW

4. The HPR of the cogeneration plant is

   $HPR = EUF/\eta_{CC} - 1 = 0.88/0.569 - 1 = 0.547$

5. The useful heat production rate is

   $Q_u = HPR \times P_{el} = 0.547 \times 3 \times (240 + 110) = 573.9$ MJ/s

6. The fuel rate per unit is

   $m_f = 1/3\,Q_f/LHV = 1/3 \times 1846.15/48.7 = 12.64$ kg/s

7. The exhaust gas flow rate per unit is

   $m_g = m_f + m_a = m_f(1 + \lambda \times AFt) = 12.64 \times (1 + 2.4 \times 17.5) = 543.5$ kg/s

8. The specific stack gas enthalpy is

   $h_{stack} = Q_f(1 - EUF)/m_g = 1/3 \times 1846.15 \times 10^3(1 - 0.88)/543.5 = 135.9$ kJ/kg

9. The exhaust gas (stack) temperature is

   $t_{stack} = h_{stack}/c_{pg} = 135.9/1.05 = 129.4°C$

GAS/DIESEL ENGINE BASED COGENERATION PLANTS

In cogeneration plants of smaller capacity the gas turbine should be replaced by an internal combustion engine, which is more efficient in the low-to-intermediate power output range (Nash, 1992). Gas engines, diesel engines, and dual-fuel gas/diesel engines are used as prime movers in these cogeneration plants. The thermodynamic basis for the gas engine is the Otto cycle, and for the diesel engine is the Diesel cycle (Wark, 1994; Heywood, 1988). They are shown on P-v diagrams in Figures 7.9a and 7.9b. Only the most important relationships for these cycles will be reviewed here.

The operation of gas engines, diesel engines, and gas/diesel engines differs in how the fuel-air mixture is formed and in the ignition. Thus the succession of processes in the lean-burn Otto gas engine includes an external mixture formation, compression, spark ignition, combustion (constant-volume heat addition), gas expansion, and exhaust rejection. The thermal efficiency of the Otto cycle is given by

$$\eta_{th} = 1 - 1/r_v^{k-1} \qquad (7.31)$$

where $r_v = v_1/v_2$ is the compression ratio of the cycle, v_1 and v_2 are the specific volumes of air at the beginning and end of the compression, respectively, and k is the isentropic exponent (for air, $k = 1.4$). The thermal efficiency of the gas engine cycle as a function of compression ratio r_v is given in Table 7.2.

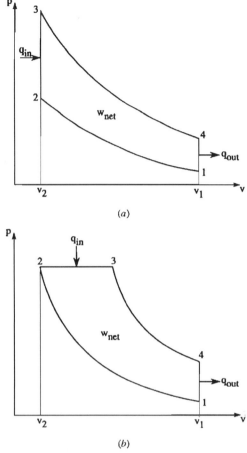

Figure 7.9. (a) Otto cycle and (b) Diesel cycle on a P-v diagram.

The diesel engine cycle consists of an air compression, fuel oil injection, internal mixture formation, autoignition, combustion (constant-pressure heat addition), gas expansion, and exhaust rejection. The thermal efficiency of the Diesel cycle is given by

$$\eta_{th} = 1 - \left(r_c^k - 1\right)/k(r_c - 1)r_v^{k-1} \tag{7.32}$$

where $r_c = v_3/v_2$ is the cutoff ratio and v_3 is the air specific volume at the end of heat addition. The thermal efficiency η_{th} of the Diesel cycle increases with an increase in the compression ratio r_v and with a decrease in the cutoff ratio r_c. Thermal efficiency of the Diesel cycle as a function

Table 7.2. Thermal efficiency of the Otto cycle

Compression ratio	Thermal efficiency
3	0.36
6	0.51
9	0.58
12	0.63

Table 7.3. Thermal efficiency of the Diesel
cycle at a compression ratio of 14

Cutoff ratio	Thermal efficiency
2	0.59
3	0.55
4	0.51
5	0.47

of the cutoff ratio r_c at a compression ratio $r_v = 14$ is given in Table 7.3. For the Otto cycle with $r_v = 14$, $\eta_{th} = 0.65$.

In the gas/diesel engine, the fuel gas/air mixture is compressed to a high pressure. Just before the end of compression, a small amount of fuel oil (3–5% of fuel gas) is injected for stable ignition.

The performance of a gas, diesel, or gas/diesel engine is characterized by mean effective pressure (MEP), indicated or effective (brake) power, specific fuel consumption, and efficiency. The MEP (Wark, 1994; Heywood, 1988) is given by

$$MEP = W_{net}/V_{dis} \quad kPa \tag{7.33}$$

where W_{net} is engine net work output in kJ and V_{dis} is engine piston displacement volume in m^3. The engine effective (brake) power is

$$P_e = V_{dis}MEPn/60s \quad kW \tag{7.34}$$

where n is speed in rpm and $s = 2$ for four-stroke engines and 1 for two-stroke engines. The specific fuel consumption of the engine is

$$b_f = 3.6 \times 10^6 m_f/P_e = 10^3/\eta_e LHV \quad g/kWh \tag{7.35}$$

where m_f is fuel consumption in kg/s, η_e is efficiency, and LHV is lower heating value of the fuel in kWh/kg.

Power output of gas engines that can be used in small-scale cogeneration (or CHP) plants is given in Table 7.4. The waste heat of a prime mover is used to generate steam or hot water.

Gas engines burning natural gas or other fuel gases are also called lean-burn engines. Gas/diesel engines typically operate on approximately 95–97% natural gas and 3–5% heavy fuel oil. Lean-burn engines and gas/diesel engines have a number of significant advantages, such as fuel flexibility, plant modularity accompanied by high cost-effectiveness, and operational flexibility. Normally, they are operated on natural gas with a small addition of fuel oil for stable combustion. The modularity of gas/diesel power plants means that they can easily be expanded to larger plants with higher output by installing additional engines as demand increases. The operational flexibility is ensured by the large number of engine modules.

Diesel engines used for cogeneration applications differ from gas turbines in the mass flow rate and temperature of the exhaust gas as well as partly in the oxygen concentration in the exhaust gas. The exhaust gas temperature in heavy-frame gas turbines reaches 600°C as compared to about 360°C for diesel exhaust. Gas turbines have mass flow rates in the range 2.5–10 kg/s per kW of power output. The exhaust gas of the gas turbine contains from 14.5% to 15.5% oxygen, whereas gas/diesel engines have up to 13% oxygen in the exhaust gas, which leaves the engine at a rate of 7–7.5 kg/s per kW of power output (O'Keefe, 1995).

Table 7.4. Power output of gas engines for small-scale combined heat and power plants

	Manufacturer								
	Ford			MAN			Caterpillar		
	BSG444	BSG666	BSG678	E2866E	E2842E	E2842LE	3508	3512	3516
Power output, kW	38	54	78	95	185	290	169	710	945

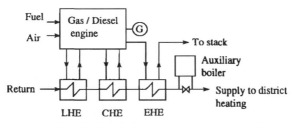

Figure 7.10. Gas/diesel engine based cogeneration plant. LHE, lubricating oil heat exchanger; CHE, jacket cooling water heat exchanger; EHE, exhaust gas heat exchanger.

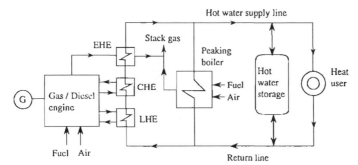

Figure 7.11. Gas/diesel engine based cogeneration plant with peaking boiler and thermal storage.

Figures 7.10 and 7.11 show two configurations of gas, diesel, and gas/diesel engine based cogeneration plants. The waste heat of these engines can be recovered at three temperature levels. The low temperature level is that of the lube-oil cooling system, which produces water at about 80°C, and that of the jacket cooling water, which is available at approximately 90°C. The exhaust gas of the gas/diesel engines has the highest temperature, e.g., around 360°C. Three heat exchangers—the lubricating oil heat exchanger (LHE), cooling water heat exchanger (CHE), and exhaust gas heat exchanger (EHE)—must be employed to recover the waste heat.

The exhaust gas heat from gas/diesel engines can be used to partially replace the feedwater heaters in steam power plants. As indicated in Chapter 4, the feedwater temperature reaches 300–320°C in advanced steam power plants. The steam consumption of the feedwater heaters is rather high, and this results in a substantial reduction of the work production per kg steam entering the turbine because of the reduced steam flow through the intermediate-pressure and low-pressure turbines. An essential improvement of performance of an existing steam power plant can be achieved through the integration of gas/diesel engines into a hybrid power plant repowering configuration (Shelor, 1995). The low-temperature waste heat from the LHE and CHE is appropriate for partial replacement of steam in low-pressure feedwater heaters. The exhaust gas heat from the EHE can be used to heat feedwater in the economizer of a boiler.

To reduce NO_x emissions from the engine to approximately 5 ppm, a selective catalytic reduction unit will be installed in the gas duct upstream of the economizer prior to an air heater. The exhaust gas contains up to 13% oxygen. Hence the exhaust gas after the economizer, retaining a temperature suitable for combustion air, may be introduced into the furnace of a boiler and thus replace a significant portion of the combustion air. An optimum exhaust gas to air ratio of 1 to 1 (by volume) can be maintained, and thus an appropriate oxygen concentration in the combustion air of about 17.5% will be achieved. The coal can be efficiently burned in specially designed low-NO_x burners at this oxygen level. The remaining combustion air is introduced into the furnace as overfire air. This technique is environmentally superior compared to the flue gas recirculation used to reduce NO_x emissions in boiler furnaces. Due to a reburning atmosphere, the NO_x emissions from the diesel engine exhaust gas will be further reduced (Shelor, 1995).

Repowering an existing steam power plant by means of a number of diesel engines with simple-cycle efficiency up to 40% can significantly enhance the overall efficiency of the combined-cycle plant. Thus an incremental heat rate of about 6400 kJ/kWh, which relates to roughly 25% of the total power output, can be achieved. Thereby, the heat rate for the combined diesel engine–steam plant may be improved by about 10%. In addition, using diesel engine waste heat for feedwater heating increases system output by about 25% (Shelor, 1995). For example, lean-burn and gas/diesel engines are available from Wärtsilä Diesel, USA, in six models from 150 to 460 mm bore, with outputs from 220 kW to 16 MW. They can be employed for economical and reliable power generation predominantly on a capacity scale of up to 40 MW as well as in distributed energy supply systems. For electric power output above 40 MW, gas turbine plants are more advantageous.

Diesel engines can also be beneficial for repowering smaller steam power plants up to ~40 MW of electrical power. Incremental power enhancement up to 55% and incremental heat rate reduction by about 35% are predicted, in particular, for plants with micronized coal boilers. Heat rates of about 9500 kJ/kWh or 9000 BTU/kWh could then be achieved (O'Keefe, 1995).

Examples 7.5 and 7.6 present the calculation of electrical efficiency, the energy utilization factor, and the fuel energy savings for a gas/diesel engine cogeneration plant.

Example 7.5

Calculate the electrical efficiency and the energy utilization factor of a gas/diesel engine cogeneration plant with a useful heat output of 23 MW and a heat-to-power ratio of 1.3. The engine consumes 0.95 kg/s of fuel mixture, which consists of 97% natural gas with LHV $= 49$ MJ/kg and 3% fuel oil No. 2 with LHV $= 45.26$ MJ/kg.

Solution

LHV value of the fuel mixture

$$LHV = (m_{gas}LHV_{gas} + m_{oil}LHV_{oil}) = 0.97 \times 49 \text{ MJ/kg} + 0.03 \times 45.26 \text{ MJ/kg}$$
$$= 48.89 \text{ MJ/kg}$$

Rate of heat input

$$Q_{in} = m_f LHV = 0.95 \text{ kg/s} \times 48.89 \text{ MJ/kg} = 46.445 \text{ MW}$$

Electrical power of the cogeneration plant

$$P_{el} = Q_u/\sigma = 23 \text{ MW}/1.3 = 17.69 \text{ MW}$$

Electrical efficiency of the cogeneration plant

$$\eta_{el} = P_{el}/Q_{in} = 17.69/46.445 = 0.381 = 38.1\%$$

Energy utilization factor

$$EUF = (P_{el} + Q_u)/Q_{in} = (17.69 + 23)/46.445 = 0.876 = 87.6\%$$

Example 7.6

Calculate the fuel energy savings of a gas/diesel engine cogeneration plant in comparison with the separate production of power and useful heat in a conventional steam power plant and boiler. Over a specified period of time, 100 MWh of electrical energy and 130 MWh of useful heat are produced in the energy systems of both types.

Assume that the steam power plant and boiler efficiencies are $\eta_{PP} = 39\%$ and $\eta_B = 90\%$, respectively, and that the energy utilization factor of the cogeneration plant is EUF $= 89\%$. The LHV of the fuel is 43 MJ/kg.

Solution

Fuel energy requirements of the cogeneration plant

$E_{CG} = (E_{el} + Q_u)/EUF = (100 + 130)/0.89 = 258.4$ MWh

Fuel energy requirements for separate production of power and useful heat

$E_{sep} = E_{el}/\eta_{PP} + Q_u/\eta_B = 100/0.39 + 130/0.9 = 400.85$ MWh

Fuel consumption for the cogeneration plant over the specified period of time

$m_{CG} = E_{CG}/LHV = 258.4$ MWh $\times$ 3600 s/h/43 MJ/kg $= 21,633.5$ kg

Fuel consumption for separate production of power and useful heat

$m_{sep} = E_{sep}/LHV = 400.85$ MWh $\times$ 3600 s/h/43 MJ/kg $= 33,559.9$ kg

Fuel savings of the cogeneration plant in comparison with separate production

$\Delta m_f = m_{sep} - m_{CG} = 33,559.9 - 21,633.5 = 11,926.4$ kg

or

$\Delta m_f/m_{sep} = 2.9/30 = 0.355 = 35.5\%$

The main incentive of cogeneration is the production of process steam and/or useful heat for hot water and space heating. In general, very low ratios of electric power generation to total useful energy produced over the same period of time are not considered economical for cogeneration. It should be stressed that cogeneration plants are economically viable if the cost of electricity generated by the plant in addition to the useful heat is less than that purchased from a utility. Hence the economics of a cogeneration plant are sharply influenced by the additional cost of generating electricity.

CLOSURE

Cogeneration plants present a technology that allows significant fuel energy savings due to combined power generation and useful heat production. For process heat and power supply in industry, cogeneration plants with back-pressure steam turbines are most suitable, whereas for power supply and district heating, extraction-condensing steam turbines are engaged. In the power output range up to 20–40 MW, gas/diesel engine based cogeneration plants can be more efficient than gas turbine based cogeneration plants, which are more suitable for higher power outputs, e.g., above 40 MW.

The recent development trend is to employ combined-cycle cogeneration plants both in newly constructed power plants and for repowering. Operational flexibility is enhanced by using supplementary coal firing in the steam power plant portion of a repowering project. Diesel engines can also be favorably used for repowering of steam power plants. Maximum EUF values of 85–90% can be attained in cogeneration plants based on gas/diesel engines.

PROBLEMS

7.1. A back-pressure turbine in a cogeneration plant that delivers power and process steam is operated under the following conditions: main steam at turbine throttle is 180 bars and 560°C, back-pressure is 6 bars, main steam mass flow in the turbine is 230 kg/s, and turbine/generator efficiency is 0.9. The exhaust steam is used for industrial heat supply, and the condensate is returned to the power plant at a temperature of 50°C. Calculate (a) the power output of the plant in MW, (b) the electrical efficiency, (c) the heat-to-power ratio, and (d) the energy utilization factor of the plant.

7.2. Consider a cogeneration plant consisting of a 150-MW gas turbine for electric power generation and a waste-heat recuperator producing useful heat for district heating. The gas turbine exhaust gas and stack gas temperatures are 520 and 180°C, respectively, and its flow rate is 500 kg/s. The water supply and return temperatures in the district heating system are 110 and 70°C, respectively. Calculate (a) the heating water flow rate in kg/s, (b) the useful heat output in kJ/s, (c) the heat-to-power ratio, and (d) the energy utilization factor of the plant. Assume a specific heat of exhaust gas and water of 1.05 and 4.19 kJ/(kg K), respectively.

7.3. Calculate the electrical efficiency and the energy utilization factor of a gas/diesel engine cogeneration plant with a useful heat output of 4.8 MW and heat-to-power ratio of 1.2. The engine consumes 0.22 kg/s of natural gas with LHV = 51 MJ/kg.

7.4. Calculate the electrical efficiency and energy utilization factor of a gas engine cogeneration plant with a useful heat output of 770 kW and heat-to-power ratio of 1.26. The gas engine burns 0.031 kg/s of natural gas with LHV = 49.5 MJ/kg.

7.5. Calculate the fuel energy savings for a gas/diesel engine based cogeneration plant as compared with the separate power generation in a steam power plant and useful heat in a heating plant. Over a specified period of time, the two sets of energy production facilities have the same energy generation, namely, 40 MWh of power and 52 MWh of useful heat production. Assume that the steam power plant and heating plant efficiencies are 38% and 90%, respectively, the energy utilization factor of the cogeneration plant is 89%, and the LHV of the fuel is 49.5 MJ/kg.

REFERENCES

Adlhoch, W., and Bolt, N. 1994. Möglichkeiten zur Weiterentwicklung der Kombi-Kraftwerkstechnik. *VGB Kraftwerkstechnik* 74(7):609–612.

Anon. 1995. Kombiheizkraftwerk Mitte-saubere Elektrizität und Fernwärme für Berlins Stadtzentrum. *ABB Technik* 1:1–15.

Bohn, T. 1992. *Gasturbinen, Kombi-, Heiz- u. Industriekraftwerke.* Gräffelfing: Resch.

Brown, W. R., Moore, R.-B., and Klosek, J. 1989. Coproduction of electricity and methanol. In *Eighth Annual EPRI Conference on Coal Gasification, Proceedings*, pp. 4.1–4.15. Palo Alto, Calif.: Electric Power Research Institute.

Haywood, R. W. 1991. *Analysis of engineering cycles.* New York: Pergamon.

Heywood, J. B. 1988. *Internal combustion engine fundamentals.* New York: McGraw-Hill.

Horlock, J. H. 1987. Cogeneration-combined heat and power. In *Thermodynamics and economics.* New York: Pergamon.

Kehlhofer, R. 1991. *Combined-cycle gas and steam turbine power plants.* Lilburn, Ga.: Fairmont Press.

Li Kam, W., and Priddy, A. P. 1985. *Power plant system design.* New York: Wiley.

Nash, F. 1992. The development and first commercial application of an innovative Diesel engine based cogeneration system. *Proc. Inst. Mech. Eng. Part A J. Power Energy* 206(A3):197–207.

O'Keefe, W. 1995. Engine/generators reconfigured to compete in the next century. *Power* 139(10):52–62.

Peraiah, K. C. 1993. Design of steam turbines in cogeneration systems. *Power Int.* 39(10):10–12.

Plohberger, D. C., Fessl, T., Gruber, F., and Herdin, G. R. 1995. Advanced gas engine cogeneration technology for special applications. *Trans. ASME J. Gas Turbine Power* 117:826–834.

Roy, G. K. 1993. Selecting gas turbines for power cogeneration: Supplement mechanical requirements with these efficiency-related guidelines to better match the turbine to the process. *Hydrocarbon Proc.* 72(3):115–118.

Schnitz, K. W., and Koch, G. 1995. *Kraft-Wärme-Kopplung.* Düsseldorf: VDI Verlag.

Shelor, F. M. 1995. Repower with diesel engines to bolster feedwater heating. *Power* 139(6):96–101.

Steimle, F. 1995. *Kraft-Wärme-Kopplung mit Gasturbinen.* Essen: Vulkan-Verlag.

Taft, M. 1991. A comprehensive classification of combined cycle and cogeneration plant. Part 2: Introducing the integrated steam cycle in practice. *Proc. Inst. Mech. Eng. Part A J. Power Eng.* 205(A3):145–159.

Wark, K. 1994. *Advanced thermodynamics for engineers.* New York: McGraw-Hill.

Weston, K. C. 1992. *Energy conversion.* New York: West Publishing.

Wills, R.A., Johnson, N. H., and Knox, H. L. 1989. *Standard handbook of powerplant engineering*, ed. T. C. Elliott. New York: McGraw-Hill.

Chapter Eight

FUEL CELL AND MHD-BASED POWER PLANTS

Fundamentals of and progress in fuel cell and magnetohydrodynamic (MHD) power generation technologies are reviewed in this chapter. The principal advantages of fuel cell technology are high efficiency, independent of the Carnot cycle limitation, and very low environmental impact. A thermodynamic analysis is provided for an H_2-O_2 fuel cell with a reversible cycle efficiency of 83%. The major types of fuel cells are phosphoric acid fuel cells (PAFCs), molten-carbonate fuel cells (MCFCs), and solid-oxide fuel cells (SOFCs), which are described along with their component electrodes and electrolytes. Combined cycles based on high-temperature fuel cells (650°C for MCFCs and about 1000°C for SOFCs) promise to attain overall efficiencies of more than 60%. Because of some technological material problems and primarily due to high initial costs, large-scale fuel cell based combined-cycle power plant projects are predicted to be realized only in the first decade of the next century. Current fuel cell facilities include some pilot plants in the power output range of 2–11 MW.

The MHD generator is another technology suitable for application in advanced combined-cycle power plants. This chapter describes the current state of development of MHD technology. It should be emphasized that there are critical technological problems that prevent a rapid realization of large-scale combined-cycle plants. The critical issue is a high-temperature air heater. One promising technology is the externally fired combined-cycle (EFCC) MHD generator configuration.

FUEL-CELL FUNDAMENTALS

The energy chemically stored in a fuel may be converted directly into electric energy by means of an electrochemical process in a fuel cell. A fuel cell is an electrochemical device for the direct conversion of fuel energy into a direct current (dc) of low voltage. Most often, hydrogen is used as the fuel in fuel cells. Hydrogen production technology is described in this chapter. Natural gas or product gas of coal gasification may also be used.

The principles of the fuel cell were discovered in 1839. A fuel cell (see Figure 8.1) consists of two electrodes, an anode, and a cathode, which are separated by an electrolyte. The fuel, e.g., hydrogen, is fed to one electrode, the anode, and the oxidant, i.e., oxygen gas or air, is fed to the other electrode, the cathode. The electrolyte must be highly permeable to an ion that is a charge carrier traveling through the electrolyte. Depending upon the type of fuel cell, either positive or negative ions, for example, a negatively charged ion CO_3^{-2} or a positively charged ion H^+, may result from the fuel oxidation reaction. A second charge carrier is the electron that travels through the external (load) circuit.

In general, the overall reaction in a fuel cell may be written as follows

$$\text{Fuel} + \text{Oxidant} \rightarrow \text{Oxidation products} + \text{Useful work} + \text{Heat rejected} \qquad (8.1)$$

The useful work in fuel cells is the electrical work. When the oxidant is oxygen, the appropriate

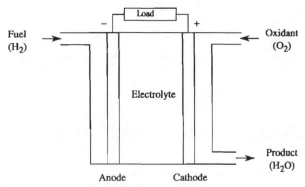

Figure 8.1. Principle of fuel-cell operation.

reaction of complete combustion of a fuel may be written, on a molar basis, as

$$\text{Fuel} + nO_2 \rightarrow xCO_2 + yH_2O \tag{8.2}$$

If air is used as the oxygen source, then the corresponding quantity of the inert gas nitrogen N_2 is added on both sides of Eq. (8.2).

Let us consider a hydrogen-oxygen fuel cell consisting of hydrogen electrode, oxygen electrode, and electrolyte. The charge carrier traveling in an alkaline electrolyte is a positively charged hydrogen ion H^+ (proton). The direct conversion of heat to electric energy occurs in the electrochemical reactions on electrodes, which may be written as follows (Sutton, 1966; Spring, 1965).

The anode reaction in the H_2-O_2 fuel cell is

$$H_2 \rightarrow 2H^+ + 2e^- \tag{8.3}$$

The electrons flow through the external load to the positive electrode, i.e., cathode, where they interact with oxygen. Thus the cathode reaction is

$$2H^+ + 2e^- + 1/2O_2 \rightarrow H_2O \tag{8.4}$$

Thus the overall reaction in the H_2-O_2 fuel cell is

$$H_2 \text{ (gas)} + 1/2O_2 \text{ (gas)} \rightarrow H_2O \text{ (liquid)} \tag{8.5}$$

Taking into account the energy balance, we obtain the full expression

$$H_2 + 1/2O_2 \rightarrow H_2O + 2\,\text{eV} + Q - 3/2RT \tag{8.6}$$

where 2 eV is the electrical energy output, Q is the heat rejected, $3/2RT$ is the mechanical energy absorbed when liquid water is formed, R is the gas constant, and T is the absolute temperature.

The Gibbs free energy is defined as (Sutton, 1966; Spring, 1965)

$$G = H - TS \quad \text{J} \tag{8.7}$$

where H is the enthalpy, T is the temperature, and S is the entropy. Consequently, for the Gibbs free energy change at constant temperature, e.g., for a chemical reaction,

$$\Delta G = \Delta H - T\Delta S \tag{8.8}$$

where ΔH is the enthalpy change of the overall reaction and $T\Delta S$ is a measure of heat absorbed by the system during a reversible process. In addition,

$$\Delta G = nF E_{\text{rev}} \tag{8.9}$$

Table 8.1. Standard electrode potential for reactions
relevant to fuel cells at 25°C

Element	Reaction	E°, V
Hydrogen	$2H^+ + 2e^- = H_2$	0.000
Oxygen	$O_2 + 4H^+ + 4e^- = 2H_2O$	1.229

Source: Sutton (1966) and Spring (1965).

where ΔG is the Gibbs free energy change of the reaction, n is the number of electrons participating in the reaction per mole of fuel, and F is Faraday's constant or the number of coulombs per mole of fuel, i.e., 96,487 coulombs/mol, or 96.487 kJ/(V mol), or

$$F = 96.487 \times 10^6 \text{ coulombs/(kg mol) electrons}$$

The change in the Gibbs free energy is given by

$$\Delta G = \Delta H_0 - T/T_0(\Delta H_0 - \Delta G_0) - \Delta c_p T[\ln(T/T_0) + (T_0/T - 1)] \tag{8.10}$$

where ΔH_0 and ΔG_0 are the enthalpy and the Gibbs free energy change at standard temperature $T_0 = 298$ K and 1 atm, respectively, T is the fuel cell temperature in K, Δc_p is the change in specific molar heat in the reaction Eq. (8.5) [in J/(mol K)], which is calculated from the specific molar heats of H_2O, H_2, and O_2 as follows:

$$\Delta c_p = c_p(H_2O) - c_p(H_2) - 1/2c_p(O_2) \quad J/(\text{mol K}) \tag{8.11}$$

The maximum theoretical (reversible) emf E_{rev} or voltage V_{rev} of a fuel cell can be calculated from the Gibbs free energy of the electrochemical reaction as follows (Sutton, 1966; Culp, 1991; Spring, 1965):

$$E_{rev} = -\Delta G/(nF) = -\Delta G/(4 \times 96.487n_o) \quad V \tag{8.12}$$

where n_o is the number of moles of oxygen per mole of fuel in the oxidation reaction, Eq. (8.5), e.g., $n_o = 1/2$ for hydrogen as fuel.

The standard electrode potentials for reactions relevant to fuel cells are given in Table 8.1.

The value of E_{rev} depends on temperature and pressure. Thus, for the hydrogen-oxygen cell with the gases at atmospheric pressure, E_{rev} is 1.23 V at 25°C, whereas at 200°C, it is only about 1.15 V (Sutton, 1966; Culp, 1991; Spring, 1965). The effect of temperature on the emf E_{rev} of a reversible fuel cell is taken into account by the following temperature coefficient at constant pressure:

$$(\delta E_{rev}/\delta T)_p = \Delta S/4 \times 96.487n_o \quad V/K \tag{8.13}$$

where ΔS is the entropy change in the reaction and $n_o = 1/2$ for hydrogen. The potential of the H_2-O_2 fuel cell increases with increasing pressure.

FUEL-CELL CONVERSION EFFICIENCY

The total energy of the overall reaction in a fuel cell is given by ΔH, whereas ΔG represents the maximum available energy that can be obtained in an ideal fuel cell when the reaction is being carried out reversibly. Hence $T\Delta S$ is the minimum amount of heat liberated in a reversible process at constant pressure and constant temperature.

Thus the maximum theoretical efficiency of the energy conversion in a reversible fuel cell is given by

$$\eta_{rev} = \Delta G/\Delta H = 1 - T\Delta S/\Delta H \tag{8.14}$$

The Gibbs free energy change ΔG for the formation of a mole of liquid water from hydrogen and oxygen gases is usually related to an atmospheric pressure and 25°C. The values of the enthalpy of formation and the Gibbs function at 1 atm and 25°C are given in Table 8.2. The maximum

Table 8.2. Enthalpy of formation H and the Gibbs function G at 1 atm and 25°C

Formula	Value, kJ/(kg mol)
$H_p(l) = H_f[H_2O(l)]$	$-285,830$
$H_p(g) = H_f[H_2O(g)]$	$-241,826$
$G_p(l) = G_f[H_2O(l)]$	$-237,141$
$G_p(g) = G_f[H_2O(g)]$	$-228,582$
$H_r = H_f(H_2) + 0.5H_f(O_2)$	0.0
$G_r = G_f(H_2) + 0.5G_f(O_2)$	0.0

Source: Sutton (1966) and Spring (1965).

Table 8.3. Predicted values of $-\Delta H/nF$, $-\Delta G/nF$, and the maximum efficiency, η_{rev}, of a fuel cell at various temperatures with hydrogen and methane as fuel

Fuel	$-\Delta H/nF$ 298 K	$-\Delta G/nF$ 400 K	η_{rev}, % 800 K		Energy output, kWh/kg		
					298 K	400 K	673 K
Hydrogen	1.48	1.23	1.16	1.05	83.3	78.3	14.0
Methane	1.15	1.06	1.04	1.04	92.2	90.5	6.3

(theoretical) efficiency for the conversion of heat into electrical energy in a reversible hydrogen-oxygen fuel cell at 25°C and 1 atm is $\eta_{rev} = \Delta G/\Delta H = 237/286 = 0.83$. This efficiency value is related to a hydrogen-oxygen cell regardless of the nature of the electrolyte and the individual electrode reactions. The efficiency may also be written in terms of the emf, E_{rev}, and the current I, as follows (Swanekamp, 1995):

$$\eta_{rev} = E_{rev}I\tau/\Delta H \tag{8.15}$$

where τ is the time required to consume a mole of fuel in the fuel cell. The product $E_{rev}I\tau$ in Eq. (8.15) gives the cell capacity in kWh for a given amount of fuel consumption, e.g., a mole.

Table 8.3 gives the calculated values of $-\Delta H/nF$, $-\Delta G/nF$, and efficiency of a fuel cell with hydrogen and methane as fuel at various temperatures.

The departure of a fuel cell from ideal behavior arises from several factors. One is the inherent low rate of the electrode reactions; this is dominant at low current drains. It can be reduced by an effective electrochemical catalyst and by increasing the operating temperature. At larger currents there is an additional contribution from the electrical resistance of the electrolyte (multiplied by the current strength). A low-resistance (i.e., high-conductivity) electrolyte is therefore desirable.

Even in an ideal hydrogen-oxygen cell, $100 - 83 = 17\%$ of the chemical reaction energy (enthalpy) would be liberated as heat. The proportion is increased in an actual cell because the conversion efficiency is less than the maximum of 83%. In order to avoid an excessive temperature rise, heat is removed from the fuel cell during operation. Possible ways of doing this are by the flow of excess air past the positive electrode or by circulating the electrolyte through an external cooler.

Efficiencies as high as 70% have been measured in the laboratory. For cells operating on pure hydrogen and oxygen, conversion efficiencies in the range of 50–60% can be achieved (Swanekamp, 1995; Hirschenhofer et al., 1994; Gillis, 1989). Overall efficiencies are somewhat lower when the hydrogen is derived from hydrocarbon sources or when air is used as the source of oxygen. However, they are higher than efficiencies of conventional fuel fired power plants, i.e., steam or gas turbine power plants. The following example illustrates the calculation of the conversion efficiency of an H_2-O_2 fuel cell.

Example 8.1

Compute the maximum reversible cell voltage and the maximum efficiency of an H_2-O_2 fuel cell at 800 K when the product water is in (a) vapor or (b) liquid form.

Solution

1. The specific molar heats at constant pressure are (Sutton, 1966; Spring, 1965)

$$c_p(H_2O) = 38.7 \text{ J/(mol K)} \qquad c_p(H_2) = 29.6 \text{ J/(mol K)}$$
$$c_p(O_2) = 33.7 \text{ J/(mol K)}$$

2. The change in c_p is

$$\Delta c_p = c_p(H_2O) - c_p(H_2) - 1/2 c_p(O_2) = 38.7 - 29.6 - 1/2 \times 33.7$$
$$= -7.75 \text{ J/(mol K)}$$

3. The enthalpy and the Gibbs free energy changes at standard state ($T_0 = 298$ K) are as follows.
   (a) For water in vapor form

$$\Delta H_0 = -242 \text{ kJ/mol} \qquad \Delta G_0 = -228.8 \text{ kJ/mol}$$

   (b) For water in liquid form

$$\Delta H_0 = -286 \text{ kJ/mol} \qquad \Delta G_0 = -237.4 \text{ kJ/mol}$$

4. The change in the Gibbs free energy is

$$\Delta G = \Delta H_0 - T/T_0(\Delta H_0 - \Delta G_0) - \Delta c_p T[\ln(T/T_0) + (T_0/T - 1)]$$

   (a) For water in vapor form

$$\Delta G = -242 - 800/298(-242 + 228.8)$$
$$- (-7.75 \times 10 - 3)800 [\ln(800/298) + (298/800 - 1)] = -204.3 \text{ kJ/mol}$$

   (b) For water in liquid form

$$\Delta G = -286 - 800/298(-286 + 237.4)$$
$$- (-7.75 \times 10 - 3)800 [\ln(800/298) + (298/800 - 1)] = -153.3 \text{ kJ/mol}$$

5. With $n = 1/2$ for fuel H_2, the maximum reversible cell voltage is

$$U_{rev} = -\Delta G/4nF = -\Delta G/2 \times 96.487$$

   (a) For water in vapor form, $V_{rev} = 1.06$ V, and (b) for water in liquid form, $V_{rev} = 0.8$ V.
6. The maximum cell efficiency is

$$\eta_{rev} = \Delta G/\Delta H \qquad \Delta H = \Delta H_0 + \Delta c_p(T - T_0)$$

   (a) For water in vapor form

$$\Delta H = -242 + (-7.75 \times 10 - 3)(800 - 298) = -245.9 \text{ kJ/mol}$$
$$\eta_{rev} = -204.3/(-245.9) = 0.83$$

   (b) For water in liquid form

$$\Delta H = -286 + (-7.75 \times 10 - 3)(800 - 298) = -289.9 \text{ kJ/mol}$$
$$\eta_{rev} = -153.3/(-289.9) = 0.53$$

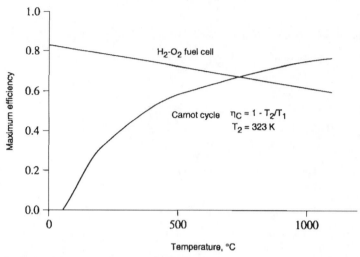

Figure 8.2. Maximum efficiency of an H_2-O_2 fuel cell in comparison with the Carnot cycle efficiency.

Fuel Cell Versus Carnot Cycle

A fuel cell is no heat engine; therefore its efficiency is not limited by the Carnot cycle efficiency, which is given by

$$\eta_C = 1 - T_c/T_h \tag{8.16}$$

where T_c is the cold sink temperature to which heat is rejected and T_h is the hot source temperature from which heat is supplied to the Carnot cycle. Figure 8.2 shows maximum efficiencies of some fuel cell reactions in comparison with the Carnot cycle efficiency.

FUEL-CELL CHARACTERISTICS

Voltage and Efficiency of a Real Fuel Cell

A fuel cell connected across an external load is an electrochemical system in which the overall reaction rate may be arbitrarily regulated by changing the current passing through the load. In real fuel cells, losses in the Gibbs free energy or useful work occur through the irreversibility. The corresponding voltage losses are called polarizations. The operation of a hydrogen-air fuel cell may be described as a series of individual steps:

- fuel (H_2) transport to the anode and dissolution in and transport through a thin film of electrolyte on the anode surface
- dissociation (the first stage of the overall electrochemical reaction) and adsorption on the surface of the anode
- transfer of an electron
- product (proton H^+) desorption and migration through the electrolyte to the cathode

While electrons pass through the anode to the terminal of the fuel cell and through the external load, they perform useful work and proceed to the cathode. There they meet protons and oxygen arriving at the surface of the cathode, and the second stage of the overall electrochemical reaction occurs. There are energy barriers at any step of the process, so that the overall polarization losses may be accounted for as a sum of the activation and concentration overpotentials for both oxidation of hydrogen and reduction of oxygen, as well as the resistance (ohmic) overpotential for the whole cell. Thus there are three types of polarization (Sutton, 1966; Hirschenhofer et al., 1994): ohmic, concentration, and activation. Ohmic polarization results from the internal resistance to the flow of electrons through electrodes and of ions through the electrolyte. Concentration polarization is due to mass transport effects (diffusion of gases through porous electrodes, and solution and dissolution of reactants and products). Activation polarization is due to the activation energy barriers for various steps in the oxidation-reduction reactions at the electrodes.

The net effect of the polarizations is a decline in terminal voltage with increasing current drawn by the load. The terminal voltage of a real fuel cell, u, is thus less than the maximum emf, E_{rev}, of the ideal (reversible) fuel cell, by the sum of the overpotentials:

$$U = E_{rev} - (E_{act, a} + E_{act, c} + E_{conc, a} + E_{conc, c} + E_R) = E_{rev} - \sum \Delta E \quad (8.17)$$

where E_{act} and E_{conc} are the activation and concentration overpotentials for oxidation of hydrogen at the respective electrodes (subscript a for anode and subscript c for cathode), $V_R = iR$ is the resistance (ohmic) overpotential, i is the current across the resistance (load) R, and $\sum \Delta E$ is sum of the overpotentials. The total polarization losses may be minimized by increasing both temperature and pressure in the fuel cell.

Voltage-Current Characteristic

The voltage-current characteristic of a fuel cell represents a relationship between the cell voltage U and the current density i in regard to the electrode surface area. It is plotted in Figure 8.3. The reversible cell voltage is 1.23 V (Hirschenhofer et al., 1994).

The efficiency drop in the fuel cell is proportional to the sum of overpotentials $\sum \Delta E$ (see Figure 8.3). The power density of a fuel cell is

$$P = Vi \quad W/m^2 \quad (8.18)$$

where U is the fuel cell voltage in volts and i is the current density in amperes (A) per m^2 of the electrode surface area.

Figure 8.4 shows a typical relationship between the power density P and the current density i of a fuel cell. The maximum power output is achieved at a fuel cell efficiency of about 50%.

The difference between the actual and theoretical voltage values of a fuel cell increases with increasing load current drawn from the cell. For the moderate currents at which fuel cells normally operate, the emf is 0.7–0.8 V.

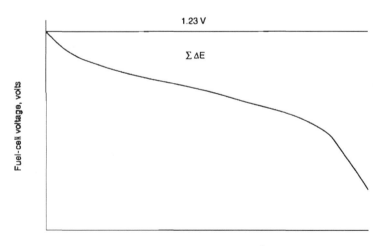

Figure 8.3. Voltage-current density characteristic of a fuel cell.

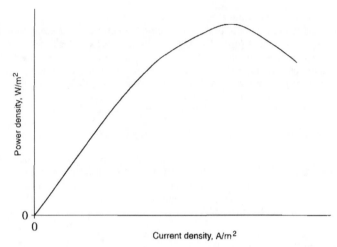

Figure 8.4. Typical power density–current density diagram of a fuel cell.

The fuel cell conversion efficiency is defined as the electrical energy output per unit mass (or mole) of fuel to the corresponding heating value of the fuel consumed. The cell conversion efficiency η_c is related to the thermal efficiency η_{th} by

$$\eta_c = \eta_{th}\eta_i\eta_v \qquad (8.19)$$

where η_v is the voltage efficiency defined as the ratio of the terminal voltage to the theoretical emf and η_i is the current efficiency defined as the ratio of the cell electrical current to the theoretical charge flow associated with fuel consumption. The voltage efficiency accounts for the various cell polarizations. Fuel cell conversion efficiencies in the range of 40–60% are possible in large-scale power plants.

Heat Generation in Fuel Cells

The rate of heat generation in fuel cells is (Sutton, 1966; Spring, 1965)

$$Q = -(T\Delta S/Fn)i + i\sum \Delta E + i^2 R \quad W \qquad (8.20)$$

The first term is based on the entropy change (ΔS) of the reaction, the second term represents the heat generated due to irreversibilities caused by slow kinetics of reactions and mass transfer limitations (where $\sum \Delta E$ is the sum of the overpotentials), and the third term is the Joule heat produced by ohmic losses (where i is the current).

Electrodes

Electrodes in fuel cells are the sites for electrochemical reactions. In addition, they serve as current collectors to carry the electrons to terminals of the fuel cell, and they maintain an interface between the electrolyte and the fuel (or oxidant). The electrodes must possess a good electrical conductivity, high resistivity to the corrosive electrolyte, and sufficient mechanical strength. Electrodes should also have good catalytic properties (if necessary with a deposition of a suitable catalyst) for specific electrochemical reactions.

Electrolytes

The electrolyte in a fuel cell should have the smallest possible resistance in order to minimize potential drops within the fuel cell. The transference number that characterizes the mobility for the ion flow should be sufficiently high, i.e., nearly 1. For use in low-temperature fuel cells (up to 200°C), aqueous electrolytes are most suitable. They include strong alkali (sodium or potassium hydroxide) solutions or strong acid (phosphoric acid or sulfuric acid) solutions. Intermediate-temperature fuel cells (500–700°C) use fused (molten) carbonates, and high-temperature fuel cells (700–1000°C) employ oxides as electrolytes.

The most efficient fuel cells operate on hydrogen gas and oxygen. For operating temperatures up to 90°C, the electrolyte is usually an aqueous solution of potassium hydroxide. In smaller cells the electrolyte is held in a porous matrix. In larger cells a free-flowing liquid may be preferred to facilitate heat removal. The negative and positive electrodes, the anode and the cathode, in this fuel cell type are usually made of porous nickel with a catalyst. Nickel or a platinum metal serves as the anode catalyst, and silver or a platinum metal serves as the catalyst for the cathode.

The gases in a hydrogen-oxygen cell must be free from carbon dioxide. Otherwise, CO_2 will react with the potassium hydroxide electrolyte to form potassium carbonate, which increases the electrical resistance of the cell and decreases its output voltage.

Hydrogen-oxygen fuel cells have been used on some manned spaceships. Both hydrogen and oxygen were stored in liquid form. Metal hydrides are also suitable for hydrogen storage.

Modified hydrogen-air cells use a gaseous or liquid hydrocarbon as the source of hydrogen. Fuel cell power plants may be operated on fossil fuels, including coal.

TYPES OF FUEL CELLS

General Classification

There are four types of fuel cells that may be used for power generation (Hirschenhofer et al., 1994; Selman, 1994; Gillis, 1989):

- alkaline fuel cell (AFC)
- phosphoric acid fuel cell (PAFC)
- molten carbonate fuel cell (MCFC)
- solid oxide fuel cell (SOFC)

Figure 8.5 illustrates the principle of operation of the high-temperature SOFC suitable for application in combined-cycle power plants, particularly, in integrated gasification fuel cell (IGFC) power plants.

Table 8.4 summarizes major characteristics (operating temperature, ionic conductor, and efficiency based on high heating value) of all four types of fuel cells.

The fuel cells may also be classified into three categories according to their operating temperatures.

Alkaline and phosphoric acid fuel cells are low-temperature (up to ~200°C), molten-carbonate fuel cells are intermediate (medium) temperature, (560–700°C), and solid oxide fuel cells are high temperature (up to ~1000°C).

Three types of fuel cells are being pursued for electric power generation—PAFC, MCFC, and SOFC (Ketelaar, 1987; Anon, 1994a, 1994b; Hirschenhofer et al., 1994; Gillis, 1989). Fuel cell

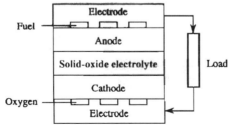

Figure 8.5. Principle of operation of a solid-oxide fuel cell.

Table 8.4. Comparison of major fuel cell types

Fuel cell type	Temperature, °C	Ionic conductor	Efficiency (HHV), %
Alkaline fuel cell (AFC)	60–90	H^+	60
Phosphoric acid fuel cell (PAFC)	160–220	O^{-2}	37–42
Molten carbonate fuel cell (MCFC)	600–650	CO_3^{-2}	50–60
Solid oxide fuel cell (SOFC)	800–1000	H^+	60–65

plants produce dc (direct current) and require an inverter for the conversion of dc to a constant-voltage alternative current.

Low-Weight Fuel Cells

Low-weight alkaline fuel cells (AFCs) are used as the source of electricity for space and military applications. As the electrolyte, 35% KOH solution is used. These cells employ Pt on carbon as the anode catalyst and Pt or Ag on carbon as the cathode catalyst. Electricity is produced with an efficiency above 60%. Because of the relatively low operating temperatures of 60–90°C, utilization of the waste heat is practically impossible. Advantages of the AFCs include high power densities and fast start-up times. Disadvantages are the inability to reject CO_2 that results in carbonate formation in the porous electrode, and anode catalyst poisoning by CO, which necessitates that pure H_2 be used with electrolyte recirculation and be replenished periodically (Hirschenhofer et al., 1994; Gillis, 1989). Because of the extremely high costs and low reliability of operation, AFCs may not be used for power production on a commercial scale. At present, the specific costs per unit capacity of the AFCs are about US$70,000/kW, which are the highest costs of all known power technologies. To be economically feasible, their costs must be reduced to US$200–300/kW (Swanekamp, 1995).

Phosphoric Acid Fuel Cells

In phosphoric acid fuel cells (PAFCs) the concentrated phosphoric acid serves as the electrolyte, and processed hydrocarbon fuels such as natural gas or naphtha are used to generate electric energy. At the present time, PAFCs are the most highly developed fossil fuel cells (Appleby, 1987; Mayfield et al., 1989). The fuel is subjected to high-temperature catalytic reactions with steam, i.e., to steam reforming and to the water-gas shift reaction.

The hydrogen-rich gas is produced by steam reforming or partial oxidation of hydrocarbons (including methane, natural gas, or naphtha) according to the following reaction equations:

$$CH_4 + H_2O + Heat \rightarrow CO + 3H_2 \tag{8.21}$$

$$C_mH_n + m/2O_2 \rightarrow mCO + n/2H_2 \tag{8.22}$$

The water-gas shift reaction is

$$CO + H_2O \rightarrow CO_2 + H_2 \tag{8.23}$$

Thus the product gas contains around 80 vol.% of hydrogen; the rest is mainly CO_2 with a small proportion of CO. The CO_2 has practically no effect on the phosphoric acid electrolyte, and this tolerance of PAFCs to CO_2 is their most attractive feature. In order to prevent poisoning of the catalysts on the fuel cell electrodes, the fuel must be sulfur free. PAFCs are less sophisticated and more reliable, but less efficient than the two other types of fuel cells—MCFC and SOFC. With platinum electrodes on both sides of a silicon carbide matrix, the phosphoric acid electrolyte generates only 2/3 V per cell (Appleby, 1987; Mayfield et al., 1989).

The electrolyte is retained in an asbestos or carbon sheet, which is placed between fibrous–carbon sheet electrodes. The electrodes have a platinum or platinum alloy catalyst deposited on their outer surfaces. Thus the hydrogen-rich gas is supplied to the anode, and air to the cathode.

Operating temperatures between 150 and 220°C and discharge voltages of 0.7–0.8 V are common. A large number of fuel cells, each only a few millimeters thick, are stacked in a package of certain size to produce the desired voltage and power.

For electric and heat supply to small users, such as dwellings, hospitals, and small industrial enterprises, PAFCs with capacities from some kW up to 10 MW may be employed. They may also be applied for small-scale on-site and distributed power production. The PAFC technology is considered viable today.

Fuel cell plants with power output in the MW range are now being studied in some demonstration projects. Several MW fuel cell plants are being planned for construction in Japan and the United States. An 11-MW PAFC plant has been in operation for over 5 years in Goi, Japan, and current Japanese projects include 5-MW plants (Appleby, 1987; Selman, 1994; Homma et al., 1994). In the 200-kW capacity range, the fuel cell plant specific capital costs of US$3170/kWe with natural gas as fuel are at present approximately 3 times higher than those required in order to achieve economic feasibility. The specific capital costs must be reduced to approximately US$900/kWe, and the service life of the fuel cell should be increased from its present 15,000 hours to approximately 40,000 hours (Appleby, 1987; Ketelaar, 1987; Anon, 1994a, 1994b; Whitaker and Lueckel, 1994).

PAFC Operating Experience

The production-model fuel cell offered by ONSI is called the PC25. Since 1992, ONSI sold almost 60 of the units for distributed and on-site power generation. ONSI's third-generation fuel cell, the PC25C, released in January 1995, achieved a 30% reduction in weight and footprint, a 38% reduction in number of components, and an impressive 50% reduction in capital cost, compared to the PC25A. Corrosion problems and other technical concerns appeared at the Goi plant, including reduced cell life, unit derating because of cell voltage decays, phosphoric acid leakage, and degraded stack performance because of aging of the platinum catalyst (Whitaker and Lueckel, 1994).

Molten-Carbonate Fuel Cells

MCFCs and SOFCs are being developed for utilization in the high-temperature region. MCFCs employ a molten-carbonate mixture as the electrolyte and can use a variety of fossil fuels, including coal, as the primary energy source. They oxidize CO to CO_2 as well as hydrogen to water during their operation. Therefore coal gases such as the synthesis gas containing hydrogen and carbon monoxide can be used to fuel the MCFCs; CO_2 in the fuel gas has a minor effect. However, the gas must be free of sulfur compounds, as they poison the electrodes. The maximum theoretical emf of the CO-O_2 cell is about 1 V at 700°C. The synthesis gas can be produced by coal gasification with air and steam.

The anode reaction in the MCFC is

$$H_2 + CO_3^{-2} \rightarrow H_2O + CO_2 + 2e^- \tag{8.24}$$

Then the cathode reaction is

$$2e^- + CO_2 + 1/2 O_2 \rightarrow CO_3^{-2} \tag{8.25}$$

The electrolyte in the MCFCs is a mixture of molten alkali-metal (lithium, sodium, and potassium) carbonates at temperatures of 560–680°C (Myles and Krumpelt, 1989). The electrolyte is retained in a porous matrix placed between two porous nickel electrodes. There is no need for a catalyst at the high operating temperatures of MCFCs. The fuel (mixture of hydrogen and carbon monoxide) is supplied to the anode, and oxygen (from the air) to the cathode. The discharge voltage of the cell is about 0.8 V.

MCFCs are particularly suited to electric-utility needs because they have higher efficiencies and potentially lower initial cost, require a smaller plant footprint (square meters per kilowatt), and operate at temperatures of 560–680°C, particularly suitable for heat recovery from exhaust gases in steam bottoming cycles or cogeneration applications. MCFCs are highly modular, with the power output from commercial-scale fuel-cell stacks ranging from 125 kW to some hundred kilowatts. Plants of this type have minimal environmental impact and can achieve efficiencies of 50–60% (based on the lower heating value) without a bottoming cycle.

MCFCs have several important advantages over PAFCs, such as high power densities at higher cell voltages and the capability to oxidize hydrocarbons including natural gas directly without any external reformation. Due to higher temperatures of the exhaust gas of MCFCs, above 550°C, a combined-cycle plant can be employed for the production of additional power and heat for industrial processes and district heating. Thereby, in a waste-heat boiler (WHB), steam will be raised to drive a steam turbine. An overall efficiency of 80–85% will thus be achieved (Monn, 1995; Ray, 1994).

Solid-Oxide Fuel Cells

High-temperature fuel cells operating in the range of 800–1000°C may used for electrolytes certain solid ceramic oxides such as zirconium dioxide containing a small amount of another oxide to stabilize the crystal structure. This material is able to conduct oxygen ions (O^{-2}) at high temperatures. Due to the high temperatures, electrochemical catalysts are not required. SOFCs can utilize the same fossil fuels with the same processing as the MCFCs. SOFCs are as yet in the early stages of development.

There are two types of SOFC design: tubular and planar (Bates, 1989; Drenckhahn et al., 1994). The tubular SOFC features an air electrode made of strontium-doped lanthanum. The tubular geometry limits thermal gradients and allows free thermal expansion. In the planar design of high-temperature SOFCs, metallic and ceramic materials are employed, which enables high power densities to be achieved. Facilities for the mass production of these materials already exist, ensuring cost-effective fuel-cell fabrication. Large-volume stacks comprise a large number of individual cells.

A fuel-cell stack consists of two metallic end plates and several bipolar plates, which direct the process gases, i.e., fuel gas and air, to the electrochemically active elements. An active cell element comprises an electrolyte sandwiched between two electrodes—the anode and the cathode. Thus the electrolyte and electrodes form a thin, self-supported planar cell. The electrolyte membrane is an approximately 150-mm-thick substrate produced by tape casting and sintering a thin film of material. The approximately 50-mm-thick electrode layers are applied to both sides of the substrate by means of a screen-printing process, and then sintered to form a permanent bond with the electrolyte. The SOFC stack operating temperature lies at around 950°C.

The bipolar plate as the structural element of the fuel cell must meet strict requirements with regard to physical properties such as high electrical and thermal conductivities and good mechanical strength at operating temperatures of about 1000°C. Also the electrode surface layers must remain stable in the environment of process gases and comply with thermal expansion behavior.

In the past, only ceramic materials could have met all of these requirements. A new metallic, chromium-based alloy, developed by Siemens, may also be used as the material for bipolar plates (Monn, 1995; Drenckhahn et al., 1994). Its mechanical strength and electric and thermal conductivities exceed those of ceramic materials. Hence it is well suited for high current densities and ensures a uniform temperature distribution. The fuel gas, i.e., hydrogen or a fuel gas containing hydrogen, carbon monoxide, and water vapor, is supplied to the anode, while oxygen or air is supplied to the cathode. The oxygen ions, O^{-2}, migrate through the gas-tight electrolyte from the cathode to the anode. At the three-phase interfaces between electrodes, electrolyte, and the process gases, corresponding oxidation and reduction electrochemical reactions take place. Thus at the anode interface, the oxygen ions react with hydrogen to form water. The electrons are drawn from the anode through the external load to the cathode, thus producing the electric current.

Miscellaneous Fuels

Alternative energy sources that can be conveniently stored and transported in liquid form, such as methanol, ammonia, and hydrazine, have been proposed for fuel cells (Culp, 1991). Methanol (CH_3OH) can be catalytically reformed with steam at about 200°C to yield a mixture of hydrogen

(75 vol.%) and carbon dioxide. This gas can be supplied to the anode of a fuel cell, and air can be supplied to the cathode. The fuel cell can use aqueous phosphoric acid or potassium hydroxide solution as the electrolyte. The silver electrodes supply a finely divided palladium/platinum electrochemical catalyst for the anode (fuel electrode) and silver for the cathode (air electrode).

In the ammonia-oxygen (air) fuel cell, ammonia gas, obtained from the stored liquid, is decomposed catalytically into hydrogen and nitrogen. Part of the hydrogen is burned in air to provide the heat required for the decomposition. The bulk of the hydrogen is then supplied to the negative electrode of a hydrogen-air fuel cell. The most suitable electrolyte would be potassium hydroxide (KOH) solution.

A compact fuel cell for a mobile power source utilizes the liquids hydrazine (N_2H_4) and hydrogen peroxide (H_2O_2) or air as the energy source. Hydrazine is injected as required into the electrolyte (KOH) to provide the active material at the negative electrode. The oxygen for the positive electrode is obtained either by the catalytic decomposition of hydrogen peroxide or from the ambient air. The electrodes are made of a nickel matrix with the electrochemical catalyst of nickel (anode) and silver (cathode). The overall cell reaction is the oxidation of hydrazine to water and nitrogen, but the discharge emf is similar to that of the hydrogen-oxygen cell.

Ceramic Electrodes and Electrolytes for SOFCs

The electrodes are made of porous ceramic materials with an ionic and electronic conductivity greater than 10 S/cm. The best electrode materials at present are a perovskite solid solution (Sr doped $LaMnO_3$) as the cathode material, and an NiYSZ cermet (40% Ni, 60% YSZ) as the anode material (Drenckhahn et al., 1994). The porosity of electrodes is about 0.4, and the thermal expansion coefficient is approximately the same as that of the electrolyte, i.e., about $11 \times 101/K$. With improvements in microstructure, Sr and Co doping, porosity, three-phase interfaces, and contact surfaces between electrolyte and electrodes, Siemens optimized these materials. This has allowed increasing the current density to 1000 mA/cm² and the cell output to 0.7 W/cm² at a cell voltage of 0.7 V (Drenckhahn et al., 1994).

The solid-oxide electrolyte must be extremely resistant to active corrosive media at the cathode (oxidizing medium) and at the anode (reducing environment). At present, as the electrolyte membrane of SOFCs, yttria-stabilized zirconium oxide (YSZ) with a cubic crystalline structure is used. Its ionic conductivity of 0.10–0.16 S/cm at operating temperatures of 900–1000°C is about 10 times lower than that of liquid electrolytes (Drenckhahn et al., 1994). In order to reduce resistance losses, the electrolyte thickness should be less than 200 mm.

The Siemens high-temperature SOFC set a record for power output using hydrogen and oxygen fuel in 1994. It achieved a 1.8 kW electric power output, whereas the previous record was 1.3 kW. This fuel cell attained a 0.6 W/cm² power density. It is anticipated that a 5-kW SOFC with an enhanced long-term stability will soon be developed, and a 100-kW cogeneration plant is to be designed by Siemens (Monn, 1995; Drenckhahn et al., 1994).

FUEL CELL BASED POWER PLANTS

Advantages and Limitations of Fuel Cells

Hydrogen-fueled cells have the following advantages (Hirschenhofer et al., 1994): (1) high efficiency (above 50%) both in the part load and in the full load operation modes, and (2) practically no pollutant emissions because the reaction product is water. These advantages are balanced against certain limitations such as high capital costs and a short service life.

Fuel cells may be considered as an ideal source of electricity. They achieve an electrical efficiency of at least 50% in power plants and an overall efficiency of more than 80% in cogeneration plants. With the fuel-cell technology, energy savings of 40–60% may be achieved. The NO_x emissions of fuel cell based power plants are up to 50–90% less in comparison with conventional power plants. CO_2 emissions may thereby be decreased by 50%.

Fuel cells are often described as continuously operating batteries or electrochemical engines. By reacting hydrogen with air, they produce electric power without using combustion or rotating machinery. Unlike batteries, fuel cells can be operated continuously using natural gas, coal gas, or other fuel gases to produce electricity. The world's total capacity of fuel cell based plants in 1994 was only 50 MW (over 40 MW are PAFC plants, most in the 50- to 200-kW range) (Swanekamp, 1995; Mayfield et al., 1989).

Due to high efficiency, modularity, quiet operation, low environmental impact, small capacity, and short lead time, fuel cells could be the ideal technology option for on-site and distributed power generation. High-temperature fuel cells are very well suited for use in cogeneration plants, as they produce significant quantities of waste heat. Depending on the system size and operating temperature, the waste heat of fuel cells may be used to produce steam or hot water for thermal loads, or to generate additional electricity in gas and steam turbines. Benefits of distributed energy generation may be estimated as more than $100/kW (Swanekamp, 1995).

If fuel-cell reliability would increase, fuel cells could achieve the goal of unmanned power plants. Fuel-cell companies are pushing their technology to be ready. Dozens of 100-kW fuel cells are already on-line in the United States. In the international arena, some 14 fuel cells are currently generating electric energy throughout Europe, and there are more than 80 units in Japan.

Fuel Cell Based Power and Cogeneration Plants

The fuel cell based power plant (see Figure 8.6) should have the following main components:

- fuel processor consisting of a gasifier, reformer, and gas cleaning unit, used to convert the fossil fuel, e.g., coal, natural gas, oil, or naphtha, to a hydrogen-rich gas
- power section consisting of a fuel-cell stack generating dc (direct current)
- inverter for changing the dc into alternating current (ac)

High-temperature fuel cells are being developed especially for application in power plants. They are characterized by exceptionally low environmental impact and high efficiency. In the power production mode, efficiencies up to 60% can be achieved. Unfortunately, the direct electricity generation in high-temperature fuel-cell power plants has not been economically feasible so far. Using the waste heat and the chemically bound energy of the flue gas increases the overall performance of fuel-cell power plants. If hydrogen is present in the exhaust gas, it should be electrochemically separated from the flue gas and recirculated to the fuel cell. The high-temperature waste heat may be used either for power production only or for cogeneration of useful heat and power.

The Santa Clara Demonstration Project (SCDP) in the United States is the world's first demonstration of natural gas fueled MCFC technology in a size module suitable for designing plants for distributed power generation (Swanekamp, 1995; O'Shea et al., 1994). The primary objective of the project is demonstration of molten carbonate fuel-cell technology at full scale. It must also demonstrate the specific advantages to power plants using carbonate technology, including superior efficiency, low emissions, high reliability and availability, and capability for unattended operation. The plant includes a dc power system comprising sixteen 12-kW fuel-cell stacks; process systems necessary for handling the fuel, steam, oxidant, and exhaust gas streams; and electrical systems required to convert dc power to ac and to interconnect with the local grid. Table 8.5 summarizes the SCDP design specifications (O'Shea et al., 1994).

Another example is the 200-kWe fuel cell based cogeneration plant PC25 that has supplied heat and electricity in Buena Park, California, since June 1993 at an overall efficiency above 80%. The PC25 produces 200 kWe of electric power as well as 800 MJ/h (222 kWth) of thermal energy and

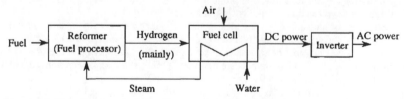

Figure 8.6. Flow diagram of a fuel-cell power plant with a fuel processor.

Table 8.5. Design performance data of the Santa Clara Demonstration Project

Plant parameter	Data
Nominal capacity, MW	2
Net plant rating (ac), MW	1.8
Heat rate (LHV) at rated power, kJ/kWh	7230
Estimated availability, %	90
Expected start-up time (cold to rated power), hours	40
Fuel-cell stack life (at operating temperature above 540°C), hours	10,000
Maximum emissions at rated power, lb/MWh (g/MWh)	
SO_x	0.0003 (0.14)
NO_x	0.0004 (0.18)
CO_2	5 (2260)

Source: O'Shea and Leo (1994).

has a very high reliability (Whitaker and Lueckel, 1994). It has negligible pollutant emissions, making it the cleanest fossil-fueled electricity generation plant.

The so-called direct fuel cell technology using MCFCs provides internal reforming of hydrocarbon fuels within the fuel cells, thus eliminating any external reformer and simplifying the plant configuration. The heat required for reforming is supplied directly from the exothermic fuel-cell reactions. This significantly reduces heat rejection requirements and the amount of waste heat generated. Such power plants can be operated on most hydrocarbon fuels, thus providing enhanced fuel flexibility.

Figure 8.7 shows a schematic diagram of a fuel cell based power plant with a reformer to produce hydrogen, an SOFC, a WHB to generate steam for the reforming process, and a heat exchanger to use the exhaust heat. An MCFC-based power plant is now being developed by MTU Friedrichshagen GmbH and RWE AG in Germany. It is planned for a commercial operation in 2010. The project includes the following components:

- high-temperature gasifier
- steam generator using raw gas heat
- unit for coal gas cleanup and preparation
- gas turbine
- MCFC stacks
- waste-heat exchanger for utilization of heat rejected by the anode
- units for preheating the fuel gas and air or oxygen to be supplied to the anode and to the cathode, respectively
- waste-heat exchanger for hot water production

In comparison to conventional power plants, this concept of the fuel-cell power plant has important advantages such as very low emissions and power production at the highest efficiency (up to 65%).

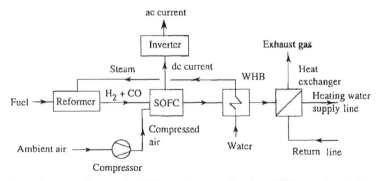

Figure 8.7. Flow diagram of a SOFC-based cogeneration plant. WHB, waste heat boiler.

Due to rather complicated and efficient coal gas cleanup and preparation required for fuel-cell operation, the resulting emissions are extremely low.

The development of power plant concepts on the basis of medium- and high-temperature fuel-cell (MCFC and SOFC) technologies is ongoing. The wide commercial application of these fuel cells at the present time is prohibited mainly because of their high capital costs. However, due to their high efficiency and low pollutant emissions, the power plants based on MCFCs and SOFCs may achieve economic feasibility when their capital costs decrease to about US$1500/kW of installed capacity. These capital costs would be achieved if the total capacity of fuel-cell power plants of 200–300 MW would be installed annually. Japan plans to set in commercial operation fuel-cell power plants with a total capacity of 2 GWe by the year 2000. In the future, fuel-cell power plants would become a viable option to conventional thermal power plants.

HYDROGEN PRODUCTION AND UTILIZATION

Conventional Methods for Hydrogen Production

Hydrogen is at present produced from natural gas, mineral oil (naphtha), or coal. Thereby, a hydrogen-rich product gas is produced from these fuels either by steam reforming of natural gas or naphtha (roughly at 900°C) or by coal gasification (partial oxidation with oxygen roughly at 1400°C). When natural gas or naphtha is used as the initial fuel, the product gas has the following composition (in % by volume): H_2 (68%), CO (10%), CH_4 (10%), CO_2 (10%), and N_2 (2–3%). The coal gas contains, in % by volume, H_2 (30–40%), CO (50–55%), CH_4 (1–3%), CO_2 (3–12%), and N_2 (1–2%). In a two-stage catalytic conversion process at 300–500°C (first stage) and 200°C (second stage), CO in the raw product gas is converted to CO_2, which is removed in a downstream scrubber, and thus the resulting product gas is hydrogen.

The hydrogen production process from natural gas or mineral oil as fuel may be described as follows. Hydrogen production by steam reforming of hydrocarbons is written

$$C_mH_n + mH_2O = mCO + (m + n/2)H_2 \tag{8.26}$$

Steam reforming of carbon monoxide is

$$CO + H_2O = CO_2 + H_2 \tag{8.27}$$

Similarly, for steam reforming of methane (catalytic reactions),

$$CH_4 + H_2O = CO + 3H_2 - 205 \text{ kJ/mol} \tag{8.28}$$

$$CH_4 + 2H_2O = CO_2 + 4H_2 - 164 \text{ kJ/mol} \tag{8.29}$$

The energy content of the product gas hydrogen is approximately 75–80% of that of the fuel when natural gas is used and only 55–60% when coal is used. The hydrogen production costs depend on the fuel and the production process used. Some typical hydrogen production cost values related to 1 GJ of the energy content of the product gas are given in Table 8.6. These cost values may be favorably compared with the electricity price of approximately US$20/GJ.

Electrolysis

The processes of water dissociation into hydrogen and oxygen by electrolysis, photolysis, or biophotolysis represent alternative methods for hydrogen production. In the electrolysis of water,

Table 8.6. Typical hydrogen production costs in relation to the fuel used

Fuel	Costs, US$/GJ
Natural gas	16
Lignite	17
Hard coal	20

hydrogen is produced directly:

$$H_2O + Energy = H_2 + 1/2O_2 \tag{8.30}$$

The reaction enthalpy ΔH is approximately 3–5 kWh/m^3 hydrogen.

The electrolysis efficiency is defined as the ratio of the theoretical to actual energy requirements. It is approximately 80%. By decreasing energy losses, the electrolysis efficiency can be increased to approximately 90%. They may be coupled with fuel cells. There are several electrolyzer types such as alkaline, membrane, and high-temperature steam electrolyzers.

Hydrogen Use for Power and Heat Production

In energy-related technologies, hydrogen may be used for the following:

- heat production by (1) high-temperature combustion with oxygen or air or (2) flameless catalytic low-temperature combustion (practically pollution free)
- power production in systems based on (1) high-temperature fuel cells and membrane fuel cells or (2) the so-called direct steam production process
- power and heat cogeneration in combined plants based on gas/diesel engines or gas and steam turbines
- energy storage in metal hydrides and pressurized H_2 storage
- pollution-free automobiles

Conventional and Catalytic Combustion of Hydrogen

Hydrogen may be burned in conventional high-temperature combustors as well as in low-temperature flameless catalytic combustors. Unlike to natural gas, hydrogen as a fuel has a very high combustion velocity and high flame temperature. Because of the high flame propagation velocity for hydrogen (2.37 m/s in comparison with 0.42 m/s for natural gas), combustion instabilities can occur. High flame temperatures lead to higher NO$_x$ emissions.

The low-temperature flameless catalytic combustion of hydrogen can be conducted at temperatures as low as 500°C. Catalytic burners operate practically pollution free and with high efficiencies. Such burners with capacities up to 50 kW can be particularly applied in heating and absorption cooling plants. Through a combination of high-temperature flame combustion and low-temperature flameless catalytic combustion in the temperature range between 800 and 1500°C, high capacities in the MW capacity range with low NO$_x$ emissions may be achieved.

Cogeneration Plants Based on Gas Turbine and Diesel Engine

For power and heat cogeneration, combined-cycle plants using gas turbines or diesel engines may be employed. At the present time, combined-cycle power plants based on natural gas fired combustion turbines and diesel engines have the lowest capital costs and, simultaneously, the highest (up to 58%, HHV) efficiencies among power plants of all types in the capacity range up to 500 MW. Without any essential technical modifications, these combined-cycle power plants may be switched over to firing hydrogen.

The estimated capital and operating costs and efficiencies of gas turbine and diesel based combined-cycle plants are presented in Table 8.7. The service life of these plants is estimated to

Table 8.7. Estimated costs and overall efficiency of a combined-cycle power plant based on a hydrogen-fired gas turbine (GT), and energy utilization factor (EUF) of a cogeneration plant based on a diesel engine (DE)

Parameter	GT combined cycle power plant	DE cogeneration plant
Power output, MW	400	3.1
Plant capital cost, US$/kW	570	570
Annual operating cost, US$/kW	18	35
Overall efficiency (EUF), %	52	84 (electric 25, thermal 59)

be 30 years with 8000 hours of operation per year. The time required for further development of this technology to the commercial plant level is estimated to be 5–10 years.

Direct Steam Production

A steam generator based on the direct oxidation of hydrogen with oxygen may be used to supply steam to a high-pressure, high-temperature steam turbine. This process may appear to be a viable alternative to gas and steam turbine combined-cycle power plants, provided that its efficiency of at least 50% is achieved.

An effective method called HYDROSS for direct steam production from hydrogen and oxygen is now being developed. The principle of this method is simple: stoichiometric amounts of hydrogen and oxygen are introduced into a combustor by means of a supply device. The water is first used for cooling the combustor wall and then introduced into the combustor through a number of nozzles. The hot products of combustion are cooled from an initial temperature above 3000°C down to a temperature of 500–1000°C that is required for steam production. An efficiency of at least 50% is achieved in the part-load range and up to nearly 100% at the full load. The method should be implemented in economically feasible power plants. Steam generator concepts with a capacity of 30–100 MW are being developed. With their exceptionally short start-up time, such steam generators may be incorporated into instant reserve power plants. It should be stressed that in the area of the hydrogen technology, there are still great R and D needs.

MAGNETOHYDRODYNAMIC ENERGY CONVERSION

General Principles

MHD power generation is based on the Faraday effect, by which an electric current is induced in a conductor moving through a magnetic field. An MHD generator is a device for the direct conversion of heat into electrical energy without a conventional electric generator. Figure 8.8 shows the principle of the MHD generator. Like other heat engines, the MHD generator converts part of the heat supplied by burning fuel at a high temperature into electrical power and rejects the remainder to the ambient. Therefore the thermal efficiency of an MHD generator is increased by increasing the temperature of heat supply and by reducing the temperature of heat rejection.

Principally, an ionized gas or liquid metal can be used as the working fluid in an MHD generator. The two types of MHD generators have advantages and limitations. Thus the liquid-metal MHD generator can easily produce ac power directly. The power density of the liquid-metal system is about an order of magnitude higher than that of an ionized gas generator. However, the ionized gas MHD generator easily achieves a high gas velocity using a nozzle, can produce relatively high dc voltages, and can have a higher conversion efficiency than the liquid-metal system. Therefore an ionized gas is used in the present MHD generator projects.

According to the Carnot principle, an MHD generator intrinsically has high efficiency because it operates at very high temperatures, about 2550–2700°C. Coal-burning MHD/steam power plants

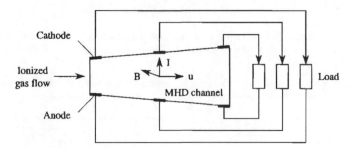

Figure 8.8. Principle of the MHD generator. B, magnetic-field strength; T, output current; u, fluid velocity.

promise to generate power at up to 50% efficiency, which is much higher than in conventional coal-fired power plants.

Operation of the MHD generator is based on the Faraday effect. If an electrically conducting fluid (an ionized gas or liquid metal) is forced through a perpendicular magnetic field at high velocity, as shown in Figure 8.8, an electric current is induced in the fluid. The fluid flows with a high (sonic or supersonic) velocity and has a very high temperature. The thermal energy of the fluid as well as its kinetic energy are partially converted into electric energy in form of a dc current. Thus electric power may be extracted from the MHD channel at its terminals.

The voltage generated in the MHD channel depends on the fluid velocity, magnetic field strength, and interelectrode gap width (Sutton, 1966; Culp, 1991; Spring, 1965). Thus it is given by

$$U_g = uBd \quad V \tag{8.31}$$

where d is the interelectrode gap width in m, u is the fluid velocity in m/s, and B is the magnetic field strength in Wb/m^2. Equation (8.31) is valid when the values of u and B are constant in the interelectrode gap.

The short-circuit current density is

$$J_{sc} = uB/\rho = uB\sigma \quad A/m^2 \tag{8.32}$$

where ρ is the average electrical resistivity of the MHD fluid in Ω m and σ is the average conductivity of the MHD fluid in $1/(\Omega \text{ m})$.

The internal voltage drop in the MHD generator is

$$\Delta U_g = iR_g = i\rho d/A = J\rho d \quad V \tag{8.33}$$

where i is the current in A, $R_g = \rho d/A$ is the internal resistance of the MHD generator in Ω, and A is the electrode area in m^2.

The external voltage drop on the load is then given by

$$U_L = U_g - J\rho d \quad V \tag{8.34}$$

where J is the current density in A/m^2.

The loading factor K of the unit is the ratio of the external load voltage to the generated voltage (Culp, 1991):

$$K = U_L/U_g = U_L/uBd \tag{8.35}$$

The generator current density is defined as

$$J = i/A = (U_g - U_L)/\rho d = uB(1 - K)/\rho \quad A/m^2 \tag{8.36}$$

The output current is

$$i = AJ \quad A \tag{8.37}$$

The output electrical power of the MHD generator is

$$P = iU_L \quad W \tag{8.38}$$

The output electrical power of an MHD generator per unit volume is the output power density

$$P/u = U_L i/Ad = KU_g J/d = KuBJ = Ku^2 B^2(1 - K)/\rho \quad W/m^3 \tag{8.39}$$

The maximum power density of an MHD converter is a function of the loading factor. It corresponds to an optimum value of K_{opt} of 1/2, which is found by setting $d(P/u)/dK = 0$. Substituting $K_{opt} = 1/2$ into Eq. (8.39) yields the maximum power density of an MHD converter as

$$(P/u)_{max} = u^2 B^2/4\rho \quad W/m^3 \tag{8.40}$$

There are a number of power losses in an MHD generator. For example, the joule-heating power loss per unit volume of the MHD generator (Culp, 1991) is given by

$$Q_j/u = i^2 R_g/Ad = J^2 A^2 (d\rho/A)/Ad = J^2 \rho = u^2 B^2(1 - K)^2/\rho \quad W/m^3 \tag{8.41}$$

Ignoring all other energy losses, the conversion efficiency for maximum power density (Culp, 1991) is

$$\eta_{max} = P/(P + Q_j) = 1/(1 + Q_j/P) = K_{max} = 1/2 \tag{8.42}$$

Thus the maximum possible conversion efficiency of an MHD generator is 0.5.

In addition to the joule-heating power loss, an MHD generator has the following power losses: Hall-effect loss, end loss, and friction and heat transfer losses. The Hall-effect loss is caused by the Lorentz force acting on the electrons in the interelectrode gap. A force called the Lorentz force acts on any charged particle that passes through a perpendicular magnetic field. For an electron this force is given by (Sutton, 1966; Culp, 1991)

$$F_L = euB \tag{8.43}$$

This force makes the electron travel in a circular arc, and thus all the electrons try to travel to one end of the collecting electrode. This produces very large currents in the collecting electrode with the accompanying resistance loss. Using an insulated, segmented collecting electrode, such as that shown in Figure 8.8, helps to solve this problem. The current can be reduced by increasing the MHD channel aspect ratio, i.e., length/interelectrode gap, by permitting the magnetic field poles to extend beyond the end of the electrodes, and/or by using insulated vanes in the fluid duct in the inlet and outlet of the generator.

The end loss in an MHD generator is associated with the reverse flow (short circuit) of electrons through the conducting fluid around the ends of the magnetic field. It can be reduced by increasing the aspect ratio (L/d) of the generator, by permitting the magnetic field poles to extend beyond the end of the electrodes, and/or by using insulated vanes in the fluid duct in the inlet and outlet of the generator.

In the MHD system, high friction and heat transfer losses occur. The high fluid turbulence also increases the convection heat transfer rate from the gas or liquid metal to the containment walls.

The electrode loss is another loss that may be significant in an MHD generator.

The specific power output per unit volume of an MHD generator increases with increase

- in the electrical conductivity of the working fluid
- in the magnetic field strength
- in the fluid flow velocity

Because of relatively low conductivity of the working fluid, MHD generators operate practically with high flow velocities and high magnetic field strengths.

The theoretical power output of an MHD generator is proportional to the square of the magnetic field strength. Hence a strong magnetic field of innovative superconducting magnets is necessary.

The efficiency of the MHD generator may approximately be estimated as follows:

$$\eta = R_e/(R_e + R_i) \tag{8.44}$$

where R_e and R_i are external and internal resistance of the ionized gas (plasma), respectively. It has been shown that a conversion efficiency of 50% and higher may be achieved in advanced MHD systems that are now being developed.

MHD-Based Combined Cycle

The MHD converter can be used as a topping system for a conventional steam or gas turbine plant that serves as a bottoming plant. Thus a combined-cycle conversion system with a high overall thermal efficiency will be formed. Similar to the gas and steam turbine combined cycle, the overall conversion efficiency of such a system is given by

$$\eta_o = \eta_{MHD} + \eta_{steam} - \eta_{MHD}\eta_{steam} \tag{8.45}$$

where η_{MHD} and η_{steam} are the individual conversion efficiencies of the MHD generator and the steam power plant, respectively.

The working fluid in an MHD generator usually consists of a carrier gas, i.e., combustion gas. The carrier gas is rendered electrically conducting either by injecting a solid seed material such as potassium carbonate (about 1% of the total flow rate) or a liquid-alkali metal such as cesium into a flowing carrier gas. When heated to high temperatures, the seed substance ionizes and its atoms split off electrons, thus making the carrier gas electrically conducting. Temperatures required in the MHD generator are about 3000 K. At such high gas temperatures resulting from the combustion of a fossil fuel with oxygen or highly preheated air, high thermal efficiencies are possible. For example, by using an MHD generator as a topping unit in a combined-cycle plant with a steam turbine as a bottoming unit, an overall efficiency of about 60% is expected (Rosa, 1987).

Example 8.2

A 12-m^3 MHD generator channel with segmented electrodes has a short circuit current density of 11,000 A/m^2. The electrical resistivity of the ionized gas is 0.032 Ω m.

What is the output power if the load factor K is 0.6? What is the actual current density in the MHD channel?

Solution

Under the short circuit conditions, the load factor is zero. From the short circuit current density $J_{sc} = 11,000$ A/m^2, we obtain the product of gas velocity and magnetic field strength as follows:

$$uB = J_{sc}\rho = 11,000 \times 0.032 = 352 \text{ V/m}$$

Power output

$$P = (uB)^2 K(1 - K)V/\rho = (352 \text{ V/m})^2 0.6(1 - 0.6)12 \text{ m}^3/0.032 \,\Omega\, \text{m} = 11.15 \text{ MW}$$

Channel current density

$$J = (1 - K)uB/\rho = (1 - 0.6)352 \text{ V/m}/0.032 \,\Omega\, \text{m} = 4400 \text{ A/m}^2$$

Example 8.3

For a Faraday MHD generator, calculate the power output and the conversion efficiency if the specific resistivity ρ is 0.07 Ω m, the magnetic field strength B is 3.7 T, and the plasma velocity is 1000 m/s. Assume that the electrode area A is 1.4 m^2 and the interelectrode gap in the channel is 0.9 m.

Solution

Open circuit voltage

$$U_{oc} = uBs = 1000 \text{ m/s} \times 3.7\,\text{T} \times 0.9\,\text{m} = 3330 \text{ V}$$

Short circuit current

$$I_{sc} = A|uB|/\rho = 1.4\,\text{m}^2 \times 1000\,\text{m/s} \times 3.7\,\text{T}/0.07\,\Omega\,\text{m} = 74,000 \text{ A}$$

Current corresponding to the maximum power

$$I_m = uBA/(2\rho) = 1000\,\text{m/s} \times 3.7\,\text{T} \times 1.4\,\text{m}^2/(2 \times 0.07\,\Omega\,\text{m}) = 37,000 \text{ A}$$

Voltage corresponding to the maximum power

$$U_m = uBs - I_m\rho s/A$$

$$= 1000\,\text{m/s} \times 3.7\,\text{T} \times 37,000\,\text{A} \times 0.07\,\Omega\,\text{m} \times 0.9\,\text{m}/1.4\,\text{m}^2 = 1665 \text{ V}$$

Maximum power

$$P_m = U_m I_m = 1665 \text{ V} \times 37000\,\text{A} = 61.6 \text{ MW}$$

Conversion efficiency

$$\eta_{MHD} = U_m/(uB) = 1665\,\text{V}/(1000\,\text{m/s} \times 3.7\,\text{T}) = 0.45$$

Example 8.4

Calculate the overall efficiency of a combined-cycle plant consisting of an MHD generator with an efficiency of 0.45 and a steam turbine plant with an efficiency of 0.3 without supplementary firing.

Solution

Overall efficiency of combined-cycle plant

$$\eta_o = \eta_{MHD} + \eta_{ST} - \eta_{MHD}\eta_{ST} = 0.45 + 0.3 - 0.45 \times 0.3 = 0.615$$

Open-Cycle MHD Systems

Description

MHD energy conversion systems can operate in either open or closed cycles. In open-cycle systems, the working fluid, i.e., the seeded combustion gas, is discharged from the MHD generator to the ambient. In closed-cycle systems, heat is transferred from the combustion gas to the working fluid by means of a heat exchanger.

An open-cycle MHD system is schematically presented in Figure 8.9. The system consists of an MHD generator, an air compressor, and a high-temperature air heater (HTAH). The carrier gas is obtained by burning a fossil fuel, e.g., coal, in a suitable combustion chamber. The seed material, generally potassium carbonate, K_2CO_3, is injected into the combustion chamber, where the potassium is ionized by hot combustion gases at temperatures of about 2700–3000 K (Tompson et al., 1993).

In application to electric utility power generation, MHD is combined with steam power generation in a binary cycle. The MHD generator is used as a topping unit to the steam bottoming plant, as illustrated in Figure 8.9. Starting with combustion products at a pressure of 5–10 bars

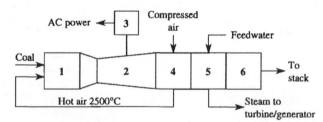

1 - Combustor with
    nozzle
2 - MHD generator
3 - Inverter
4 - High-temperature
    air heater
5 - Steam generator
6 - Flue gas cleaning
    and seed material
    recovery

Figure 8.9. Advanced coal-fired MHD/steam plant combined-cycle configuration with a high-temperature air heater.

and a temperature sufficiently high (about 2550°C) to produce a working fluid or adequate electric conductivity, when seeded with an easily ionizable salt such as potassium carbonate, the hot ionized gases flow through the MHD generator at approximately sonic velocity. The MHD generator duct, or channel, extracts energy from the gas, and the flow is expanded so that it can maintain its velocity against the decelerating forces resulting from its interaction with the magnetic field. Development and use of high-temperature refractory heat exchangers in future advanced MHD plants would allow realization of the full efficiency potential of MHD and, correspondingly, improved fuel utilization and even lower energy costs.

Advanced MHD power cycles have also been proposed in which the heat energy of the MHD generator exhaust gas is utilized in a bottoming gas turbine plant. The gas turbine working fluid in such a plant is clean air heated by the MHD generator exhaust gas. Such MHD power cycles do not need cooling water for steam condensation and heat rejection. Furthermore, any efficiency advantage offered by the use of a high-temperature gas (air) turbine instead of a steam turbine in the bottoming plant will improve the overall MHD power cycle accordingly.

The MHD topping cycle components are the magnet, coal combustor, nozzle, MHD channel, associated power conditioning equipment, and diffuser. The magnetic field required in a commercial plant is typically 4.5–6 T. Therefore, only superconducting magnets can be employed.

Emissions of SO_x, NO_x, and particulates below present NSPS standards are anticipated in commercial MHD power plants.

To burn fuel, compressed air, oxygen-enriched air, or pure oxygen can be used. To attain the combustion gas temperature of about 3000 K, the air must be preheated to at least 1400 K. The oxygen-enriched air requires less preheating, the pure oxygen practically no preheating.

The pressurized high-temperature gas from the combustion chamber passes through a Laval nozzle and enters the divergent channel of the MHD generator at a supersonic flow velocity. The MHD channel is made of a heat-resistant alloy and is water cooled. In the electrically conducting fluid flowing across the magnetic field, dc is generated, which can be converted into ac by means of an inverter. To conduct the dc to an external load, electrodes are arranged on the oppositely located channel walls. The so-called Hall effect causes the energy losses arising from the Hall current as the result of the interaction of the magnetic field on the generated (Faraday) current, which produces a voltage in the fluid flow direction. Various electrode connections can be used to utilize the Faraday current and to minimize energy losses arising from the Hall effect. An alternative is utilizing the Hall current only by short-circuiting each electrode pair outside the generator and connecting the load between the electrodes at the two ends of the MHD generator (Rosa, 1987).

As the working gas travels along the MHD channel and heat is converted into electricity, the gas temperature falls. However, the lower limit of the operating temperature of the MHD system lies by about 2170–2200 K (Rosa, 1987). At lower temperatures, the extent of ionization is insufficient to maintain an adequate electrical conductivity.

The thermal efficiency of the MHD generator is typically about 30%. The working gas leaves the MHD channel at a temperature of about 1900°C. Its heat is further utilized to preheat the combustion air in an HTAH and to raise steam in a heat recovery steam generator (HRSG) to produce additional power in a steam turbine plant. The MHD system is thus converted into a combined-cycle power plant. An overall efficiency up to 60% can be achieved in advanced MHD-based combined-cycle plants (Rosa, 1987). In coal-fueled MHD systems the fly ash is removed from the flue gas leaving the HRSG, and the seed material is recovered from the ash.

The scientific feasibility of MHD generation has been demonstrated in several experimental MHD conversion systems. Some technical problems affecting the efficiencies attained and the life of the equipment have still to be resolved for achieving the economic feasibility of the MHD simple- and combined-cycle systems. One of the main problems is the short life of the equipment caused by corrosion and erosion. The highly corrosive environment of combustion gases at very high temperatures loaded with ash (or slag) residue from the burning coal requires the utilization of corrosion-resistant materials for the combustor, MHD channel, electrodes, and air preheater. In coal-fired MHD plants the separation of the seed material from the fly ash and its reconversion into the carbonate form is another problem. One possible remedy to these problems may be the utilization of an ash-free and sulfur-free fuel gas derived from coal, rather than coal itself, in the combustor. Burning the fuel gas in preheated air would provide adequate operating temperatures.

Another advanced concept includes the use of hydrogen gas produced from coal. Burning the hydrogen gas in high-pressure oxygen produces high-temperature steam. Being seeded, it can be used as a working fluid in the MHD generator and, after removing the seed material, as working fluid in the steam turbine.

SO_x and NO_x Emissions

Unless the sulfur in the coal has been removed, the original potassium carbonate seed is converted in the MHD channel into potassium sulfate, K_2SO_4, which must be recovered from the ash, reconverted into potassium carbonate, and further utilized as seed. The removal of potassium sulfate makes the desulfurization of the flue gas unnecessary. When combustion of fuel is conducted with oxygen alone, no thermal NO is formed. However, at high operating temperatures a large amount of thermal NO is formed from nitrogen of air in the combustion chamber unless NO_x control methods such as staged combustion are employed. By this method, the combustion air is split into two portions. The primary air supplied to the combustion chamber is not sufficient for complete fuel burning, and the secondary air is introduced into the gas flow downstream of the MHD generator in order to complete the combustion of unburned constituents in the fuel gas. Owing to lower combustion temperature and deficit of oxygen, the nitrogen oxide concentration in the flue gas is decreased.

Closed-Cycle MHD Systems

Two types of closed-cycle MHD generators are possible, with electrical conductivity of the working fluid provided by ionization of a seed material in one type, and with a liquid metal in the other type. The carrier, usually a chemically inert gas, is circulated in a closed loop and is heated by the combustion gases using a heat exchanger.

The closed-cycle plant operates on the Joule cycle (see Chapter 5). The gas is compressed, heated in a heat exchanger by an external source, and expanded in the MHD generator, while its pressure and temperature drop. After leaving the generator, heat is removed from the gas by a cooler, and the gas is recompressed and returned for reheating.

In a nonequilibrium ionization MHD converter, the monotomic carrier gas such as argon or helium is compressed and heated by combustion gases in a heat exchanger, and the seed (cesium metal) is injected into the hot gas. In the carrier gas with a cesium seed, nonequilibrium ionization occurs at a relatively low temperature. Thus electrical conductivity required for the MHD generator can be attained at a temperature below 2000°C. The ionized working fluid is injected into the generator at high speed. Upon leaving the MHD generator, the hot fluid enters an HRSG, which supplies steam to a turbine generator. While the working fluid is cooled, the seed is condensed and separated for reuse, and the gas is recompressed and reheated in the heat exchanger. The flue gases are used to preheat the incoming combustion air and then cleaned to remove pollutants prior to discharge through a stack to the atmosphere.

Because of lower operating temperatures, the closed-cycle MHD generator permits a wider choice of materials but has a lower thermal effieiency than an open-cycle system.

High-Temperature Air Heater

The combination of energy extraction and flow expansion causes the gas temperature to drop. If the gas temperature becomes too low (about 2050°C), the energy extraction ceases. The gases exhausting from the generator contain significant heat energy. This energy can be further used in the bottoming plant to raise steam to drive a steam turbine and generate additional power, and also to preheat the combustion air to a temperature of 1375–1650°C in a ceramic regenerative heat exchanger. The required high combustion temperature can also be achieved if the oxygen-enriched combustion air will be preheated to a lower temperature in a metal tubular heat exchanger.

A HTAH has long been recognized as a requirement for the most efficient MHD plants in order to reach high operating temperatures (Tompson et al., 1993). Ceramic composite materials enable the use of recuperative air heaters for heating air to temperatures well above that feasible with metal tube recuperators. The application of direct coal fired open-cycle gas turbines in combined-cycle plants has been impeded because of the problems in hot gas cleanup required for reliable turbine operation. The utilization of an HTAH eliminates the need for hot gas cleanup and thus makes closed-cycle air turbine plants more feasible.

Consider a combined cycle consisting of an MHD generator topping plant and a steam turbine bottoming plant. The first prerequisite for achieving high overall efficiency of a combined-cycle power plant is the enhanced efficiency of the topping plant. The efficient power production in an open-cycle MHD generator is possible with the flame temperature in the coal-fired combustor of about 2700–3000 K, which is required for adequate ionization of the working fluid. This temperature can be achieved by burning coal in oxygen-enriched air and by preheating the air to a high temperature of about 1400°C using the waste heat of the MHD generator.

To achieve high conversion efficiency, the advanced MHD/steam combined-cycle power plant needs a coal-fired air heater to heat air to about 1400°C. Whereas earlier MHD concepts were based on a regenerative air heater with their material problems and high capital costs, recent MHD systems include using direct combustion of coal-derived clean fuel in oxygen-enriched air.

Most economical system design in the power output range of 1000 MWe can provide an overall efficiency of about 45% (heat rate 8000 kJ/kWh).

The full conversion efficiency potential of the MHD/steam power plant requires that an HTAH be directly fired by the coal combustion products exiting the MHD generator. This configuration concept is shown in Figure 8.10.

The performance improvement that is realized in the plant efficiency depends on the air temperature achieved, but can approach 50% (heat rate 7200 kJ/kWh) for modest technology systems and the mid-50% range for advanced technology systems, such as steam cycles at higher pressures and temperatures and higher field magnet strengths.

From a thermodynamic viewpoint, it is apparent that the utilization of the gases exiting the MHD diffuser at around 2050°C, heating air to about 1400°C, greatly improves the Second Law efficiency. An attractive MHD power generation system is the closed-cycle nonequilibrium MHD generator. This system uses helium as a working fluid in the MHD generator. Helium can be kept in an ionized state at temperatures around 2000°C. A high-temperature heat exchanger is required to transfer the heat from the coal combustion products to helium. The current test facility under consideration at the Tokyo Institute of Technology utilizes a regenerative, pebble bed heat exchanger (Tompson et al., 1993).

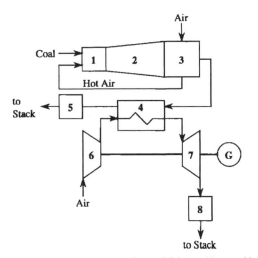

1 - Combustor
2 - MHD generator
3 - High-temperature air heater
4 - Heat exchanger
5 - Flue gas cleaning and seed material recovery
6 - Compressor
7 - Turbine
8 - Waste-heat boiler

Figure 8.10. Advanced coal-fired MHD/gas turbine combined-cycle configuration with a high-temperature air heater.

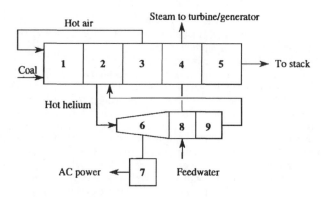

1 - Coal combustor
2 - High-temperature helium heater
3 - Intermediate-temperature air heater
4 - Steam generator
5 - Flue gas clean-up and seed material recovery
6 - MHD generator
7 - Inverter
8 - Economizer
9 - Compressor

Figure 8.11. Externally fired combined-cycle MHD generator configuration.

Externally Fired Combined Cycle

The externally fired combined cycle (EFCC) MHD generator configuration shown schematically in Figure 8.11 consists of a coal combustor that burns coal with hot air heated in an HTAH placed downstream of a high-temperature helium heater and upstream of a heat recovery steam generator (HRSG) with a subsequent unit for flue gas cleaning and seed material recovery. The helium gas is pressurized by a compressor, heated to a high temperature in the helium heater, and introduced via a nozzle into the MHD generator, where it is used as the working fluid. The helium gas stream exiting from the MHD channel is then cooled down passing the economizer of the HRSG.

A conceptual design of an advanced direct pulverized coal-fired MHD/steam plant with a net power output of 960 MWe and net efficiency of 58.5% has been developed. The plant includes a direct-fired regenerative refractory heater to heat air up to 1700°C (Tompson et al., 1993). The thermal efficiency of the MHD power plant is therefore limited predominantly by the minimum temperature T_{min} at which a sufficient gas ionization is still maintained.

Advanced coal-burning MHD/steam power generation plants can achieve efficiencies of up to 60% with less environmental impact than from conventional coal-burning steam plants. Compared with other direct coal-burning power generation technologies, MHD technology potentially has such advantages as

• higher efficiency
• lower fuel usage
• lower generating costs
• less environmental intrusion

The use of MHD generators will significantly reduce emissions of carbon dioxide, heat rejection, and solid-waste production.

CLOSURE

The two technologies, fuel cell and MHD, described in this chapter have a potential of achieving efficiencies of over 60% in combined-cycle power plants. Fuel cells have an additional benefit of low environmental impact. High-temperature fuel cells—MCFC and SOFC—require substantial improvement in material stability and essential reduction in costs to be economically viable for efficient combined-cycle plants based on fuel cells. The state-of-the-art MHD technology is described in this chapter along with some novel concepts.

The main difficulties in the development of MHD-based combined-cycle power plants that have a potential of over 60% efficiency are caused by temperatures of over 2700 K required in the MHD channel. Complex material and design problems must be solved in order to realize MHD-based combined-cycle power plant projects. Significant progress is anticipated in the first decade of the 21st century.

PROBLEMS

8.1. Evaluate the maximum cell voltage, the maximum work, and the theoretical (maximum) conversion efficiency for a hydrogen-oxygen fuel cell at the standard reference conditions of 1 atm and 25°C. Consider the two cases when the product water is in the liquid and vapor phases.

8.2. A molten-carbonate fuel cell (MCFC) stack operates at a temperature of 650°C and consumes 25 kg of hydrogen per hour. Calculate the maximum values of the voltage, efficiency, and power output of the cell when the product is water vapor.

8.3. Find the output voltage and conversion efficiency of an oxygen-hydrogen solid-oxide fuel cell (SOFC) operating at 950°C. Assume that the hydrogen is supplied at 1.08 atm and the oxygen is supplied from air at 1.12 atm, and assume that the water vapor product is at 1 atm.

8.4. An 8-m^3 MHD generator channel with segmented electrodes has a short circuit current density of 1.2 A/cm^2. The electrical resistivity of the ionized gas is 0.048 Ω m. What is the actual current density in the MHD channel and its power output if the load factor k is 0.55?

8.5. For a Faraday-MHD generator, the magnetic field strength is 4 T, the specific resistivity of ionized gas is 0.09 Ω m, and the ionized gas velocity is 970 m/s. Assuming the electrode area of 1.2 m^2 and the interelectrode gap in the channel of 0.8 m, calculate (*a*) the short circuit current in A, (*b*) the open-circuit voltage in V, (*c*) the maximum power output in W, and (*d*) the conversion efficiency corresponding to maximum power output.

8.6. Calculate the overall efficiency and the heat rate of a combined-cycle plant consisting of MHD generator with an efficiency of 0.48 and a steam turbine plant with an efficiency of 0.32 without supplementary firing.

REFERENCES

Anon. 1994*a*. SOFC sets performance record. *Power Eng.* 98(10):5.

Anon. 1994*b*. Fuel cell power plant sets reliability record. *Power Eng.* 98(10):8.

Appleby, A. J. 1987. Phosphoric acid fuel cells. In *Fuel cell trends in research and applications*, ed. A. J. Appleby. Washington, D.C.: Hemisphere.

Bates, J. L. 1989. Solid oxide fuel cells: A materials challenge. In *Proceedings of the Sixteenth Energy Technology Conference*, pp. 205–219. Government Institutes, Inc.

Culp, A. W. 1991. *Principles of energy conversion*. New York: McGraw-Hill.

Drenckhahn, W., Greiner, H., and Ivers-Tiffee, E. 1994. Materials for solid-oxide high-temperature fuel cells. *Siemens Power J.* 4:36–38.

Gillis, E. 1989. Fuel cells. *EPRI J.* (September):34–36.

Hirschenhofer, J. H., Stauffer, D. B., and Engleman, R. R. 1994. *Fuel cells: A handbook*, rev. 3. Reading, Mass.: Gilbert/Commonwealth Inc.

Homma, T., Mori, S., and Nakaoka, A. 1994. Current status of fuel cells in Japan. Presented at the Fuel Cell Seminar, San Diego, Calif., November 1994.

Ketelaar, J. A. 1987. Molten carbonate fuel cells. In *Fuel cell trends in research and applications*, ed. A. J. Appleby. Washington, D.C.: Hemisphere.

Mayfield, M. J., Beyma, E. E., and Nelkin, G. A. 1989. Update on U.S. Department of Energy's Phosphoric Acid Fuel Cell Program. In *Proceedings of the Sixteenth Energy Technology Conference*, pp. 184–196. Government Institutes, Inc.

Monn, M. 1995. Project brings fuel cells closer to utility user station: Show-case study. *Power.* 139(4):76–78.

Myles, K. M., and Krumpelt, M. 1989. Status of molten carbonate fuel cell technology. In *Proceedings of the Sixteenth Energy Technology Conference*, pp. 197–204. Government Institutes, Inc.

O'Shea, T. P., and Leo, A. J. 1994. Santa Clara Demonstration Project: The first 2-MW carbonate fuel cell power plant. Presented at the Fuel Cell Seminar, San Diego, Calif., November 1994.

Ray, E. R. 1994. Westinghouse solid oxide fuel cell technology status update. Presented at the Fuel Cell Seminar, San Diego, Calif., November 1994.

Rosa, R. J. 1987. *Magnetohydrodynamic energy conversion*. Washington, D.C.: Hemisphere.

Selman, J. R. 1994. Fuel cell development and commercialization in Japan and Europe. In *American Power Conference*. South Windsor, Conn.: ONSI Corp.

Spring, K. H., ed. 1965. *Direct generation of electricity*. London: Academic.

Sutton, G. W., ed. 1966. *Direct energy conversion*. New York: McGraw-Hill.

Swanekamp, R. 1995. Fuel cells inch towards mainstream power duties. *Power*. 139(6):82–90.

Tompson, T. R., Boss, W. H., and Chapman, J. N. 1993. Application of high temperature air heaters to advanced power generation cycles. In *Proceedings, American Power Conference*, pp. 460–466.

Whitaker, F. L., and Lueckel, W. J. 1994. The phosphoric acid PC25 fuel cell power plant—and beyond. In *American Power Conference*. South Windsor, Conn.: ONSI Corp.

Chapter Nine

CLEAN COAL POWER GENERATION TECHNOLOGY

One of the most challenging problems in the area of power engineering is efficient utilization of coal to power generation with reduced environmental impact. This includes combustion with an objective of diminishing the inert and harmful constituents such as sulfur and ash. Innovative technologies that convert coal and solid wastes into liquid and gaseous fuels are now being developed or improved.

Fluidized bed technology is applied both to coal combustion and gasification. The currently used technology is circulating fluidized bed combustion (CFBC) technology. It has been implemented in power plants with power output up to 175 MW. A further step in the development of CFBC boilers is 300-MW power plants. The hydrodynamic and heat transfer considerations presented in this chapter create a basis for the design of fluidized bed combustion boilers. Therefore this issue is covered in greater detail.

As an addition to Chapter 2 and 6, this chapter continues the discussion of the coal gasification technology in application to integrated gasification combined-cycle (IGCC) plants. The IGCC technology combines a coal gasification plant and a gas turbine and steam turbine combined-cycle power plant into one unit. Thus the overall energy conversion efficiency is improved, and the environmental impact is reduced. The most critical technological problem is the development of reliable hot gas cleanup (HGCU). This and other problems of the IGCC technology are discussed in this chapter. The major advantages of fluidized bed technology are its intrinsic capability of controlling the sulfur and nitrogen oxide emissions by sorbent addition and by keeping the fuel combustion temperature low. The emphasis is put on the pressurized fluidized bed combustion technology, which has advantages as compared to the atmospheric fluidized bed technology (AFBT) because it can be directly applied to combined-cycle power plants.

COAL BENEFICIATION, LIQUEFACTION, GASIFICATION

The clean coal technologies include (Longwell et al., 1995; Maude, 1993; Lozza et al., 1996)

- coal upgrading processes (beneficiation)
- coal gasification for the production of clean fuel gas
- coal liquefaction for production of liquid fuels (gasoline and light diesel oil)
- atmospheric and pressurized FBC systems
- integrated gasification combined-cycle (IGCC) power plants
- integrated gasification fuel-cell (IGFC) power plants

Other advanced coal-based power cycles are (Longwell et al., 1995; Anand et al., 1996): (1) externally fired gas turbine combined cycle (EFCC), (2) coal-fired diesel engine cycle, and (3) hybrid system including partial (mild) gasification, pressurized fluidized bed combustion (PFBC) and combined-cycle power plant.

Coal Upgrading Processes: Beneficiation

Coal beneficiation is the treatment of as-mined coal to reduce the content of the mineral matter (ash) and sulfur. Coal usually contains 3–15% extraneous mineral matter (ash). The sulfur is present in coal in organic and inorganic (pyritic) forms. Coal beneficiation results in a substantial decrease in the amount of solid wastes remaining after the coal is burned and, in particular, in a 30–50% decrease in the pyritic sulfur content.

Physical methods based on the difference in density or surface properties of coal itself (density 1.2–1.4) and inert matter (density 1.8–5.0) are used for coal beneficiation in the jig, dense media, concentrating table, and froth flotation methods (Longwell et al., 1995). In the jig method the coal is fed onto a perforated submerged screen and is subjected to the action of pulsating motion of the water. As a result, the coal ascends to the top of the screen and is carried off by the water flow, while the mineral matter is removed from the bottom of the screen. Froth flotation is used for cleaning coal fines suspended in water. Air is passed through the water, and air bubbles adhere to the fine coal particles and carry them to the surface. The clean coal is recovered from the froth, and the mineral matter is discharged from the bottom. In the concentrating table method, the surface of a slightly tilted, grooved table oscillates in a direction perpendicular to the water flow. The suspension of finely ground coal in water is fed onto the upper end of the table. The inert particles are trapped and removed at the side of the table, whereas the coal is washed over the end of the table. Dry cleaning methods use air instead of water. Since coal itself is nonmagnetic and iron sulfides (pyrites) are magnetic, a high-gradient magnetic field can be used to separate iron sulfides from a coal-water slurry in a magnetic separator.

Precombustion Coal Desulfurization

Certain physical, chemical, or biological processes can be used to remove some of the sulfur from the coal. As stated earlier, physical cleaning methods can remove 30–50% of the pyritic sulfur, which is only 10–30% of the total sulfur in the coal. Through chemical and biological treatment, some of the organic sulfur can be removed from the coal. Thus, up to 90% of total sulfur content can be removed (Longwell et al., 1995).

Another approach is to convert the high-sulfur coal into a low-sulfur gaseous fuel. The most suitable method of chemical coal cleaning is leaching both the organic sulfur and mineral matter with hot sodium-based or potassium-based caustic. In the biological method the coal is exposed to bacteria or fungi that have an affinity for sulfur. Sulfur-digesting enzymes may be added (Minchener et al., 1991). However, the most effective method of precombustion coal treatment is coal gasification. The sulfur in coal is converted to H_2S, which can be relatively easily removed from the producer gas. Yet another possibility is the conversion of coal into combustible liquids.

Coal Liquefaction

The conversion of coal into a mixture of liquid hydrocarbons by adding hydrogen is called coal liquefaction. The product may be equivalent to a synthetic crude oil (syncrude), which differs from petroleum crude oil in that it has a higher content of aromatic hydrocarbons, such as benzene. Coal chlorine content is usually not higher than 0.2%. To prevent corrosion from chlorides, its content in coal is limited to a maximum of 0.1%. For the same reason, low-sulfur coals are desired. The yield of liquid fuel is approximately 0.44–0.53 m^3 per 1000 kg of coal.

There are four basic processes for coal liquefaction (Longwell et al., 1995) indirect (Fischer-Tropsch process), rapid pyrolysis, hydrogenation, and solvent extraction. Figure 9.1 shows the classification of coal liquefaction processes.

Coal Liquefaction by Fischer-Tropsch Process

This process includes the following two steps: (1) complete coal gasification to produce a synthesis gas (syngas) and (2) catalytic conversion of syngas to methanol by the Fischer-Tropsch process.

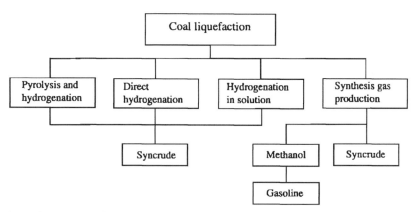

Figure 9.1. Coal liquefaction processes.

The syngas is a mixture of carbon monoxide (CO) and hydrogen (H_2). The syngas conversion to methanol occurs at pressures above 100 bars according to the following reaction (Longwell et al. 1995)

$$CO + 2H_2 \rightarrow CH_3OH \qquad (9.1)$$

Methanol (CH_3OH) may be further converted into gasoline.

Rapid Coal Pyrolysis

The main product of rapid coal pyrolysis is light oil with a high content of aromatic hydrocarbons, with char and coal gas as by-products. The faster the complete pyrolysis is accomplished, the greater the yield of liquid hydrocarbons. Char can be used as a boiler fuel. The process can be improved by combining pyrolysis with hydrogenation using a hydrogen-rich gas. The two processes may be conducted simultaneously or in separate stages.

Direct Coal Hydrogenation

A slurry of pulverized coal with some product oil content is heated and fed into a catalytic reactor. There, it is bombarded with high-temperature hydrogen gas at about 200 bars and 450°C to produce liquid hydrocarbons and some gaseous by-product. The latter is cleaned of H_2S and NH_3 and is available for sale. Higher hydrogen gas pressures and longer reaction times yield lighter oils. At lower pressures and shorter times, heavier components predominate in the liquid product. A coal-oil paste is used in a similar process.

Solvent Extraction

Coal is slurried in hydrogen-rich liquid solvent and is heated with a small amount of hydrogen to about 430°C at 70 bars. The solvent is made by hydrogenating one of the liquid products of the process. In another process, the coal is partly dissolved in a product oil in the presence of hydrogen-rich gas at high temperature and pressure.

Coal Gasification

Advanced coal combustion systems that reduce the sulfur content before fuel burnout include integrated gasification combined-cycle (IGCC) systems, hybrid systems with mild coal gasification, and coal liquefaction. The fundamentals of coal gasification are discussed in Chapter 2.

IGCC systems involve the following four basic steps (McAllester, 1991; Richards, 1994; Robertson and Horazak, 1993; Schellberg and Kuske, 1992; Smith, 1992; Stambler, 1993; Verweyen, 1994; Tait and McDonald, 1994):

1. coal gasification with steam and oxygen (or air)
2. gas cleanup to remove sulfur and alkali metal
3. combustion of the clean gases with the hot exhaust used to drive a gas turbine to generate electricity
4. residual heat recovery by steam generation, which is used to power a conventional steam-driven generator.

During gasification, most of the sulfur is converted to H_2S rather than SO_2. Several alternatives are available for efficient removal of H_2S. In some gasification combined-cycle systems, the hot gases are passed through a bed of zinc ferrite, which absorbs H_2S. Commercial-grade sulfur is produced when the ferrite bed is regenerated. In other applications, limestone sorbent injection is used for H_2S capture.

However, the main advantage of IGCC systems is that pollutant reduction is accomplished before air addition and combustion. Thus the volume of gas is much less, and the pollutants have significantly higher concentrations and therefore are easier to remove.

In the hybrid systems a mild gasification process produces char and fuel gas. The char is burned in a boiler furnace, and the clean gas is burned in a topping gas turbine combustor. Both IGCC and hybrid systems are discussed in more detail below.

FLUIDIZED BED COMBUSTION TECHNOLOGY

Fluidized bed technologies for coal combustion in power plants can be classified as follows (Anthony, 1995; Minchener et al., 1991; Minchener, 1992; Robertson and Horazak, 1993; Schellberg and Kuske, 1992; Smith, 1992; Stambler, 1993):

- atmospheric bubbling fluidized bed combustion (AFBC)
- atmospheric circulating fluidized bed combustion (CFBC)
- pressurized bubbling fluidized bed combustion (PFBC)
- pressurized circulating fluidized bed (PCFB) combustion

Basic concepts of atmospheric bubbling, atmospheric circulated, and pressurized fluidized bed combustors are shown in Figure 9.2.

An AFBC differs from other fluidized bed systems through a stable operating zone of fluidized material with a distinct upper surface. Because of gas bubbles formed in the AFBC, this type of fluidized bed is referred to as the bubbling fluidized bed. Crushed coal and limestone are fed into the fluidized bed, and coarse ash particles are removed from the bed at the fluidized bed bottom, whereas fine solid particles are carried out of the fluidized bed and separated in a cyclone. Heat removal from the AFBC is accomplished by means of immersed heat transfer surfaces. Increasing the velocity of the gas ascending through the fluidized bed with a given particle size distribution, a certain upper limit of velocity is reached beyond which material tends to be carried out from the combustion chamber with the outlet gases. High rates of bed material elutriation result in loss of unburned carbon and unused limestone in AFBC systems. This lowers combustion efficiency.

Next, the fundamentals of hydrodynamics and heat transfer in both bubbling and circulated fluidized beds at atmospheric and elevated pressures are considered.

Hydrodynamics and Heat Transfer of Fluidized Bed Systems

Figure 9.3 shows pressure drop as a function of superficial gas velocity in a fixed bed and a fluidized bed. The main hydrodynamic parameters of a fluidized bed are bed porosity, pressure drop across

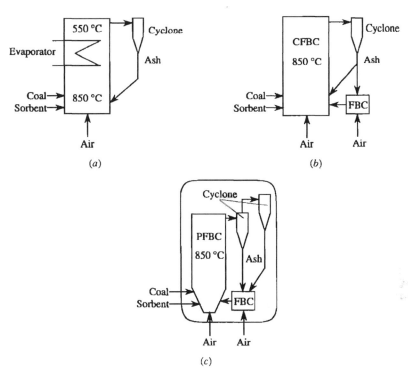

Figure 9.2. Basic concepts of (a) atmospheric bubbling, (b) atmospheric circulated (CFBC), and (c) pressurized fluidized bed combustors (PFBC).

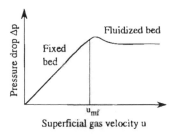

Figure 9.3. Pressure drop in a fixed bed and fluidized bed as a function of superficial gas velocity.

the fluidized bed, minimum fluidization gas velocity, and terminal (carry out) gas velocity. Bed porosity is defined as the bed void fraction:

$$\varepsilon = 1 - V_p/V \qquad (9.2)$$

where V_p is volume occupied by solid particles and V is total volume of the fluidized bed.

The pressure drop across the fluidized bed is given by

$$\Delta p = gH\rho_p(1 - \varepsilon) \quad \text{Pa} \qquad (9.3)$$

where g is acceleration due to gravity, H is bed height at the bed porosity ε, and ρ_p is solid particle density. The gas density is ignored.

The minimum fluidization velocity is the superficial gas velocity at the onset of fluidization. It can be found as follows:

$$u_{mf} = \mathrm{Re}_{mf} \, \eta/\rho d \quad \text{m/s} \tag{9.4}$$

where Re_{mf} is the Reynolds number at the minimum fluidization, η is dynamic viscosity of the fluidizing gas at the fluidized bed temperature in Pa s, ρ is density of the fluidizing gas at the fluidized bed temperature in kg/m^3, and d is the mean particle diameter in m.

The Reynolds number may be calculated as in the following empirical equation (Gogolek and Grace, 1995; Verweyen, 1994):

$$\mathrm{Re}_{mf} = (a^2 + b \, \mathrm{Ar})^{0.5} - a \tag{9.5}$$

where $a = 33.7$ and $b = 0.0408$ for $p = 1$ bar, and $a = 28.7$ and $b = 0.0408$ for $p > 1$ bar. The Archimedes number (Ar) is given by

$$\mathrm{Ar} = gd^3(\rho_p - \rho)\rho/\eta^2 \tag{9.6}$$

The transport (terminal) gas velocity for $0.4 < \mathrm{Re} < 500$ is given by (Gogolek and Grace, 1995; Verweyen, 1994)

$$u_t = \mathrm{Re}_t \, \eta/\rho d \tag{9.7}$$

with

$$\mathrm{Re}_t = (\mathrm{Ar}/7.5)^{0.666} \tag{9.8}$$

The following examples illustrate the procedures for the calculation of Δp, u_{mf}, and u_t at 1 and 16 bars.

Example 9.1

Calculate the pressure drop across an atmospheric CFBC at an average bed porosity $\varepsilon = 0.9$ for a bed height H of 10 m. The particle density is $\rho_p = 2400$ kg/m^3.

Solution

The pressure drop across the fluidized bed is given by

$$\Delta p = gH\rho_p(1 - \varepsilon) = 9.81 \times 10 \times 2400 \times (1 - 0.9) = 0.2355 \text{ bar}$$

Example 9.2

Calculate the minimum fluidization velocity and the transport (terminal) gas velocity for coal particles with a diameter of 0.3 mm and density of 2400 kg/m^3 that are fluidized in a CFBC and PCFBC boiler both operating at a temperature of 850°C and a pressure of 1 and 16 bars, respectively. Air properties at 850°C (look up in table of air properties) are

Pressure, bars	1	16
Density ρ, kg/m^3	0.31	4.94
Dynamic viscosity η, 10^{-6} Pa s	46.85	46.88

Solution

1. The Reynolds number and the minimum fluidization velocity are given by

$$\mathrm{Re}_{mf} = (a^2 + 0.0408 \, \mathrm{Ar})^{0.5}$$

where $\mathrm{Ar} = gd^3(\rho_p - \rho)\rho/\eta^2$

$$u_{mf} = \mathrm{Re}_{mf} \, \eta/\rho d$$

Thus, at 1 bar,

$$Ar = 9.81(0.3 \times 10^{-3})^3(2400 - 0.31)0.31/(46.85 \times 10^{-6})^2 = 89.77$$

$$Re_{mf} = (33.7^2 + 0.0408 \, Ar)^{0.5} - 33.7 = 0.05$$

$$u_{mf} = 0.05 \times 46.85 \times 10^{-6}/(0.31 \times 0.3 \times 10^{-3}) = 0.025 \text{ m/s}$$

Similarly, at $p = 16$ bars,

$$Ar = 9.81(0.3 \times 10^{-3})^3(2400 - 4.94)4.94/(46.88 \times 10^{-6})^2 = 1425.94$$

$$Re_{mf} = (28.7^2 + 0.0408 \, Ar)^{0.5} - 28.7 = 1$$

$$u_{mf} = 1 \times 46.88 \times 10^{-6}/(4.94 \times 0.3 \times 10^{-3}) = 0.032 \text{ m/s}$$

2. The transport (terminal) gas velocity for $0.4 < Re < 500$ is given by

$$u_t = Re_t \, \eta/\rho d$$

with $Re_t = (Ar/7.5)^{0.666}$,

$$u_t = (Ar/7.5)^{0.666} \eta/\rho d$$

Thus at $p = 1$ bar,

$$u_t = (89.77/7.5)^{0.666} \times 46.85 \times 10^{-6}/(0.31 \times 0.3 \times 10^{-3}) = 2.63 \text{ m/s}$$

and at $p = 16$ bars,

$$u_t = (1425.94/7.5)^{0.666} \times 46.88 \times 10^{-6}/(4.94 \times 0.3 \times 10^{-3}) = 1.04 \text{ m/s}$$

Heat Transfer Between Solid Particles and Gas Stream in Fluidized Bed

The heat transfer between solid particles and the gas stream in a fluidized bed proceeds very intensively due to the extended surface area of the particles. The Nusselt number for heat transfer between the gas (fluid) and the particles in a fluidized bed may be found as follows (Gogolek and Grace, 1995; Verweyen, 1994):

$$Nu_{fp} = 2 + \left(Nu_{lam}^2 + Nu_{turb}^2\right)^{0.5} \tag{9.9}$$

Thereby, for laminar and turbulent flows of a fluidizing fluid,

$$Nu_{lam} = 0.664 \, Pr^{2/3} \, Re_s^{0.5} \tag{9.10}$$

and

$$Nu_{turb} = 0.037 \, Re_s^{0.8} \, Pr/\left[1 + 2.443 \, Re_s^{-0.1}(Pr^{2/3} - 1)\right] \tag{9.11}$$

The Reynolds number Re_s is given by

$$Re_s = 18[(1 + 1/9 \, Ar^{0.5})^{0.5} - 1]^2 \tag{9.12}$$

Then the heat transfer coefficient between the gas (fluid) and particles is

$$h_{fp} = Nu_{fp} \, k/d \tag{9.13}$$

Example 9.3

Calculate the heat transfer coefficient between gas (fluid) and particles in CFBC at a temperature of 850°C and pressure of 16 bars if the bed porosity (void fraction) is $\varepsilon = 0.8$ and the Archimedes number is $Ar = 1425.94$ (see Example 9.2).

Solution

1. From Eqs. (9.12), (9.10), (9.11), and (9.9),

$$Re_s = 18[(1 + 1/9 \times 1425.94^{0.5})^{0.5} - 1]^2 = 29.46$$

$$Nu_{lam} = 0.664 \times 0.736^{2/3} \times 29.46^{0.5} = 2.938$$

$$Nu_{turb} = 0.037 \times 29.46^{0.8} \times 0.736/[1 + 2.443 \times 29.46^{-0.1} \times (0.736^{2/3} - 1)] = 0.601$$

$$Nu_{fp} = 2 + \left(Nu_{lam}^2 + Nu_{turb}^2\right)^{0.5} = 2 + (2.938^2 + 0.601^2)^{0.5} = 5$$

2. Then, from Eq. (9.13), the heat transfer coefficient between the gas and particles in CFBC is

$$h_{fp} = Nu_{fp} \, k/d = 5 \times 0.07414/0.0003 = 1236 \text{ W/(m}^2 \text{ K)}$$

Heat Transfer Between Fluidized Bed and Heat Transfer Surface

Figure 9.4 shows the heat transfer coefficient between a fixed bed, fluidized bed, or gas flow and a heat transfer surface as a function of superficial gas velocity. For a gas-solid systems—a fixed bed and a fluidized bed—the heat transfer coefficient is higher than for gas flow.

In the heat transfer between the fluidized bed and the heat transfer surface, the following three mechanisms are involved:

- heat transfer by gas (fluid) convection
- heat transfer by radiation
- heat transfer by particle convection

Hence the total heat transfer coefficient between the fluidized bed and the heat transfer surface (wall) is given by

$$h_{bw} = h_f + h_p + h_r \quad \text{W/m}^2 \text{ K} \tag{9.14}$$

where h_f is the heat transfer coefficient by gas (fluid) convection, h_p is the heat transfer coefficient by particle convection, and h_r is the heat transfer coefficient by radiation. The constituents of Eq. (9.14) are given below. The gas convective heat transfer coefficient for the atmospheric fluidized bed is (Gogolek and Grace, 1995; Verweyen, 1994):

$$h_f = 0.009 \, Pr^{1/3} \, Ar^{1/2} \, k/d \quad \text{W/m}^2 \text{ K} \tag{9.15}$$

where Pr is the Prandtl number of the fluidizing gas at the fluidized bed temperature, Ar is Archimedes number, and k is the thermal conductivity of fluidizing gas at the fluidized bed temperature in W/(m K).

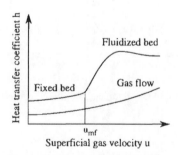

Figure 9.4. Heat transfer coefficient between a fixed bed, fluidized bed, or gas flow and a heat transfer surface as a function of superficial gas velocity.

In order to find the heat transfer coefficient by particle convection, a number of additional parameters must be calculated (Gogolek and Grace, 1995; Verweyen, 1994). The temperature dependence of the mean specific heat of coal particles is

$$c_s = 690.822 + 1.423512\, t - 0.58615 \times 10^{-3} t^2 \tag{9.16}$$

The dimensionless parameter is

$$Z = 1/6(\rho_p c_s/k)\{gd^3(\varepsilon - \varepsilon_{mf})/[5(1 - \varepsilon_{mf})(1 - \varepsilon)]\}^{0.5} \tag{9.17}$$

where ε and ε_{mf} are the fluidized bed porosities at the current gas velocity and at the minimum fluidization gas velocity, respectively. The effective free path length is

$$L = (2\pi TR)^{0.5} k/[p(2c_f - R)] \tag{9.18}$$

where T is the fluidized bed temperature in K, R is the gas constant (287 J/(kg K) for air), and p is pressure in Pa. The accommodation coefficient is

$$\gamma = 1/\{1 + \exp[0.6 - (1000/T + 1)/C_1]\} \tag{9.19}$$

Then the modified free path length is

$$l = 2L(2/\gamma - 1) \tag{9.20}$$

The second dimensionless parameter is

$$N = 4[(1 + 2l/d)\ln(1 + d/2l) - 1]/(C_2 Z) \tag{9.21}$$

The values of 2.8 and 2.6 may be assumed for C_1 and C_2 in the above equations.

Finally, the heat transfer coefficient by particle convection is given by

$$h_p = (k/d)(1 - \varepsilon)Z[1 - \exp(-N)] \tag{9.22}$$

The radiative heat transfer coefficient is approximately given by

$$h_r = 4\varepsilon_c \sigma \left(T_{fb}^4 - T_w^4\right)/(T_{fb} - T_w) \quad \text{W/(m}^2 \text{ K)} \tag{9.23}$$

where ε_c is the effective emissivity (practically, it can be taken as 0.9), $\sigma = 5.67 \times 10^{-8}$ W/(m^2 K^4) is the Stefan-Boltzmann constant, T_{fb} is the mean temperature of the fluidized bed in K, and T_w is the wall temperature of the heat transfer surface in K. The effective emissivity of the system fluidized bed wall is given by

$$\varepsilon_c = 1/[(1/\varepsilon_{gp}) + (1/\varepsilon_w) - 1] \tag{9.24}$$

With the emissivity of the fluidized bed

$$\varepsilon_{fb} = 0.5(1 + \varepsilon_p) \tag{9.25}$$

the total effective emissivity of the gas-particle system (fluidized bed) is

$$\varepsilon_{gp} = \varepsilon_{fb} + \varepsilon_p - \varepsilon_{fb}\varepsilon_p \tag{9.26}$$

Example 9.4

Calculate the heat transfer coefficient between the fluidized bed and the heat transfer surface in a CFBC boiler under the following conditions:

Pressure $p = 1 \times 10^5$ Pa
Bed temperature $t_{th} = 850°C$
Heat transfer surface (wall) temperature $t_w = 350°C$
Emissivities of the heat transfer surface and the particles $\varepsilon_w = 0.82$ and $\varepsilon_p = 0.9$
Particle diameter $d = 0.0003$ m
Particle density $\rho_p = 2400$ kg/m^3
Superficial air velocity $u = 1$ m/s
Average bed porosity (void fraction) $\varepsilon = 0.8$
Bed void fraction at minimum fluidization $\varepsilon_{mf} = 0.4$

Solution

1. From Example 9.2, the Archimedes number is 89.77 at 1 bar. The Prandtl number of air $Pr = 0.737$, the fluid (air) specific heat $c_f = 1163$ J/(kg K), the thermal conductivity of air at 1 bar and 850°C is $k = 0.07398$ W/(m K), the air gas constant $R = 287$ J/(kg K), and $C_1 = 2.8$.

2. From Eqs. (9.24)–(9.26),

$$\varepsilon_{th} = 0.5(1 + \varepsilon_p) = 0.5(1 + 0.9) = 0.95$$

$$\varepsilon_{gp} = \varepsilon_{th} + \varepsilon_p - \varepsilon_{th}\varepsilon_p = 0.95 + 0.9 - 0.95 \times 0.9 = 0.995$$

$$\varepsilon_c = 1/(1/\varepsilon_{th} + 1/\varepsilon_w - 1) = 1/(1/0.995 + 1/0.82 - 1) = 0.896$$

3. From Eq. (9.23), the radiative heat transfer coefficient is

$$h_r = 0.896 \times 5.67 \times 10^{-8}(1123^4 - 623^4)/(1123 - 623) = 146 \text{ W/(m}^2 \text{ K)}$$

4. From Eq. (9.15), the heat transfer coefficient due to gas (fluid) convection is

$$h_f = 0.009 \, Pr^{0.33} \, Ar^{0.5}k/d = 0.009 \times 0.737^{0.33} \times 89.77^{0.5}(0.07398/0.0003)$$

$$= 19 \text{ W/(m}^2 \text{ K)}$$

5. The auxiliary parameters required to find the heat transfer coefficient due to particle convection are given by Eqs. (9.16)–(9.21):

   (i) mean specific heat of coal particles at $t = 850°C$:

   $$c_s = 690.822 + 1.423512t - 0.58615 \times 10^{-3}t^2 = 1477 \text{ J/(kg K)}$$

   (ii) $Z = 1/6(\rho_p c_s/k)\{gd^3(\varepsilon - \varepsilon_{mf})/[5(1 - \varepsilon_{mf})(1 - \varepsilon)]\}^{0.5}$

   $$= 1/6(2400 \times 1477/0.07398)[9.81 \times 0.0003^3(0.8 - 0.4)/[5(1 - 0.4)(1 - 0.8)]^{0.5}$$

   $$= 106.12$$

   (iii) $L = (2\pi T R)^{0.5}k/[p(2c_f - R)] = (2\pi 1123 \times 287)^{0.5} \times 0.07398/[10^5(2 \times 1163 - 287)]$

   $$= 5.16 \times 10^{-7}$$

   (iv) $\gamma = 1/\{1 + \exp[0.6 - (1000/T + 1)/C_1]\}$

   $$= 1/\{1 + \exp[0.6 - (1000/1123 + 1)/2.8)]\} = 0.52$$

   (v) $l = 2L(2/\gamma - 1) = 2 \times 5.16 \times 10^{-7}(2/0.52 - 1) = 2.93 \times 10^{-6}$

   (vi) $N = 4[(1 + 2l/d) \ln(1 + d/2l) - 1]/(C_2 Z)$

   $$= 4[(1 + 2 \times 2.93 \times 10^{-6}/0.0003)$$

   $$\times \ln(1 + 0.0003/2 \times 2.93 \times 10^{-6}) - 1]/(2.6 \times 106.12) = 0.04$$

6. From Eq. (9.22), the heat transfer coefficient due to particle convection is

   $$h_p = (k/d)(1 - \varepsilon)Z[1 - \exp(-N)]$$

   $$= (0.07398/0.0003)(1 - 0.2)106.12[1 - \exp(-0.04)] = 205 \text{ W/(m}^2 \text{ K)}$$

7. From Eq. (9.14), the total heat transfer coefficient between the fluidized bed and the heat transfer surface (wall) is

$$h_{bw} = h_f + h_p + h_r = 19 + 205 + 146 = 370 \ W/(m^2 \ K)$$

The heat transfer has been less studied in the PFBC than in the atmospheric fluidized bed. However, the equations given above can be used for the calculation of the heat transfer coefficient in the PFBC with the appropriate gas properties at the fluidized bed pressure and temperature.

Example 9.5

Calculate the heat transfer coefficient between the fluidized bed and the heat transfer surface in a PFBC boiler if the fluidized bed pressure is 16×10^5 Pa and the temperature is 850°C. The other data are the same as in Example 9.4: heat transfer surface (wall) temperature $t_w = 350°C$, effective emissivity of the system fluidized bed-wall $\varepsilon_e = 0.896$, particle diameter $d = 0.0003$ m, particle density $\rho_p = 2400 \ kg/m^3$, superficial air velocity $u = 1$ m/s, average bed porosity $\varepsilon = 0.8$, bed porosity at minimum fluidization $\varepsilon_{mf} = 0.4$, the thermal conductivity of air at 16 bars and 850°C is $k = 0.07414 \ W/(m \ K)$, Ar $= 1425.94$ at 16 bars and 850°C (from Example 9.2).

Solution

1. The radiative heat transfer coefficient does not depend on the pressure, and thus $h_r = 146 \ W/(m^2 \ K)$.
2. From Eq. (9.15), the heat transfer coefficient due to gas convection is

$$h_f = 0.009 \ Pr^{0.33} \ Ar^{0.5} \ k/d = 0.009 \times 0.736^{0.33} \times 1425.94^{0.5} \times 0.07414/0.0003$$
$$= 76 \ W/(m^2 \ K)$$

3. From Eqs. (9.16)–(9.21), the auxiliary parameters are

   (i) $c_s = 1477 \ J/(kg \ K)$

   (ii) $Z = 1/6(2400 \times 1477/0.07414)$

   $\times [9.81 \times 0.0003^3(0.8 - 0.4)]/[5(1 - 0.4)(1 - 0.8)]^{0.5} = 105.89$

   (iii) $L = (2\pi 1123 \times 287)0.5 \times 0.07414/[16 \times 10^5(2 \times 1163 - 287)]$

   $= 3.23 \times 10^{-8}$

   (iv) $\gamma = 0.52$

   (v) $1 = 2 \times 3.23 \times 10^{-8}(2/0.52 - 1) = 1.84 \times 10^{-7}$

   (vi) $N = 4[(1 + 2 \times 1.84 \times 10^{-7}/0.0003)$

   $\times \ln(1 + 0.0003/2 \times 1.84 \times 10^{-7}) - 1]/(2.6 \times 105.89) = 0.08$

4. From Eq. (9.22), the heat transfer coefficient due to particle convection is

$$h_p = (k/d)(1 - \varepsilon)Z[1 - \exp(-N)]$$
$$= (0.07414/0.0003)(1 - 0.8) \times 105.89[1 - \exp(-0.08)] = 402 \ W/(m^2 \ K)$$

5. The total heat transfer coefficient between the fluidized bed and the surface (wall) is

$$h_{bw} = 76 + 402 + 146 = 624 \ W/(m^2 \ K)$$

In bubbling fluidized bed boilers, heat transfer surfaces used to raise steam are immersed in the fluidized bed. Higher gas and, especially, particle velocities in CFBC could drastically increase the erosion of immersed heat transfer surfaces. Therefore membrane tube-wall design is used for the evaporator in CFBC boilers instead of immersed surfaces.

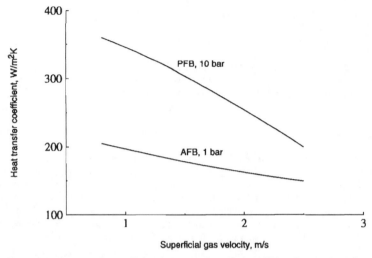

Figure 9.5. Heat transfer coefficient between atmospheric (AFB) and pressurized fluidized bed (PFB) and heat transfer surface.

The general equation that is valid for heat exchangers can be used for designing the heat transfer surfaces of fluidized bed steam generators. Thus the area required for the evaporator of a boiler placed within the fluidized bed furnace is given by

$$A = Q/U \Delta t_m \quad m^2 \tag{9.27}$$

where Q is the heat transfer rate in W, U is the overall heat transfer coefficient between the fluidized bed and the heat transfer fluid (HTF, i.e., water/steam) in W/(m^2 K), and Δt_m is the mean temperature difference between the fluidized bed and the HTF in K. The overall heat transfer coefficient is given by

$$U = 1/[1/h_{hw} + \delta_w/k_w + 1/h_i] \quad W/(m^2 \text{ K}) \tag{9.28}$$

where h_{hw} is the heat transfer coefficient from the fluidized bed to the tube outer surface in W/(m^2 K), δ_w is the tube-wall thickness in m, k_w is the thermal conductivity of the tube material in W/(m K), and h_i is the heat transfer coefficient from the tube inner surface to the HTF in W/(m^2 K). Usually, the thermal resistance of the tube wall, δ_w/k_w, may be ignored. Therefore the overall heat transfer coefficient is approximated as

$$U = (1/h_{hw} + 1/h_i) \quad W/(m^2 \text{ K}) \tag{9.29}$$

Comparison of the calculated values of the heat transfer coefficients between the fluidized bed and the heat transfer surface at 1 and 10 bars as a function of superficial gas velocity is given in Figure 9.5.

CFBC Power Generation Systems

A bubbling fluidized bed has a distinct height of fluidized material. It can operate at atmospheric pressure and higher, e.g., 12–16 bars. Coal and limestone are fed into the FBC above the gas distributing grate through which combustion air enters the bed. In a circulating FBC the superficial gas velocity should be high enough to enable the solid particles to be carried out of the furnace with the gas stream. The solid material is separated from the gas stream in a hot cyclone and recycled back to the bottom of the furnace. Spent bed material, consisting of ash, calcium sulfate, and unreacted or calcined limestone, is withdrawn from the bed at an appropriate rate.

The proportion of coal mass in the bed is only 2–3% of the total mass of the bed material. Owing to the large surface area of the particles in the fluidized bed, the rate of heat transfer between the gas stream and the particles is very high. Therefore a uniform temperature of about 850–900°C is measured throughout the bed. At this relatively low temperature, capture of SO_2 is promoted and NO formation is retarded.

Thus the most important feature of FBC boilers for power generation is the capability to capture SO_2 during combustion, by means of injected sorbent, and relatively low NO_x emissions. In addition, FBC boilers are capable of burning a range of different fuels, including low-grade lignite and solid combustible wastes. Compared to other coal combustion technologies, AFBC units may emit higher levels of the greenhouse gas N_2O and some air polluting organic compounds (Maude, 1993; Petzel, 1995).

CFBC systems use higher fluidizing velocities than AFBC boilers. This results in the circulation of the bed material, which is carried out of the bed with the combustion gases, separated from the gases in one or several hot cyclones placed at the combustor outlet, and recycled to the bottom of the bed. Small-size coal, limestone, and ash particles in the bed ensure a large gas-solid contact surface area and high heat transfer rates. Along with the recirculation of the material, this enables achieving high carbon burnouts and efficient sulfur capture with high calcium utilization rate.

Heat is extracted from the CFBC by means of heat transfer surfaces that form the membrane water tube walls of the furnace. Additional heat exchange surfaces may be installed at the cyclone wall. One or two external fluidized bed ash coolers are also used to extract the heat and to raise steam. Figure 9.6 shows the conventional design and dual leg design of circulating fluidized bed boilers with an external fluidized bed ash cooler.

Recirculation of finely ground fuel and limestone ensures high carbon burnout and high sulfur capture rates. For NO_x control in CFBC, in addition to the favorable effect of low combustion temperature, combustion air staging is used. Air staging limits NO formation by minimizing local oxygen concentrations in the combustor. Primary air is supplied through an air distributor at the bottom of the chamber, and secondary air is introduced at various height levels in the chamber.

The CFBC technology features (Minchener et al., 1991; Anon, 1995) the following:

- high fuel flexibility with a 98–99% carbon burnout
- low environmental impact, since on the one hand, over 90% of the sulfur dioxide can be removed in situ during combustion by adding small amounts of limestone or dolomite and, on the other hand, NO_x emission levels as low as 200 mg/m³ may be achieved due to low combustion temperature of about 850°C and combustion air staging
- single-unit capacities of up to 300 MWe achieved now
- compact design with small footprint of the plant
- quick start-up after relatively short still periods, e.g., over the weekend, due to high thermal capacity of the large inventory of the fluidized bed

Tables 9.1 and 9.2 contain performance data of the following CFB power plants: the Texas–New Mexico Power Company in Robertson County, Texas; the CFB power plant comprising two units with an electric capacity of 175 MWe each; the 125-MWe Emile Huchet CFB power plant, France, and the 100-MWe CFB Cogen Plant Bewag, Berlin, Germany.

The maximum unit size of CFBC boilers has not yet reached 300 MWe for hard coal. The 250-MWe rated CFBC power plant has been constructed at Gardanne in Provence, France. It has a thermal capacity of 557 MW and produces 700 t/h main steam at 163 bars and 565°C from feedwater at 249°C when firing lignite and hard coal. The steam plant uses reheat of 651 t/h of steam at 34.3 bars to 565°C (Petzel, 1995). After passing the convective portion of the boiler, the flue gases from the CFBC are cleaned of particulates in a downstream electrostatic precipitator or baghouse filter. Over 90% of the sulfur contained in the fuel is removed during combustion in the CFBC at Ca/S molar ratios of 1.6–2.0 (Foster Wheeler, 1994). The in situ desulfurization process is done with injected limestone and involves the following reaction steps.

Step 1: limestone calcination

$$CaCO_3 + Heat = CaO + CO_2 \qquad (9.30)$$

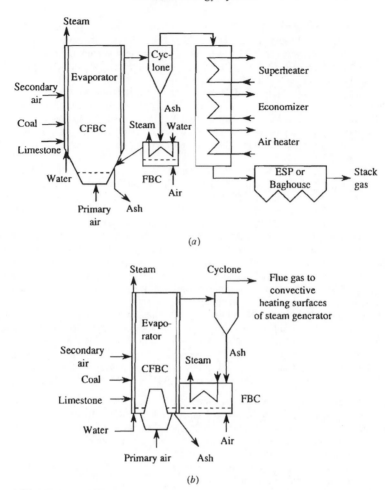

Figure 9.6. Circulated fluidized bed combustor boiler with fluidized bed cooler: (*a*) conventional design and (*b*) dual leg design.

Step 2: SO_2 capture with calcium sulfate formation

$$SO_2 + CaO + 1/2O_2 = CaSO_4 \qquad (9.31)$$

At the combustion temperature of 850°C in a CFB, the rate of the thermal NO formation is low. The fuel NO formation is inhibited by staged combustion. Therefore low NO_x emission levels (<200 mg/m³) are attained in CFBC.

PFBC Power Generation Systems

The current PFBC technology uses a bubbling fluidized bed boiler operating at 12–16 bars inside a pressure vessel in conjunction with a combined cycle. At these pressure levels, combustion is nearly 100% complete, even at low excess air levels (Karita, 1995; Stambler, 1993; Foster Wheeler, 1994).

Combustion air is compressed in a two-shaft intercooled gas turbine compressor and flows to the combustor. Coal is fed to the combustor as a coal/water paste. Some of the heat is used to

Table 9.1. Performance of the 175-MWe Texas–
New Mexico power plant

Parameter	Value
Power output, MWe	175
Thermal capacity, MWt	465
Fuel	lignite
LHV, MJ/kg	15.5
Live steam	
Mass flow, t/h	499
Pressure, bars	138
Temperature, °C	540
Reheat steam	
Mass flow, t/h	448
Pressure, bars	27
Temperature, °C	540
Feedwater temperature, °C	230
Emissions (upper limits), lbs/10^6 BTU	
Sulfur dioxide[a]	0.6/0.5
Nitrogen oxides[a]	0.6/0.2
Particulates[a]	0.03/0.01

[a] Guaranteed values/achieved values.

Table 9.2. Performance of the 125-MWe Emile Huchet CFB power plant, France,
and the 100-MWe CFB Cogen Plant Bewag, Berlin, Germany

Parameter	Emile Huchet	Bewag Berlin
Power output, MW	125	100
Thermal capacity, MW	285	222
Fuel	coal-water slurry/hard coal	lignite/hard coal
LHV, MJ/kg	10.5/20.3	16.9/28.8
Live steam		
Mass flow, t/h	367	326
Pressure, bars	126	196
Temperature, °C	542	540
Reheat steam		
Mass flow, t/h	338	269
Pressure, bars	30	42
Temperature, °C	539	540
Feedwater temperature, °C	242	300
Emissions (upper limits), mg/m^3		
Sulfur dioxide[a]	330/200	200
Nitrogen oxides[a]	300/200	200
Carbon monoxide	NA	250
Particulates	NA	20

NA, not applicable.
[a] Guaranteed values/achieved values.

generate steam in in-bed heat transfer tube bundles. The steam flows to the steam turbine, which generates approximately 80% of the plant's power output. Primary and secondary cyclones are utilized to remove most of the particulates from the flue gas before the gas enters the turbine. The blading for the gas turbine is made more rugged to accommodate a certain level of expected erosion by the particulate in the flue gas. The remaining particulates are then removed in a fabric filter before the gas is discharged out the stack.

There are two types of PFBC systems: bubbling and circulating. Figure 9.7 shows the pressurized bubbling and circulating fluidized bed combustors suitable for utilization in a combined-cycle power plant. The freeboard temperature in a bubbling bed system varies with operating load

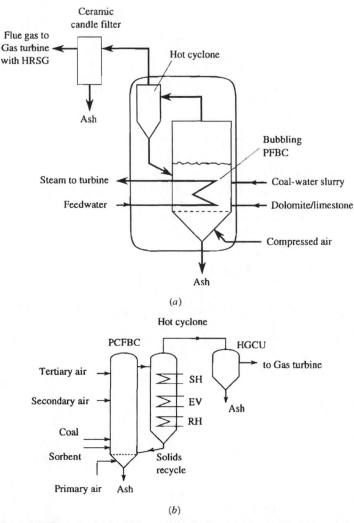

Figure 9.7. Pressurized (a) bubbling and (b) circulated fluidized bed combustor in a combined-cycle power plant.

because of the convective cooling effect of the flue gases when reducing load and therefore bed height.

PCFB technology utilizes a circulating fluidized bed boiler operating under similar conditions. PCFB may have several operating advantages over bubbling bed technology. Thermal releases are higher because of the higher velocities in the furnace, resulting in a more compact combustor for the PCFBC system.

PFBC technology is in the early stages of commercialization. Five PFBC units of less than 80 MW, two in Sweden, one in Japan, one in Spain, and one in the United States, have been operating during the last few years. PCFBC technology has the following advantages (Foster Wheeler, 1994):

- very high carbon burnout (>99%) for all coals—high volatile bituminous and subbituminous coal and chars from a pyrolizer (carbonizer) plant

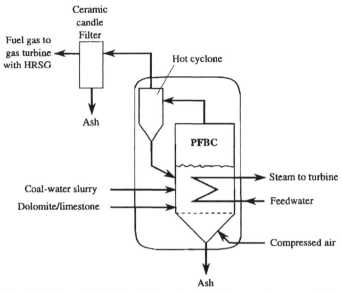

Figure 9.8. Pressurized fluidized bed combustor with hot gas cleanup consisting of a hot cyclone and a ceramic candle filter.

- very low CO emissions compared to atmospheric CFBC, i.e., <0.01 kg/GJ, which decrease by half when the combustor temperature increases from 880°C to 925°C
- sulfur capture efficiencies above 96% achieved with sorbent addition rates of Ca/S between 1.1 and 2
- excellent performance of the barrier candle filter (clay bonded SiC candles) for high-temperature gas cleanup

Its main disadvantage is relatively high NO_x emissions because of the short height of the PCFBC, which, however, is about half of the NSPS requirement. The primary air stoichiometry has the strongest effect on NO_x emissions. Alkali concentrations are several times higher than the gas turbine limit and thus a getter is required.

A combination of a hot cyclone and ceramic candle filter used for hot gas cleanup in PFBC is shown in Figure 9.8. Hot gas cleanup system of particulates in an PFBC using two-stage cyclones is shown in Figure 9.9.

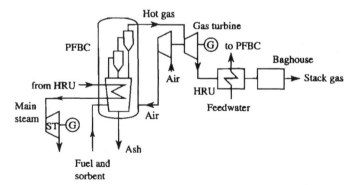

Figure 9.9. Combined-cycle power plant with pressurized fluidized bed combustor using two-stage cyclones for hot gas cleanup.

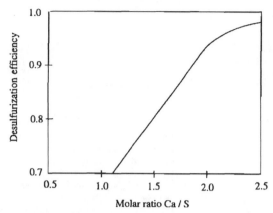

Figure 9.10. Desulfurization efficiency as a function of the molar Ca/S ratio.

The gas turbine design must allow for removal of air from the compressor for combustion and to feed hot pressurized clean flue gas for expansion in the gas turbine. The steam turbine generates about 80% of total output power, and the flue gas flowing through the gas turbine provides the drive power for the compressor and the remaining 20% of the electric output power.

The first utility-scale PFBC demonstration project in the United States, the 70-MWe Tidd plant in Brilliant, Ohio, has been in operation since 1991. The main early operating problems experienced at the Tidd plant are attributed to coal preparation and cyclone ash removal. Essential improvements in these systems are needed for commercialization of the technology. The requirement of at least 95% SO_2 removal at a Ca/S molar ratio of less than 1.6 presents another technological challenge.

Owing to low combustion temperatures and staged combustion, PFBC technology exhibits low NO_x emissions. However, if emissions standards will be further lowered, the utilization of the selective catalytic reduction (SCR) process described in Chapter 3 may become necessary. Reburning in the freeboard zone of the combustor may be required at part load.

Two larger commercial PFBC plants were put under construction in 1994: a 350 MWe in Japan and a cogeneration plant in the Czech Republic, which will produce 690 MWe of power, 135 MWt heat for district heating, and 250 t/h of process steam.

In the effort to commercialize PCFB technology by the year 2000, emphasis is being placed on the development of an efficient hot gas cleanup (HGCU) system. It consists of three cleaning stages in series: cyclone, ceramic filter, and alkali removal. In order to protect the topping combustor and gas turbine from corrosion, erosion, and deposition, the following HGCU performance requirements must be met: (1) particulates—reduction to meet 15 ppmw (parts per million by weight) limit in the stack gas, with less than 2 ppmw of particles larger than 5 microns for turbine protection; and (2) control of alkali vapor to less than 500 ppbw (parts per billion by weight) in the gas stream entering the turbine. ABB Carbon and Babcock and Wilcox are working on improvements of the bubbling PFBC design, while Ahlstrom Pyropower, Foster Wheeler, and Lurgi Lentjes Babcock are further developing the circulating PFBC design (Karita, 1995; Petzel, 1995; Foster Wheeler, 1994; Anon, 1995). Figure 9.10 shows how the desulfurization efficiency depends on the molar Ca/S ratio in an FBC.

IGCC TECHNOLOGY

The next generation of coal-fueled power plants must reduce the emissions of such pollutants as sulfur dioxide (SO_2), nitrogen oxides (NO_x), and particulates, as well as of greenhouse gas CO_2 to protect the atmosphere. The latter can be achieved through improving the conversion efficiency, which is possible in IGCC systems. Hence the objective of IGCC technology is the

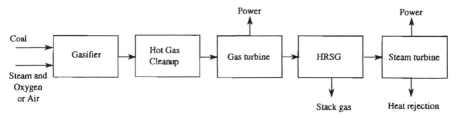

Figure 9.11. Basic flow diagram of an IGCC plant.

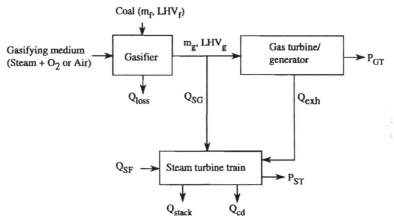

Figure 9.12. Energy flow diagram of an IGCC plant.

integration of coal gasification into a combined-cycle power plant with the goals of achieving high efficiencies of coal-to-power conversion, reducing the environmental impact, and maintaining favorable economics of power generation. Figure 9.11 shows a basic flow diagram of an IGCC plant, whereas Figure 9.12 shows its energy flow diagram.

The coal-derived fuel gas, after cleaning, will be burned like natural gas in a gas turbine. Depending on the type of gasification process used—moving fixed bed, fluidized bed, or entrained flow—the degree of integration into the combined-cycle power plant varies. To date, most IGCC plants use the entrained flow gasifier, which converts finely ground coal at temperatures of 1500–1900°C to a fuel gas. Dry coal or a coal-water slurry is fed to the pressurized gasifier. Carbon conversion is nearly 100%, and ash is removed as a stable, nonleachable slag. A large amount of sensible heat should be removed from the high-temperature raw fuel gas. This heat is used to raise steam for the IGCC.

Rather high efficiencies of IGCC plants result from the high efficiency of combined-cycle plants based on advanced second-generation combustion turbines with a gas turbine inlet temperature (TIT) of approximately 1250°C. The high combined-cycle efficiency reduces the size of the coal gasification plant and the plant capital cost. More advanced blading materials and improved cooling techniques would allow temperatures up to 1350–1450°C, and potentially as high as 1550°C. With a TIT of 1350°C, the IGCC plant net heat rate is anticipated to be improved by roughly 350 kJ/kWh (Jones, 1995; Anand et al., 1996).

Gasification Process Improvements

Table 9.3 shows the temperature conditions and the status of main gasification processes. Fluidized bed gasifiers operate at 760–1040°C. At these temperatures the raw product gas can be subjected to a

Table 9.3. Major gasification processes

Process	Exit gas temperature, °C
Entrained flow[a]	
Texaco (US), Shell (Europe/US), Destec (US)	
Prenflo (Germany), Koppers-Totzek (Germany)	1040–1540
Fluidized bed[b]	
KRW (US), HTW/Lurgi (Germany)	760–1040
Moving fixed bed	
Lurgi, dry ash (Germany),[a] British Gas Lurgi, slagging[b]	

[a] Commercial.
[b] Demonstration.

hot gas cleanup without precooling, and thus the sensible heat of the gas stream can be utilized in the gas turbine. Hence fluidized bed gasifiers potentially have a higher overall efficiency in comparison with high-temperature entrained flow gasifiers that require gas cooling prior to cleanup. Compared to moving bed gasifiers, fluidized bed units also have advantages such as higher coal throughput rates, which reduce the unit size and cost. The high-pressure, high-temperature Winkler (HTW) fluidized bed gasification process has been developed by Lurgi/Rheinbraun, Germany (Kallmeyer and Engelhard, 1992; Anon, 1995).

In a moving fixed bed gasification process, coal with coarse pieces up to approximately 50 mm moves down the reactor and passes subsequently through all the temperature zones, beginning with the drying zone, followed by the coal devolatilization, gasification, and combustion zones. Owing to the countercurrent flow, the ascending product gas leaves the gasifier at a relatively low temperature, thus contributing to a rather high overall efficiency. However, some pyrolysis products such as methane, light hydrocarbons, and tar escape oxidation, and subsequent removal of the tar is required. Also, fixed bed gasifiers are rather complex and costly plants.

There are two fixed bed gasification processes: the Lurgi dry ash process and the British Gas/Lurgi slagging process. In the former, the dry ash is removed through the slots in a rotating grate, whereas in the latter, the slag is removed in a molten state. Fixed bed units have lower coal throughput than fluidized bed gasifiers. Commercial Lurgi moving bed gasifiers reach capacities of 800–1000 t of coal/d.

Overall IGCC efficiency can be improved by enhancement of gasification efficiency. In order to increase the heating value of the fuel gas, advanced gasification processes employ dry feed and/or two-stage gasifier designs. A major portion of the sensible energy content of the fuel gas leaving the gasifier can be saved by using efficient HGCU and appropriate fuel gas precooler designs. Because of energy losses inherent in gasification, the overall efficiency of an IGCC is reduced by about 15–20% in comparison with natural gas fired combined-cycle plants. The overall (electrical) efficiency of an IGCC is given by

$$\eta_o = \eta_{gas}\eta_{th} \tag{9.32}$$

where η_{gas} is the efficiency of gasification and η_{th} is the thermal efficiency of the cycle.

The efficiency of gasification η_{gas} is defined as the ratio of the chemical and sensible energy content in the clean product gas to that of the coal. It exceeds 80%. Because of energy losses inherent in gasification, the overall efficiency of an IGCC power plant is about 15–20% lower than that of natural gas fired combined-cycle plants. Values of η_o as high as 42.5–46% or even higher can be achieved in advanced IGCC power plants (see Table 9.4) (Smith, 1992; Stambler, 1993; Tait and McDonald, 1994).

The air separation unit (ASU) is capital intensive and consumes 8–10% of the plant gross power output. There are several potential ways to enhance the ASU performance and economics. Most ASU designs produce high-purity (99.5%) oxygen. For IGCC applications, lower purity (85–95%) designs can be employed, as they use less auxiliary power and cost less. The gas turbine air compressor will supply air to the ASU, oxygen is fed to the gasifier, and nitrogen to the gas turbine, where it is used for NO_x control in place of steam diluent. The nitrogen lowers the

Table 9.4. Efficiency of IGCC plants with fluidized bed and entrained flow gasifiers

| | Fluidized bed/KRW | | Entrained flow/ABB CE | | |
| | Air in situ + CGCU | HGCU | Oxygen HGCU | Air | |
Efficiency				CGCU	HGCU
Gasification efficiency (based on HHV), %	81.3	86	85	79	84.5
Overall (electrical) efficiency (based on HHV), %	44.5	46	43	42.5	44.5

KRW, Kellogg-Rust-Westinghouse; ABB CE, ABB/Combustion Engineering; CGCU, cold gas cleanup; HGCU, hot gas cleanup; HHV, high heating value.

gas turbine combustor flame temperature and thus reduces NO formation. This improves cycle efficiency and has a slight benefit in capital cost. Using compressed nitrogen for NO_x control in place of steam diluent also significantly reduces process make-up water.

Two options are used for cooling high-temperature product gas: (1) quench process with cooling by means of water sprays, and (2) cooling by means of a waste heat boiler (WHB) with production of high-pressure steam. The quench process is simple and reliable, but thermal energy of the hot gas is not recovered owing to direct-contact water spraying of the gas. In the WHB process, heat of the high-temperature product gas is used to raise high-pressure steam, and thus it provides higher overall efficiency. However, the WHB process is more complicated and costly.

Gasification is carried out at high pressures (up to 30–80 bars), and therefore large-capacity IGCC systems can use smaller equipment with reduced specific capital costs per kW electric power output. High-pressure operation provides an additional advantage, since it enables the utilization of physical solvents in place of chemical solvents for H_2S removal.

Hot Gas Cleanup

The raw coal derived gas contains particulate matter, sulfur compounds, and other harmful constituents, such as alkali metals (sodium and potassium), which can cause high-temperature corrosion of gas turbine blades if the gas will be burned in a gas turbine combustor. Therefore raw gas must be cleaned before it can be burned in the gas turbine. Two cleaning methods may be employed (Gudenau et al., 1994; Ishikawa et al., 1993; Lippert et al., 1992; Minchener et al., 1991; Newby and Banniser, 1993): cold gas cleanup and hot gas cleanup. Both methods involve gas cooling with partial heat recovery to raise steam. However, cooling the gas reduces the overall efficiency because the gas must then be reheated before combustion. Conventional cold gas cleanup designs effectively control all kinds of air pollutant emissions and also remove alkali metals, but they involve removal of large quanities of sensible heat while cooling the gas to a low temperature of gas scrubbing. Providing gas cleanup at the high temperature and pressure makes the pressurized gas compatible with the gas turbine while meeting fuel specifications. HGCU of sulfur dioxide and particulates at elevated pressures greatly reduces the volume of gases to be cleaned. HGCU occurs in ceramic candle filters for particulate removal and in a barium titanate system for sulfur removal at about 550°C or higher, and thus plant net heat rates below 8000 kJ/kWh can be achieved.

IGCC systems based on the Destec and Texaco entrained flow gasifiers have a low overall efficiency of 38 and 40%, respectively (Longwell et al., 1995). The Kellogg-Rust-Westinghouse (KRW) air-blown process is scheduled for demonstration at the Sierra Pacific Power Company, and the two-stage ABB/Combustion Engineering process is scheduled for demonstration at City Water, Light and Power in Springfield, Illinois. The first stage of the entrained flow system operates at 1480–1650°C and produces a molten slag. The second-stage gas leaves at 1070°C and is then cooled to a temperature of 540–590°C, suitable for HGCU with zinc titanate/zinc ferrite sulfur removal and candle filters (Longwell et al., 1995).

For the air-blown systems, use of HGCU in place of cold gas cleanup results in an overall efficiency gain of about two percentage points (Maude, 1993). Thus the most efficient system now is the air-blown fluidized bed gasifier with HGCU plus in-bed sulfur removal. A gain of over three

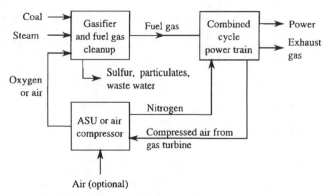

Figure 9.13. Block diagram of an IGCC plant with air-blown or oxygen-blown gasifier.

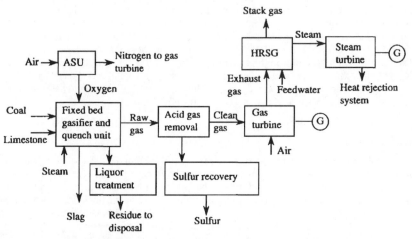

Figure 9.14. IGCC plant with moving fixed bed oxygen-blown gasifier, HGCU, gas turbine, HRSG, and steam turbine.

percentage points in net electrical efficiency is achieved compared to the oxygen-blown entrained flow gasifier with cold gas cleanup. However, carbon dioxide emissions increase by 4.5% owing to the calcination of limestone in the fluidized bed gasifier.

Oxygen-blown systems produce about half the gas volume of an air-blown system but require energy for oxygen manufacture in an ASU. Therefore air-blown systems appear to be more attractive for IGCC power generation plants. A block diagram of an IGCC plant with an air-blown or oxygen-blown gasifier is shown in Figure 9.13. Figure 9.14 shows a flow diagram of an IGCC plant with moving fixed bed oxygen-blown gasifier, HGCU, gas turbine, HRSG, and steam turbine.

The gasifer throughput should be increased to reach a goal of one gasifier per combustion turbine. In most current commercial gasifiers, oxygen rather than air is used as the oxidant. However, air-blown gasifiers with HGCU can potentially improve both efficiency and cost of IGCC systems, particularly in the 80–150 MW plant capacity range. Using air-blown gasifiers in IGCC systems would avoid the high cost of an ASU.

The net efficiency of advanced combined-cycle power plants that are operated on natural gas approaches 58% (HHV) or 60% (LHV). It is impossible for IGCC plants to achieve these efficiency levels, especially at comparable emissions values.

Table 9.5. IGCC projects in Europe (all systems have cold gas cleanup)

Project/site	Power, MW	Gasification technology	Efficiency, %	Startup
Demkolec	253	Shell entrained flow[a]	41	1994
Elcogas	300	Prenflo entrained flow[b]	43	1997
KoBra, RWE	312	HTW fluidized bed[c]	43	2000

[a]Oxygen-blown, Siemens gas turbine V94.2.
[b]Oxygen-blown, Siemens gas turbine V94.3.
[c]Air-blown, Siemens gas turbine V94.3.

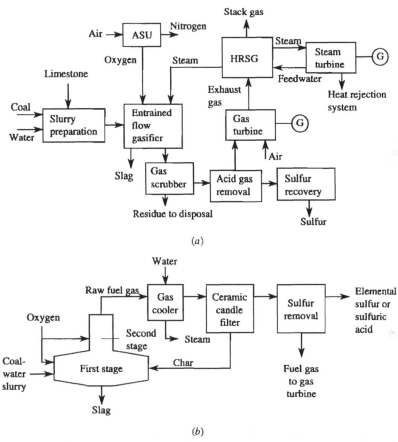

Figure 9.15. Two-stage entrained flow gasifier with (a) cold and (b) hot gas cleanup for a combined-cycle power plant.

Figure 9.15a shows a basic flow diagram of an IGCC based on an entrained flow gasifier with cold cleanup. An IGCC using a two-stage entrained flow gasifier with HGCU is shown schematically in Figure 9.15b. Figure 9.16 shows an IGCC plant with fluidized bed oxygen-blown gasifier, HGCU, gas turbine, HRSG, and steam turbine.

The general characteristics of three IGCC projects in Europe—Demkolec (SEP) at Buggenum, the Netherlands; Elcogas at Puertollano, Spain; and KoBra at Huerth, Germany—are presented in Table 9.5 (Longwell et al., 1995). The performance and economic goals of the U.S. Department of Energy's combined-cycle technology (CTT) (DOE's) Program related to IGCC, integrated

Table 9.6. Performance and economic goals for near- and mid-term integrated
gasification-based systems

	IGCC	IGAC	IGFC
Efficiency, %	45	≥50	≥60
SO₂/NO$_x$ emissions, fraction of NSPS	1/10	1/10	1/10
Predicted capital cost, $/kW		1200	1050 1100

Source: U.S. Department of Energy.
  IGCC, integrated gasification combined cycle; IGAC, integrated gasification advanced cycle;
IGFC, integrated gasification fuel cell; NSPS.

Table 9.7. Pulverized coal systems: Performance and economic goals

	PCLEB	PCCG	EFCC/HIPPS	DCFGT	DCFD
Efficiency, %	42	70	50	40	45
Capital cost, $/kW	1400	NA	1200	1400	1300

Source: Longwell et al. (1995).
  PCLEB, pulverized coal, low-emission boiler; PCCG, pulverized coal, cogeneration plant; EFCC/HIPPS, externally fired combined cycle/high-performance power system; DCFGT, direct coal-fired gas turbines; DCFD, direct
coal-fired diesel; NA, not applicable.

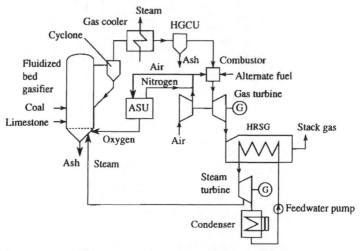

Figure 9.16. IGCC plant with fluidized bed oxygen-blown gasifier, HGCU, gas turbine, HRSG, and steam turbine: ASU, air separation unit.

gasification advanced cycle (IGAC), and integrated gasification fuel-cell (IGFC) power generation technologies are presented in Table 9.6. The performance and economic goals of the DOE's
CCT Program related to pulverized coal power generation technologies are presented in Table 9.7.
The following systems are included: advanced pulverized coal low-emission boiler (PCLEB),
pulverized coal fired cogeneration (PCCG), externally fired combined-cycle/high-performance
power system (EFCC/HIPPS), direct coal-fired gas turbines (DCFGT), and direct coal-fired diesels
(DCFD).

Thus IGCC offers a coal-based power technology with low emissions, high thermal efficiency,
and the potential for phased construction, that is, building simple-cycle natural gas fired combustion
turbines first, then converting to combined-cycle, and finally adding coal gasification as gas prices
increase or gas availability deteriorates. Future advances in gasification-based power production
are linked to increases in gas turbine firing temperature, HGCU of the fuel gas, coproduction
of both chemicals and electricity, improved gasifier designs, and integration of gasification with
advanced gas turbine cycles and fuel cells.

Table 9.8. Emissions from advanced lignite-fired power plants

Pollutant	Emissions, mg/m^3		
	PFBC	PCFBC	IGCC
SO$_2$	150	<100	<25
NO$_x$	130	<120	<150
CO	10	<50	<3.7
Particulates	15	<5	7.5

PFBC, pressurized fluidized bed combustion; PCFBC, pressurized circulating fluidized bed combustion; IGCC, integrated gasification combined-cycle plant.

Construction of the 300-MW IGCC Elcogas demonstration plant is under way at Puertollano in Spain (Siemens, 1995b). The project sponsored within the European Thermie Program features an oxygen-blown Krupp Koppers Prenflo gasifier (Germany) with an ASU of Air Liquide, France, a gas cleanup system, and a Siemens combined-cycle power train. The combined-cycle plant consists of a gas turbine generator, a triple-pressure reheat HRSG of the combined-circulation drum type, and a steam turbine generator with auxiliaries. Nitrogen from ASU is used in the gas turbine to augment the power output. The Siemens V94.3 gas turbine with two horizontally opposed silo-type combustors is used. High efficiency and low environmental impact are expected to be achieved. The NO$_x$ emissions control involves steam injection.

The 1000-MWe IGCC power plant at Tapada, Portugal, features state-of-the-art natural gas fired, highly efficient power technology (Siemens, 1995a). The plant consists of three single-shaft combined-cycle units, each with a Siemens V94.3A gas turbine, a natural-circulation heat-recovery steam generator, and a reheat condensing steam turbine driving a common generator. The gas turbine and generator are connected by a rigid coupling, whereas the steam turbine is provided with a synchronous clutch. The modular design with three identical single-shaft combined-cycle units enables completion of the plant construction within a short specified period of time at a low capital cost. Data on the major air pollutant emissions from advanced lignite-fired power plants are presented in Table 9.8 (Longwell et al., 1995).

Cogeneration IGCC plants that generate power and produce steam, heat, or chemicals have a potential of improving the cost-effectiveness of the system. The most efficient combined cycles in the long term would be IGFC systems. Fuel cells can be easily integrated with coal gasifiers. Such IGFC systems are potentially the most efficient and least polluting method to generate electricity from coal. Molten-carbonate fuel cells (MCFCs) are the most attractive near-term systems for utility applications. Fuel cells are discussed in Chapter 8 in more detail.

Since 1995, several demonstration fuel-cell power plants ranging in size from 200 kW to 2 MW have been operating in the United States and Japan. These natural gas fueled plants are expected to enter the utility market as 1- to 5-MW units with electrical efficiency greater than 50% by the year 2000. In larger sizes, with steam or gas bottoming cycles, efficiency will be 60% or higher.

Solid-oxide fuel cells (SOFCs) operate at temperatures of about 950–1000°C. The largest SOFC-based system now is the Westinghouse tubular design usable for units of up to 20-kW capacity. Characteristics of emerging IGFC technologies based on the three types of fuel cells integrated with coal gasifiers are presented in Table 9.9. Such systems based on MCFC or SOFC fuel cells

Table 9.9. Integrated gasification fuel-cell system characteristics

Parameter	PAFC	MCFC	SOFC
Plant capacity, MW	150	440	300
System efficiency on coal (HHV), %	34	51	47
Capital cost, $/kW	2200	1900	2100
Demonstration on natural gas	1993	1998	1998

PAFC, phosphoric acid fuel cell; MCFC, molten-carbonate fuel cell; SOFC, solid-oxide fuel cell.
Emissions (NO$_x$ and SO$_x$) are less than 0.5 kg/MWh.

would comprise the following units (Longwell et al., 1995):

- coal gasification plant to produce syngas
- reformer to convert the coal gas to hydrogen
- fuel-cell stacks for electrochemical energy conversion and generation of direct current
- current conditioning unit
- bottoming gas turbine or steam turbine plant
- a process steam or district heat unit

IGFC combined-cycle plants could achieve heat rates below 7500 kJ/kWh. However, to achieve this, fuel-cell stack production costs must be drastically reduced.

HYBRID COMBINED CYCLES

First-generation PFBCs have relatively low efficiency because of the low gas turbine inlet temperature of 850°C. Therefore hybrid cycles including PFBC integrated with air-blown gasification and high-temperature gas turbines will achieve higher overall efficiencies. The general process in a hybrid power generation system involves the following major steps: coal carbonization, char combustion in a PCFBC, coal gas combustion in a topping gas turbine combustor, and combined-cycle power generation (Maude, 1993; Takahashi et al., 1994; Nakabayashi, 1994; Fujita, 1994; Goto, 1995; Matsuda, 1995).

Thus the hybrid power generation technology incorporates key elements of PFBC and coal gasification processes. Coal is first devolatilized/carbonized. Then the LHV fuel gas produced by this process is burned in a gas turbine topping combustor, and the char is injected into the PCFBC furnace. Gas turbine inlet temperatures above 1150°C are achieved due to this combination. The carbonizer, combustor, and particulate-capturing HGCU systems operate at 850–875°C. Sulfur capture is accomplished by injection of lime-based sorbents. It is anticipated that the efficiency of the hybrid cycle will be about four percentage points higher than the efficiency of the first-generation PFBC systems.

A major portion of the heat transfer surfaces is placed in an external fluidized bed heat exchanger. Since the fluidizing velocity gas there is low, e.g., 0.15 m/s, the tube erosion is negligible. Before entering the gas turbine topping combustor and the gas turbine, the flue gases leaving the carbonizer and the char PCFB combustor must be cleaned in an HGCU system to a solid particle concentration below 20 ppm. In both pressurized combustion and gasification, sulfur capture at high pressure is more efficient than at 1 bar, provided that the sulfation time is long enough. Although this concept could lead to a higher efficiency than integrated coal gasification combined-cycle (IGCC) power plants (because steam conditions are often subcritical), the environmental performance is the same as that for a conventional fluidized bed power plant. SO_2 emissions cannot be as low as for those for an oxygen-blown IGCC with cold gas cleanup. The ash particles, which are not molten in the process, contain varying proportions of $CaSO_4$ and unreacted limestone. A hybrid IGCC power generating system is shown in Figure 9.17. It comprises a pressurized circulating fluidized bed (PCFB) coal carbonizer for partial coal gasification producing fuel gas and char, PCFB char combustor, two hot gas cleanup units (HGCU) for fuel gas and products of char combustion, respectively, fuel gas-fired gas turbine topping combustor and combined cycle power plant consisting of gas turbine train, heat recovery steam generator (HRSG) and steam turbine train.

Two other varieties of advanced coal cycles that are being developed now are the direct coal combustion system (DCCS) and the so-called clean coal combustion system (CCCS) for gas turbine based combined-cycle plants (Jones, 1995).

Pressurized slagging combustors are being developed that would operate at 1370–1650°C and remove the ash and harmful constituents during the combustion process. This is a long-term development. At the present time, most of these concepts include a gasifier as the first stage, an HGCU system, and a staged combustion system. The CCCS comprises three stages of combustion. The first stage is fuel-rich combustion. All the available O_2 will be used only for the oxidation of carbon and hydrogen. Sulfur in the fuel will form calcium sulfide, and fuel nitrogen will convert to molecular N_2. The next stage is also fuel-rich. Calcium sulfide will be encapsulated in molten slag and removed from the combustor. The carbon burnout will be completed in the last stage

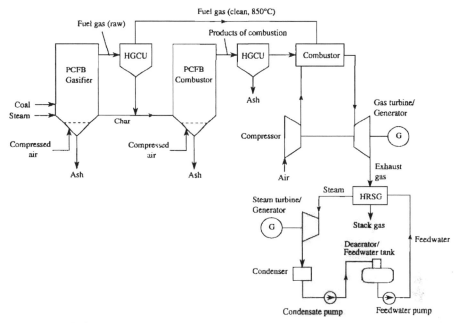

Figure 9.17. IGCC system consisting of pressurized circulating fluidized bed (PCFB) coal gasifier (carbonizer), PCFB char combustor, gas cleanup units (HGCU), gas turbine topping and gas turbine-based combined cycle power plant including gas turbine, heat recovery steam generator (HRSG) and steam turbine.

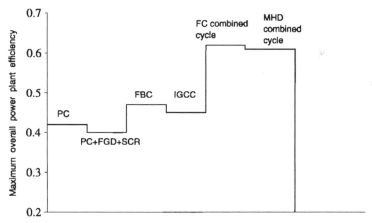

Figure 9.18. Efficiency comparison of advanced power plant concepts.

under excess-air conditions. A portion of the flue gas will be recirculated to prevent the formation of thermal NO_x.

Efficiencies of advanced power plant concepts are compared in Figure 9.18. The general performance comparison of advanced coal-based power generation systems is presented in Table 9.10. Figure 9.19 shows efficiencies and operational temperature ranges of advanced energy systems of all types. Capital costs of utility power plants and cogeneration plants are compiled in Table 9.11. Table 9.12 presents a general overview of CO_2 abatement techniques that are applied in the field

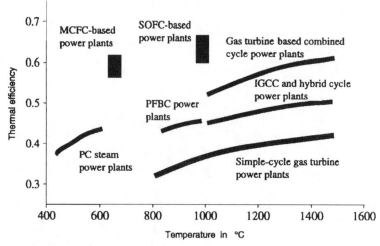

Figure 9.19. Efficiencies and operational temperature ranges of advanced power plants and IGCC plants.

Table 9.10. Performance comparison of advanced coal-based power generation systems

Technology	Efficiency, %	Coal conversion technology	Power generation technology
PC	38–42	C	240–300 bars, ST
PCLEB	42	C	300 bars, ST
PFBC	40	PFBC	130 bars, CC
IGCC	40	Entrained bed gasifier	1300°C, CC
EFCC	45	Slagging combustor + 1260°C heat exchanger	1300°C, CC
PFBC-2	45	Carbonizer + PCFBC	1300°C, CC
IGCC-2	45	Fluidized bed gasifier	1370°C, CC
HIPPS	50	HTAF	1400°C, CC
Improved PFBC-2	>50	PCFBC	1450°C, GT + 300 bars, ST
IGAC	>50	Fluidized bed gasifier	1450°C, HAT
IGFC	>60	Fluidized bed gasifier	MCFC (650°C) + HRSG/ST

PC, pulverized coal power plant; C, conventional coal combustion; ST, steam turbine; GT, gas turbine; CC, combined cycle (gas turbine + HRSG + steam turbine); HRSG, heat recovery steam generator; PCLEB, pulverized coal, low-emission boiler; PFBC, pressurized fluidized bed combustion; IGCC, integrated gasification combined cycle; EFCC, externally fired combined cycle; HIPPS, high-performance power system; HTAF, high-temperature air furnace; IGAC, integrated gasification advanced cycle; PCFBC, pressurized circulating fluidized bed combustion; IGFC, integrated gasification fuel cell; HAT, humidified air turbine; MCFC, molten-carbonate fuel cell.

Table 9.11. Capital cost of various types of power plants

Technology	Capital cost, $/kW
Steam power plant with FGD	1000–1500
Simple-cycle gas turbine power plant	350–550
Combined-cycle power plant	450–670
Gas turbine based cogeneration plant	450–670
Extraction steam turbine based cogeneration plant	1000–1600
IGCC plant	1200–1650
Nuclear power plant	2000–2700

FGD, flue gas desulfurization; IGCC, integrated gasification combined cycle.

Table 9.12. CO_2 abatement techniques for power generation

Technology	Steam conditions	Net plant efficiency (LHV), %	CO_2 emissions, g C/kWh
PC + FGD	Sb	35	262
AFBC	Sb	37	257
PC + FGD	Sc, 25 MPa/560°C/0.003 MPa	45	210
PC + FGD	Sc, 30 MPa/600°C/0.003 MPa	48	200
PC + natural gas firing	Sb	36–38	233–242
PC, coal-oil mixture	Sb	35	252
IGCC plant	Sb	43	213
PFBC plant	Sb	42	230
PFBC gasification cycle	Sc, 25 MPa	42–45	196–200
Pressurized PC plant	Sb	48	195
PC − natural gas + FGD, $deNO_x$	Sb	40	202
PC − natural gas + FGD, $deNO_x$	Sc, 30 MPa	49	166
MHD + steam plant	Sb	45–55	193–217
Fuel cells	Sb	45–58	147–193
Cogeneration	Sb	77–85	110–119

PC, pulverized coal power plant; FGD, flue gas desulfurization; Sb, subcritical steam parameters; Sc, supercritical steam parameters; AFBC, atmospheric fluidized bed combustion; IGCC, integrated gasification combined cycle; $deNO_x$, NO_x-removal plant; MHD, magnetohydrodynamic generator; PFBC, pressurized fluidized bed combustion.
Source: Longwell et al. (1995), Maude (1993), and Fujita (1994).

of coal-based power generation technologies (Longwell et al., 1995; Maude, 1993; Petzel, 1995; Jones, 1995).

CLOSURE

This chapter deals mainly with fluidized bed combustion and IGCC technologies. The largest CFBC plants built so far have a power output of 175 MW, whereas the required capacity is 300 MW. Both PFBC and IGCC technologies are still in the early stages of commercialization. Because of the limited combustion temperature (around 850°C), combined-cycle power plants with PFBC have rather low efficiency. The efficiency of the IGCC should also be increased, particularly by employing HGCU. To increase the efficiency of both PFBC and IGCC power plants, reliable HGCU is required. Efficiencies of around 45% can be achieved. Some pilot plant projects with PFBC and IGCC technology are now being realized. Major coal gasification process options, both air blown and oxygen blown, discussed in this chapter, are being used in current IGCC technology.

PROBLEMS

9.1. Calculate the pressure drop across an atmospheric circulating fluidized bed combustor (CFBC) with an average bed porosity of 0.87 at a bed height of 8 m. Assume a paricle density of 2300 kg/m^3.

9.2. Calculate the minimum fluidization velocity for coal particles with a diameter of 1 mm and density of 2300 kg/m^3 that are fluidized in an atmospheric circulating fluidized bed combustor (CFBC) and a pressurized circulating fluidized bed combustor (PCFBC) boiler both operating at a temperature of 850°C and a pressure of 1 bar and 15 bars, respectively.

9.3. Calculate the heat transfer coefficient between the fluidized bed and heat transfer surface in a pressurized fluidized bed combustor (PFBC) boiler if the fluidized bed pressure is 1.5 MPa and teperature is 850°C. The heat transfer surface (wall) temperaure is 420°C, effective emissivity of the system fluidized bed wall is 0.9, particle diameter is 1 mm, particle density is 2300 kg/m^3, particle specific heat is 1477 J/(kg K), superficial air velocity is 3 m/s, average bed porosity is 0.8, and bed porosity at minimum fluidization is 0.4. Assume the following air properties at 1.5 MPa and 850°C: thermal conductivity 0.075 W/(m K), density $\rho = 4.65$ kg/m^3, dynamic viscosity 46.87×10^{-6} Pa s, and specific heat 1153 J/(kg K). Prandtl number for the air is 0.736, and the air gas constant is 287 J/(kg K). Additional constants are $c_1 = 2.8$ and $c_2 = 2.6$.

REFERENCES

Anand, A. K., Cook, C. S., Corman, J. C., and Smith, A. R. 1996. New technology trends for improved IGCC system performance. *Trans. ASME J. Eng. Gas Turbine Power* 118:732–736.

Anon. 1995. CFB. Lurgi AG, Germany.

Anthony, E. J. 1995. Fluidized bed combustion of alternative solid fuels: Status, successes and problems of the technology. *Prog. Energy Combust. Sci.* 21:239–267.

Foster Wheeler. 1994. Second generation PFBC.

Fujita, M. 1994. Start-up of the Wakamatsu PFBC. Presented at the EPRI FBC Conference, Atlanta, Ga., May 1994.

Gogolek, P. E. G., and Grace, J. R. 1995. Fundamental hydrodynamics related to pressurized fluidized bed combustion. *Prog. Energy Combust. Sci.* 21:419–451.

Goto, H. 1995. Operating experience of the 71 MW Wakamatsu PFBC demonstration plant. Presented at the 13th ASME FBC Conference, May 1995.

Gudenau, H. W., Hoberg, H., and Mayerhofer, A. 1994. Hot gas cleaning for combined cycle based on pressurized coal combustion. ASME paper 1993, 94-GT-417, pp. 1–8.

Ishikawa, K., Kawamata, N., and Kamei, K. 1993. Development of a simultaneous sulfur and dust removal process for IGCC power generation system. In *Gas cleaning at high temperatures*, vol. 2, pp. 419–435. London: Blackie Acad.

Jones, C. 1995. New coal-based power cycles aim to compete with gas. *Power* 139(3):52–58.

Kallmeyer, D., and Engelhard, J. 1992. KoBra-Kombikraftwerk mit integrierter HTW-Braunkohlevergasung. *Brennstoff-Wärme-Kraft* 44:388–391.

Karita. 1995. 360 MWe PFBC will be the first P800. *Mod. Power Syst.* 15(2):33–35.

Lippert, T. E., et al. 1992. Development of hot gas cleaning systems for advanced, coal based gas turbine cycles. ASME paper 92-GT-431, pp. 1–9.

Longwell, J. P., Rubin, E. S., and Wilson, J. 1995. Coal: Energy for the future. *Prog. Energy Combust. Sci.* 21:269–360.

Lozza, G., Chiesa, P., and DeVita, L. 1996. Combined-cycle power stations using "Clean coal technologies": Thermodynamic analysis of full gasification versus fluidized bed combustion with partial gasification. *Trans. ASME. J. Eng. Gas Turbine Power* 118:737–748.

Matsuda, T. 1995. Conversion to 350 MW AFBC at Takehara-2. Presented at the 13th ASME FBC Conference, May 1995.

Maude, C. 1993. *Advanced power generation.* London: IEA Coal Research.

McAllester, P. 1991. Pinon pine set to demonstrate IGCC. *Mod. Power Syst.* 11(12):19–23.

Minchener, A. J. 1992. Next generation coal power targets 1998 demonstration. *Colliery Guardian* 240(3):105–109.

Minchener, A. J., et al. 1991. Advanced clean coal technology for power generation. In *FBC technology and the environmental challenge, Proceedings of the Institute of Energy's 5th International Fluidized Combustion Conference*, pp. 331–341.

Nakabayashi, Y. 1994. Development of high-performance coal-fired power generation technologies, EPDC. Presented at the International Clean Coal Technology Symposium on PFBC, Kitakyushu, Japan, July 1994.

Newby, R. A, and Banniser, R. L. 1993. Advanced hot gas cleaning system for coal gasification processes. ASME paper 93-GT-338, pp. 1–10.

Newby, R. A., Domeracki, W. F., McGuigan, A. W., and Banniser, R. L. 1994. Integration of combustion turbine systems into pressurized fluidized bed combustion combined cycles. ASME paper 94-GT-333, pp. 1–11.

Petzel, H.-K. 1995. Die Wirbelschichtfeuerung auf dem Weg zur betriebsgewährten Großfeuerung? *VGB Kraftwerkstechnik* 75(4):380–385.

Richards, P. C. 1994. Gasification: The route to clean and efficient power generation with coal. *Can. Min. Metall. Bull. CIM Bull.* 87:135–140.

Robertson, A., and Horazak, D. 1993. Effect of PFB and AFB combustors on carbonizer-based power plant efficiencies. In *FBC and AFBC Projects and Technology, 1993 International Joint Power Generation Conference*, pp. 33–38. New York: ASME.

Schellberg, W., and Kuske, E. 1992. The qualities of PRENFLO coal gas for use in high-efficiency gas turbines. ASME paper 92-GT-263, pp. 1–5.

Siemens. 1995a. A 1000-MW single-shaft combined-cycle (GUD) power plant.

Siemens. 1995b. The Puertollano integrated coal gasification combined-cycle (IGC-GUD) power plant in Spain.

Smith, D. J. 1992. Commercialization of IGCC technology looks promising. *Power Eng., Barrington* 96(2):30–32.

Solomon, P. R., et al. 1996. A coal-fired heat exchanger for an externally fired gas turbine. *Trans. ASME J. Gas Turbine Power* 118:22–32.

Stambler, I. 1993. Second generation PFBC coal plants target 50% HHV efficiency. *Gas Turbine World* 23(6):22–27.

Tait, K. M., and McDonald, M. M. 1994. Recent development in IGCC technology. *Can. Min. Metall. Bull. CIM Bull.* 87:27–35.

Takahashi, Nakabayashi, Y., Fujita, M., and Kimura. 1994. Aktueller Stand der 350-MW-Wirbelschichtfeuerung Takehara und der 71-MW-Druckwirbelschichtfeuerung Wakamatsu der EPDC sowie der fortschrittlichen Stromerzeugung in Japan. *VGB Kraftwerkstechnik* 11:1003–1009.

Verweyen, N. 1994. *Zur Modellierung von stationären Kohlewirbelschichten*. Düsseldorf: VDI.

Chapter Ten

ADVANCED ENERGY STORAGE SYSTEMS

The efficiency and economics of conventional power plants may be improved by using energy storage. In solar energy or wind energy systems, the application of energy storage is required for a reliable energy supply. This chapter deals with advanced energy storage technologies. For utility power plants the most promising energy storage technologies are pumped hydro power storage, compressed-air energy storage, and electric batteries. These technologies are discussed in the first half of this chapter.

The emphasis in the second half of the chapter is on thermal energy storage for low-, intermediate-, and high-temperature applications, particularly in solar energy systems. Both sensible and latent heat storage technologies can be utilized. Equations for calculation of energy storage capacity and of mass and volumes of energy storage media required to store a certain amount of energy are presented. In addition, energy storage technologies for solar power plants are discussed.

ENERGY STORAGE PERFORMANCE CHARACTERISTICS

Energy storage is very important component of some power-generating systems. Solar and wind energy power systems require either an energy storage system or an alternative source of energy to supply energy when power production is low. Energy storage systems can also be used by electric utilities. An energy storage system converts electric energy (alternating current) into an energy form suitable for storage, e.g., potential, kinetic, or chemical energy. This energy is then stored in a storage reservoir. The stored energy can be recovered during periods of peak energy demand. This increases the reliability of the power supply and reduces the need for the less cost-efficient peaking facilities. The utilization of energy storage is beneficial whenever there is a temporal mismatch between energy supply from the energy conversion plant and demand by the energy user. The general principle of energy storage is schematically shown in Figure 10.1.

The First Law of Thermodynamics may be applied to the energy storage. The change in the storage energy content per unit time is determined by the rates of energy input and output as well as by the energy loss rate. Thus

$$dE_s/d\tau = E_{in} - E_{out} - E_l \tag{10.1}$$

where E_s is energy content of storage in Joules (J), τ is time in seconds (s), E_{in} and E_{out} are time-dependent rates of energy input and output, in watts respectively, and E_l is rate of energy loss of the storage in watts.

A storage operational cycle is made up of charging, storing, and discharging periods with durations τ_i, τ_s, and τ_0, respectively:

$$\tau_c = \tau_i + \tau_s + \tau_0 \tag{10.2}$$

For a storage cycle, change in stored energy E_s is zero. Therefore, integrating over τ_c yields the

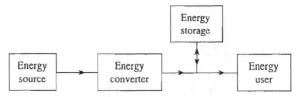

Figure 10.1. Schematic diagram of an energy system with energy storage.

corresponding energy quantities

$$\int E_{out}\, d\tau = \int E_{in}\, d\tau - \int E_1\, d\tau \qquad (10.3)$$

Specific energy density is the amount of energy stored in a unit of mass or volume of storage medium:

$$e_s = E_s/m \quad \text{J/kg} \qquad (10.4)$$

and

$$e'_s = E_s/V \quad \text{J/m}^3 \qquad (10.5)$$

Storage media with high densities require less mass and volume to store a given amount of energy. Fuels have the highest specific energy densities, of the order of 30–50 MJ/kg.

The storage efficiency (or recovery efficiency) is the total energy output divided by the total energy input over a cycle:

$$\eta_s = E_{out}/E_{in} \qquad (10.6)$$

Energy storage systems may be classified according to the form in which energy is stored, as follows (Beckman and Gilli, 1984; Duffie and Beckman, 1991; Bitterlich, 1987; Dinter, 1992; Khartchenko, 1995; Garg et al., 1985): mechanical, thermal, electrical and electromagnetic, and chemical. For example, chemical energy storage systems include storage batteries, hydrogen storage, and thermochemical energy storage based on reversible endothermic/exothermic chemical reactions.

MECHANICAL ENERGY STORAGE

Mechanical energy storage systems include gravitational energy storage or pumped hydro power storage (PHPS), compressed-air energy storage (CAES), and flywheels. The PHPS and CAES technologies along with electrical batteries can be used for large-scale utility energy storage, while flywheels are more suitable for intermediate storage.

Storage charging is carried out when inexpensive off-peak power is available, e.g., at night or on weekends. The storage is discharged when power is needed because of insufficient supply from the base-load plant.

Pumped Hydro Power Storage

PHPS is a potential-energy storage system. It represents the most economical means to store energy presently available to electric utilities. It is the most developed and used of all mechanical energy storage systems. The turbine/generator can be brought on-line from standstill within 90 s and to full power within 120 s. Changeover from pumping to generating can be done in 180–240 s. The overall efficiency of a pumped storage plant is approximately 80%. However, PHPS plants are very expensive, require suitable sites and long lead times, and need to be built in large size (Elliot, 1995).

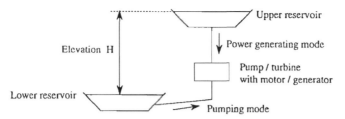

Figure 10.2. Pumped hydro power storage (PHPS).

The amount of potential energy PE that can be stored in a uniform gravitational force field is given by

$$PE = mg\Delta z \quad J \tag{10.7}$$

where m is the mass in kg, g is the acceleration due to gravity (9.81 m/s^2 at zero elevation), and Δz is the change in elevation in m.

In order to store 1 kWh of energy with a weight system, a 1000-kg mass would have to be raised 367 m above the reference level. Despite the small amount of energy that can be obtained from a unit mass, this system is used to store large amounts of energy by moving large quantities of water through reasonable distances in the PHPS system.

The potential energy of a mass of water m_w in kg, at an elevation H in m, above a reference plane is given by

$$PE = gm_w H \quad J \tag{10.8}$$

To store large quantities of energy, a large mass of water must be elevated to sufficiently large heights. A PHPS system is depicted in the schematic diagram of Figure 10.2. It consists of two reservoirs (upper and lower), pump turbine, motor-generator set, and connecting lines. During off-peak periods, electricity generated by base-load plants is used to pump water into the upper reservoir. When the power demand exceeds the supply, the water is allowed to flow back down through a hydraulic turbine, which drives an electric generator.

The operating heads on the pump turbine in pumping mode, H_p, and in power generating mode, H_t, are different because of different flow rates and head losses. Thus

$$H_p = H + \Delta H_p \qquad H_t = H - \Delta H_t \tag{10.9}$$

where H is static head (elevation differential between the upper and lower reservoirs) in m and ΔH_p and ΔH_t are head loss in pumping and generation mode, respectively, in m.

The power input in pumping mode and the power output in generating mode are given by

$$P_p = g\rho V_p H / \eta_p \eta_m \eta_l \tag{10.10}$$

and

$$P_t = g\rho V_t H \eta_t \eta_g \eta_l \tag{10.11}$$

where P is power in W, ρ is density of water in kg/m^3, V is volumetric flow rate in m^3/s, H is elevation differential between the upper and lower reservoirs (static head) in m, and η is efficiency. Subscripts denote p, pump; t, turbine; g, generator; m, motor; and l, water channel (line).

The PHPS turnaround efficiency is the total energy output during the generation period divided by the total energy input during the pumping period:

$$\eta_s = E_t / E_p \tag{10.12}$$

It is of the order of 75–80% (Elliot, 1995).

Site selection is more complex for PHPS than for hydroelectric plants. The suitable topography will enable the construction or selection of two reservoirs (one of which is usually a river or lake).

Advanced Energy Systems

Table 10.1. Characteristics of large PHPS systems

| Plant | Power, MW | | Head, | Number of pump-turbine sets |
	Turbine	Pump	m	
Bath County, U.S.	2100	2280	360	6
Grand Maison, France	1220	1260	955	8
Oku-Yoshino, Japan	1210	1280	515	6
Piastra, Italy	1280	1430	1070	9
Raccoon, U.S.	1530	1600	320	4
Zagorsk, Russia	1200	1250	1120	6

Source: Burnham (1993).

Underground PHPS systems reduce visual impact of storage on the environment (Elliot, 1995). Data of some large PHPS systems are given in Table 10.1.

Example 10.1

Calculate the pumping and generation powers and the energy storage efficiency for a PHPS system under the following conditions:

Elevation difference between the two reservoirs	$H = 300$ m
Volumetric flow rate in pumping mode	$V_p = 50$ m^3/s
Pumping period	8.5 h/d
Pump efficiency	$\eta_p = 0.8$
Volumetric flow rate in generating mode	$V_t = 85$ m^3/s
Generating period	6 h/d
Turbine efficiency	$\eta_t = 0.9$
Head loss in water channel (line)	5% ($\eta_l = 95\%$)
Generator/motor efficiency	$\eta_g = \eta_m = 0.97$

Solution

Generation power output

$$P_t = g\rho V_t H \eta_t \eta_g \eta_l = 9.81 \text{ m/s}^2 \times 1000 \text{ kg/m}^3$$
$$\times 85 \text{ m}^3/\text{s} \times 300 \text{ m} \times 0.9 \times 0.97 \times 0.95 = 207.5 \text{ MW}$$

Daily energy generation

$$E_t = t_t P_t = 6 \times 207.5 = 1245 \text{ MWh}$$

Pumping power input

$$P_p = g\rho V_p H / \eta_p \eta_m \eta_l = 9.81 \text{ m/s}^2 \times 1000 \text{ kg/m}^3$$
$$\times 50 \text{ m}^3/\text{s} \times 300 \text{ m} / (0.8 \times 0.97 \times 0.95) = 199.6 \text{ MW}$$

Daily energy input

$$E_p = t_p P_p = 8.5 \times 199.6 = 1696.6 \text{ MWh}$$

Turnaround efficiency of the energy storage

$$\eta_s = E_t / E_p = 1245/1696.6 = 0.734$$

Compressed-Air Energy Storage

Compressed-air energy storage (CAES) is potential-energy storage. In CAES, electrical energy in excess of the demand is used to compress air, which is stored in a reservoir for later use in a gas turbine to generate electricity. Compressed-air storage can be used by electric utilities for load leveling and for storing electrical energy generated in large-scale solar and wind plants. The

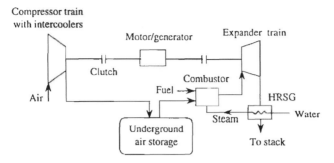

HRSG Heat recovery steam generator

Figure 10.3. Schematic diagram of a compressed-air energy storage (CAES) plant.

overall recovery efficiency defined as the ratio of recovered energy to the energy input over a storage cycle is about 75–80% (Elliot, 1995).

CAES is environmentally benign and has a low specific capital cost. It has larger energy densities per 1 m^3 of volume in comparison with PHPS plants.

Figure 10.3 shows a schematic diagram of an advanced CAES plant. The CAES store is an underground reservoir such as a cavern or salt dome. The machinery includes the compressor train, turbine/expander train, and motor-generator with clutch. The clutch disconnects the compressor in generation mode and the turbine in the storage charging mode. To charge the storage, the motor-generator, working as a motor, drives the compressor, which delivers the high-pressure air into the store. The electric energy required to compress the air is approximately 0.8 kWh per 1 kWh of power produced during storage discharging. In storage discharging, the expander is connected to the motor-generator, which acts now as a generator. The compressed air from the cavern is expanded in the expander, driving the generator. The air may be preheated in a recuperator using the expander exhaust heat, and further heated in a combustor firing natural gas or fuel oil. The fuel energy requirements are approximately 1.2 kWh per 1 kWh generated by the plant. A two-stage expander with reheat cycle is used in a CAES plant. The exhaust from the high-pressure stage is reheated before entering the low-pressure stage (Elliot, 1995).

Mass and energy conservation equations must be considered simultaneously. They may be written as follows:

$$d(\rho V)/dt = m_i - m_e \tag{10.13}$$

and

$$d(\rho V c_v T)/dt = P_i - P_o \tag{10.14}$$

where ρ is air density, V is volume of the reservoir, t is time, c_v is isochoric specific heat of air, m_i and m_e are incoming and exiting mass flow rate of air, respectively ($m_e = 0$ in compression mode, $m_i = 0$ in generation mode), and P_i and P_o are power input and output, respectively. Heat loss is ignored in Eq. (10.14).

According to Chapter 5, the power input is given by

$$P_i = m_i c_p T_a \left[\beta_c^{(k-1)/k} - 1 \right] / \eta_{ic} \eta_m \tag{10.15}$$

where c_p is isobaric specific heat of air, T_a is atmospheric air temperature, β_c is compression pressure ratio, k is isentropic exponent, η_{ic} is compressor isentropic efficiency, and η_m is motor efficiency. Similarly, the power output is given by

$$P_o = m_e c_p T_{t,i} \left[1 - 1/\beta_t^{(k-1)/k} \right] \eta_{it} \eta_g \tag{10.16}$$

where $T_{t,i}$ is the gas turbine inlet temperature, β_t is turbine pressure ratio, η_{it} is turbine isentropic efficiency, and η_g is generator efficiency.

Table 10.2. Performance data of CAES plants

Data	Huntorf, Germany	McIntosh, Alabama	Sunagawa, Japan
Capacity, MW	290	110	35
Generation, hours	2	26	6
Compression, hours	4	1.6	1.2
Volume, 10^3 m^3	311	538	30
Cavern temperature, °C	35	35	50
Expander train			
High pressure			
Inlet pressure, bars	46	45	40
Inlet temperature, °C	540	540	800
Low pressure			
Inlet pressure, bars	11	15	15
Inlet temperature, °C	670	670	1250
Expander mass flow, kg/s	415	154	47
Recuperator	no	yes	yes

The fuel heat input in the combustor is

$$Q_{in} = m_c c_p (T_{t,i} - T_s) \tag{10.17}$$

where T_s is the storage temperature. The storage efficiency for the storage cycle is given by

$$\eta_s = \int P_o \, dt \Big/ \left(\int P_i \, dt + \int Q_{in} \, dt \right) \tag{10.18}$$

The 290-MW CAES plant, which stores 4 generation hours, has been in operation since 1979 in Huntorf, Germany, and the 110-MW CAES, which stores 26 generation hours, has been in operation since 1991 in McIntosh, Alabama. Table 10.2 contains the main performance data of these plants along with the design data of a CAES plant in Sunagawa, Japan, scheduled for operation in 1997.

The capital cost of a CAES plant depends on the power rating of the plant and on the required storage time. Specific capital costs (in $/kW) of pumped hydro electric plants are 2–3 times higher than for CAES. Performance of the CAES technology can be significantly improved by using

- advanced gas turbines with higher turbine inlet temperature and higher cycle efficiency
- steam injection into the combustor
- humid air turbine (HAT) cycle with a saturator to humidify the air
- heat recovery steam generator (HRSG) instead of recuperator

As was discussed in Chapters 5 and 6, steam injection into the combustor is used to increase power output and reduce NO_x emissions, and the use of air humidification prior to combustion reduces the amount of air needed and lowers NO_x emissions.

Example 10.2

Calculate the CAES storage volume required for a 100-MWe peaking compressor/gas turbine as well as the average air flow rate during a 5-hour cavern charging period.

Compressor and peaking gas turbine isentropic efficiencies η_{ic} and η_{it} are 85 and 80%, respectively. The duration of power generation period is 8 hours. The compressor intake is at 15°C and 100 kPa and discharge is at 6.4 MPa. Supplementary firing raises the compressed-air temperature to a turbine inlet temperature (TIT) of 1050°C by burning natural gas with a heating value (HV) of 49 MJ/kg. Find the fuel rate also.

Assume a constant specific heat for the air of 1.05 kJ/(kg K) in the entire temperature range.

Solution

The required storage capacity is equal to the total energy input to the gas turbine train during the power generation period of 8 hours. Thus, taking into account the energy losses,

$$Q_s = P\tau_g/\eta_{it} = 100 \times 8/0.8 = 1000 \text{ MWh}$$

Compressor discharge air temperature, reversible and irreversible, respectively, are

$$T_{2s} = T_1\beta^{(k-1)/k} = (15 + 273)(6400/100)^{(1.4-1)/1.4} = 945 \text{ K}$$

$$T_2 = T_1 + (T_{2s} - T_1)/\eta_{ic} = 288 + (945 - 288)/0.85 = 1061 \text{ K}$$

Fuel requirements per kg air

$$FA = c_p (TIT - T_2)/HV = 1.05(1050 + 273 - 1061)/49000 = 0.00561 \text{ kg/kg}$$

Mass of air required

$$m_a = Q_s/c_p\Delta T = 1 \times 10^6 \text{ kWh} \times 3600 \text{ s/h}/1.05 \text{ kJ/(kg K)}(1050 + 273 - 288) \text{ K}$$
$$= 3,312,629.4 \text{ kg}$$

Total fuel requirements

$$m_f = FAm_a = 0.00561 \times 3,312,629.4 = 18,583.8 \text{ kg}$$

Fuel flow rate per hour of generation

$$18,583.8 \text{ kg}/8 \text{ h} = 2323 \text{ kg/h}$$

Air density at 1061 K

$$\rho = p/RT = 6.4 \times 10^6 \text{ Pa}/[287 \text{ J/(kg K)}1061 \text{ K}] = 21.02 \text{ kg/m}^3$$

Cavern volume required

$$V = m_a/\rho = 3,312,629.4 \text{ kg}/21.02 \text{ kg/m}^3 = 157,594.2 \text{ m}^3$$

Average air flow to cavern during 5 hours of compression

$$157,594.2 \text{ m}^3/5 \text{ h} = 31,518.8 \text{ m}^3/h = 8.755 \text{ m}^3/s$$

It is more reasonable to employ a two-stage intercooled compression and two-stage reheat expansion.

Flywheel Energy Storage

Flywheel storage systems convert electrical energy into kinetic energy stored in a flywheel rotating at high speeds. Figure 10.4 shows a schematic of a flywheel energy storage.

Flywheels store kinetic (rotational) energy. They could be used for electric utility peaking units, for storage of solar and wind energy, and for vehicle propulsion. Energy recovery efficiency up to 90% can be achieved (Garg et al., 1985; Jensen and Sorensen, 1981).

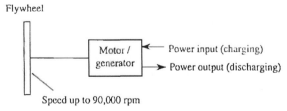

Figure 10.4. Flywheel energy storage.

A flywheel energy storage system consists of a rotating mass of flywheel, a motor-generator set, bearings, and an enclosure(s). In the charging mode, during off-peak periods, the motor adds energy to the flywheel. In the generation mode, during periods of peak demand, the energy stored in the flywheel drives the generator. The kinetic energy stored in a flywheel is given by

$$E = mR^2\omega^2/2 = 2\pi^2 mR^2 n^2 \quad J \tag{10.19}$$

where m is mass of the flywheel in kg; R is radius of gyration, i.e., the radius at which the total mass is considered to be concentrated, in m; $\omega = 2\pi n$ is angular velocity, in radians/s; and n is revolutions per second, 1/s. Note that for a disc of uniform density r, uniform thickness t, and outer radius R_0, $R = R_0/\sqrt{2} = 0.7071 R_0$.

With the moment of inertia defined as $MI = mR^2$,

$$E = MI\omega^2/2 \quad J \tag{10.20}$$

The energy absorbed or released by a flywheel between speeds of rotation n_1 and n_2 is given by

$$E_a = 2\pi^2 mR^2\left(n_2^2 - n_1^2\right) = 4\pi^2 mR^2 k_s n^2 \tag{10.21}$$

where k_s is the coefficient of speed fluctuation, i.e., the ratio of the difference $n_2 - n_1$ to the mean speed $n = (n_1 + n_2)/2$; k_s varies from 0.005 for fine to 0.2 for coarse speed regulation.

The specific kinetic energy per unit mass is

$$E/m = k_m\sigma/\rho \quad J/kg \tag{10.22}$$

where k_m is the mass-efficiency factor (maximum of 0.9 is achieved by constant-stress flywheel), σ is allowable tangential stress in the flywheel in N/m^2, and ρ is density in kg/m^3.

Proper material choice means high stress-to-density ratio σ/ρ. This ratio is 1440 kJ/kg for fused silica, 432 kJ/kg for S-glass fiber, and 324 kJ/kg for E-glass fiber, but for steel it is only 79.2 kJ/kg.

The specific energy per unit volume is given by

$$(E/V)_{max} = k_v\sigma/\rho \quad J/m^3 \tag{10.23}$$

where k_v is the volume-efficiency ratio, dimensionless. Thus, electrical energy stored by a flywheel is supplied to the motor-generator that spins up to speeds of 90,000 rpm. The energy is extracted when needed by converting the kinetic energy of the rotor to electricity when the flywheel drives the motor-generator that now serves as the generator.

Flywheel materials must have high strength, high stress-to-density ratio, and high cost-effectiveness. The battery's rotor, made of carbon fiber, is 4 times stronger than the toughest steel. It can withstand tension forces of up to 7 kN per mm^2 of cross-sectional area. Magnetic bearings keep friction losses to a minimum. One important potential application for flywheel batteries is load leveling, achieved by storing energy generated at night for use during periods of higher demand.

THERMAL ENERGY STORAGE

There are three types of thermal energy storage (TES) systems: sensible heat, latent heat, and thermochemical energy storage.

Sensible Heat Storage

Types of Sensible Heat Storage

The term sensible heat storage (SHS) refers to systems that store thermal energy by increasing the temperature of a medium without phase change, such as melting, boiling, or freezing (Beckman and Gilli, 1984; Duffie and Beckman, 1991; Bitterlich, 1987; Dinter, 1992; Khartchenko, 1995; Garg et al., 1985). Charging of SHS occurs by adding heat to the storage medium from a heat source such as a conventional power plant or alternative power plant, including solar and wind power plants. Electric energy can also be stored by means of TES. In order to recover the heat from the storage for power generation, a heat engine is required. Storage charging is accomplished during the periods of excess energy production in comparison with energy demand. While charging proceeds, temperature of the sensible heat storage medium increases in proportion to the heat input over the charging period. However, when energy demand exceeds energy production, the energy deficiency is compensated by the heat recovery from the energy storage. Thereby, the temperature of the SHS medium decreases in proportion to the heat output over the discharging period. The rate of temperature change depends on the heat input or output rate and the mass and specific heat of the SHS medium. The specific heat is the most important thermophysical property for SHS media. The higher the specific heat, the more energy can be stored per unit mass of the storage material. In other words, the storage energy density is determined by the specific heat. For example, water has a specific heat of about 4.19 kJ/(kg K) and 4.19 MJ/(m^3 K). If the temperature change in the storage is 50 K (90 F), the change in the energy content of a hot water storage per unit mass and per unit volume is 209.5 kJ/kg and 209.5 MJ/m^3, respectively.

Table 10.3 contains the specific heat of some solid and liquid energy storage materials per unit mass and per unit volume. It is seen that water has the largest specific heat both per unit mass and per unit volume. Hence water is an excellent storage medium for a temperature range between its freezing point and its boiling point, i.e., between 0°C and 100°C at a pressure of 1 bar. Therefore it is widely used to store the low-temperature heat in solar plants. If higher storage temperatures are needed, a pressurized water storage up to ~200°C can be utilized.

Some other liquid and solid materials can also be used. At higher temperatures, special hydro-carbon oils (called thermo-oils) with high boiling points and dual media (rock-oil mixture) can be used to store sensible heat. For temperatures in the range from 100 to 400°C, synthetic oils such as thermo-oil VP-1 can be used. Table 10.4 contains its physical properties (Dinter, 1992; Khartchenko, 1995).

Rock-bed heat storage is used in solar energy systems predominantly in low-temperature applications. In comparison with water, rock and pebble materials have lower specific heats and thus require larger mass and volume to store the same quantity of heat. Figures 10.5a and 10.5b show schematically a pebble-bed storage and the pebble-bed temperature profile. It is intrinsically stratified storage with a high-temperature zone in the upper part of the storage bin. The hot air manifold is placed at the top, and the hot air is supplied from the top during charging, while the cold air is passed from the bottom to the top of the storage during discharging. Owing to a large surface area per unit voume, the heat transfer rate between the air stream and pebble particles

Table 10.3. Density and specific heat of solid and liquid energy storage materials at 20°C

Material	Density, kg/m^3	Specific heat kJ/(kg K)	Specific heat kJ/(m^3 K)
Concrete	2400	1.1	2640
Iron	7800	0.5	3900
Clay	1450	1.28	1856
Ground (with coarse pebbles)	2040	0.59	1204
Rock	1600	0.84	1344
Water	1000	4.19	4190

Source: Beckman and Gilli (1984) and Khartchenko (1995).

Table 10.4. Thermophysical properties of thermo oil VP-1

t, °C	p, bars	h_l, kJ/kg	h_v, kJ/kg	c, kJ/(kg K)	k, W/(m K)	ρ, kg/m^3	η, mPa s
300	2.461	2331.2	3525.3	2.320	0.099	815	0.205
400	11.123	3381.3	4291.1	2.588	0.085	689	0.139

Here, t, temperature; p, pressure; h_l, liquid enthalpy; h_v, saturated vapor enthalpy; c, specific heat; k, thermal conductivity; ρ, density; and η, viscosity.

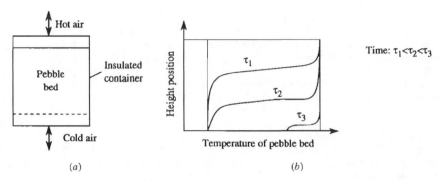

Figure 10.5. Pebble-bed sensible heat storage: (*a*) schematic and (*b*) temperature-time history by storage charging.

is very high. Also, because of low conductivity, the heat loss in the storing mode without an air stream is reasonably low. Therefore this storage technology is particularly suitable for solar systems with air used as the heat transfer fluid.

At high temperatures the selection of a proper storage medium is a rather complex problem. However, molten-salt mixtures such as Du Pont's HITEC, containing sodium nitrite and sodium and potassium nitrates, can be used to store heat up to about 540°C (Elliot, 1995). Concrete is a relatively good medium for heat storage in passively heated or cooled solar houses. It is also considered for application in intermediate-temperature solar thermal plants. Ceramic pebbles or bricks are used for heat storage on a large scale in industry. A combination of oil and rock was tested in a solar power plant project. Thermal energy storage as sensible heat in a high-temperature steam can provide utilities with a means for generating electric power to satisfy peak demands.

In summary, water with its highest specific heat is the most suitable storage medium for low-temperature SHS applications. Solid materials such as concrete, rock, and ground are also used in the low-temperature range. In the intermediate-temperature range, certain synthetic oils and dual-media (rock-oil mixture) are employed.

There are short-term and long-term TESs. The latter are of greater importance in solar space heating systems, where seasonal storages provide the only economically viable solution (Khartchenko, 1995; Dalenbäck, 1990).

Sensible Heat Storage Capacity

Performance of a TES is characterized by storage capacity, energy density, heat input and output rates while charging and discharging storage, and storage efficiency. The storage capacity of an SHS with a solid or liquid storage medium is given by

$$Q_s = mc\Delta t = V\rho c\Delta t \quad \text{J} \tag{10.24}$$

where m is mass in kg, V is volume in m^3, c is specific heat in J/(kg K), ρ is density in kg/m^3, and $\Delta t = t_{max} - t_{min}$ is maximum temperature change or difference between maximum and minimum temperatures of the storage medium in K. Equation (10.24) may be used to calculate the mass and volume of storage material required to store a given quantity of energy. Thus,

$$m = Q_s/c\Delta t \quad \text{kg} \tag{10.25}$$

$$V = Q_s/\rho c\Delta t \quad \text{m}^3 \tag{10.26}$$

For a packed bed used for energy storage, the bed void fraction ε (i.e., the volume of interparticle spaces divided by total bed volume) must be taken into account. Thus the volume of packed bed storage is given by

$$V = Q_s/\rho c(1 - \varepsilon)\Delta t \quad \text{m}^3 \tag{10.27}$$

The storage energy density per unit mass is given by

$$q = Q_s/m = c(t_{max} - t_{min}) \quad \text{J/kg} \tag{10.28}$$

Similarly, the storage energy density per unit volume is

$$q_v = Q_s/V = \rho c(t_{max} - t_{min}) \quad \text{J/m}^3 \tag{10.29}$$

Heat input rate in the TES charging mode is

$$Q_{in} = (mc_p)_s \Delta t_s = (mc_p)_f \Delta t_f \quad \text{W} \tag{10.30}$$

where m_s and m_f is mass flow rates of storage medium and heat transfer fluid in kg/s, respectively, c_{ps} and c_{pf} is isobaric specific heat of storage medium and heat transfer fluid in J/kgK, respectively, Δt_s and c_{pf} is change in temperature of storage medium and heat transfer fluid in K, respectively. Similarly, the heat output rate in the TES discharging mode may be calculated.

The energy balance of a storage given by Eq. (10.1) may be used to find the change in temperature of the storage medium of an SHS with a uniform temperature distribution. Then

$$dE_s/d\tau = Q_{in} - Q_{out} - Q_l \tag{10.31}$$

where E_s is internal energy of the storage medium in J, $d\tau$ is the time interval in s, and Q_{in}, Q_{out}, and Q_l are heat input, heat output, and heat loss flux, respectively, in W. Then,

$$dE_s/d\tau = \rho V c_v \, dt/d\tau \tag{10.32}$$

where ρ is storage medium density, V is volume, dt is change in storage temperature over the time interval $d\tau$, and $c_v =$ isochoric specific heat of storage medium. The heat loss flux from storage with a temperature t to the surroundings with temperature t_a is

$$Q_l = (UA)_s(t - t_a) \tag{10.33}$$

where $(UA)_s$ is product of storage overall heat loss coefficient times outside area in W/K.

Let us first consider the case when $Q_{in} = Q_{out} = 0$. Substituting Eq. (10.33) into Eq. (10.31) and rearranging yields

$$dt/(t - t_a) = -[(UA)_s/(\rho V c_v)] \, d\tau \tag{10.34}$$

Integrating Eq. (10.1) for a time interval from 0 to τ results in

$$\ln [(t - t_a)/(t_i - t_a)] = -[(UA)_s/(\rho V c_v)] \, \tau \tag{10.35}$$

where t_i and t are temperature in storage at $\tau = 0$ and τ, respectively, in °C. Thus

$$t = t_a + (t_i - t_a) \exp \{-[(UA)_s/(\rho V c_v)] \, \tau\} \tag{10.36}$$

In general, when Q_{in} and Q_{out} are not nil, Eq. (10.31) can be approximately solved using the finite difference method. Thus for an interval of 1 hour, the storage temperature is given by

$$t_{i+1} \approx t_i + [Q_{in} - Q_{out} - 3600(UA)_s(t_i - t_a)]/m_s c_p \tag{10.37}$$

Hot Water and Water/Steam Storages

A TES system consists of a storage medium, storage container with insulation, and facilities (heat exchangers) for energy addition and withdrawal during the charging and discharging periods. Figure 10.6a depicts a schematic of a TES for energy storage in a vessel with hot water. Here the heat exchange surfaces for storage charging and discharging can be arranged either inside or outside the storage vessel. For an efficient operation of the TES, thermal insulation is a critical issue. This kind of TES is applicable up to 95°C in the nonpressurized version and up to about 150–180°C in pressurized facilities (Dinter, 1992; Khartchenko, 1995). Figure 10.6b shows the cycle of operation of a sensible heat TES. It consists of charging, storing, and discharging.

Water/steam accumulators are often used by utilities to store heat in the temperature range up to about 180°C. In the storage charging mode of this TES, steam is introduced into the storage

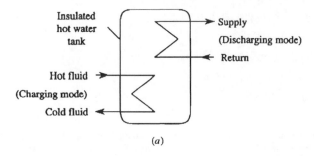

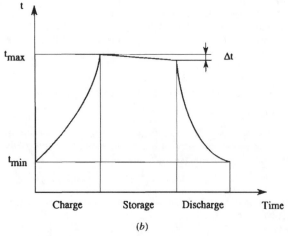

Figure 10.6. (a) Hot water sensible heat storage and (b) cycle of
operation of a sensible heat storage.

vessel through steam nozzles. The pressure in the vessel rises from p_1 to p_2. Thereby, the steam
condenses in direct contact with water, and as a result, the water temperature and enthalpy rise
from t_1 to t_2 and from h_1 to h_2, respectively. The heat of steam condensation is stored as sensible
heat of hot water. The steam mass added during storage charging is found from the energy balance
on the hot water storage as follows:

$$m_s = m_{w1}(h_1 - h_2)/(h_s - h_2) \quad \text{kg} \tag{10.38}$$

where m_{w1} is initial mass of the water in the storage vessel at p_1 and t_1, in kg; h_1 and h_2 are initial
and final water enthalpy in storage in kJ/kg; and h_s is steam enthalpy in kJ/kg. In the fully charged
state, the hot water storage contains a mass of $m_{w2} = m_{w1} + m_s$ of saturated water at pressure p_2
and temperature t_2.

In the storage discharging mode, steam is extracted from storage through a throttling valve.
Throttling is a constant-enthalpy process. Therefore the storage pressure constantly drops from p_2
to $p = p_2 - \Delta p$ during storage discharging. A portion of the water vaporizes until the temperature
in the storage vessel is equal to the saturation temperature at a new pressure p. The mass of extracted
saturated steam is found from the energy balance on the storage for the discharging process.

Typically, the steam is extracted from a steam turbine at a pressure of about 20–30 bars with a
saturation temperature of 212–234°C for charging the accumulator. Storage discharging continues
until the pressure is reduced to 2–5 bars (the saturation temperature is 120–152°C). Thus the
temperature of saturated water recovered from storage varies within these limits during discharging.
The quantity of heat that must be stored in the accumulator to generate elelctric power during

discharging of the TES is given by

$$Q_s = P_{el} \tau_d / \eta_p \quad \text{kJ} \tag{10.39}$$

where P_{el} is electric power output of a peaking power plant in kW, τ_d is duration of storage discharging in s, and η_p is efficiency of the peaking power plant. The mass of water needed to be recovered and flashed to steam is given by

$$m = Q_s / (h_{f1} - h_{f2}) \quad \text{kg} \tag{10.40}$$

where Q_s is heat recovery during discharging of the TES in kJ, h_{f1} is saturated water enthalpy in the initial state in kJ/kg, and h_{f2} is saturated water enthalpy in the final state in kJ/kg.

Under the assumption of a uniform temperature distribution in the storage, the efficiency of a water/steam accumulator can be approximately found from

$$\eta_s = 1 - [1 - \exp(-\tau_d / \tau_c)(t_1 - t_a)/(t_1 - t_2) \tag{10.41}$$

where τ_d is storage discharging duration in s, τ_c is storage time constant in s, t_1 is initial saturated water temperature at p_1 in °C, t_2 is final saturated water temperature at p_2 in °C, and t_a is ambient temperature in °C. The time constant of the storage is approximately given by

$$\tau_c = V\rho c_p / AU \quad \text{s} \tag{10.42}$$

where V is storage volume in m³, ρ is storage medium density in kg/m³, c_p is storage medium specific heat in kJ/(kg K), A is the outside surface area of the storage in m², and U is overall heat transfer coefficient from the hot water to the surroundings in kW/(m² K). Maximum (theoretical) energy density for the following two intermediate-temperature (above 100°C) SHS media are as follows:
Hot water storage

58 kWh/m³ for $\Delta t = 50$ K   116.3 kWh/m³ for $\Delta t = 100$ K

Steam/water accumulator

13 kWh/m³ for 120–100°C   80 kWh/m³ for 180–100°C

Latent Heat Storage

Latent heat storage (LHS) is based on the heat absorption and release when a storage material undergoes a reversible phase change, usually from the solid state to the liquid state in the storage charging mode, and vice versa in the storage discharging mode. The storage material used is called phase-change material (PCM). LHS systems have certain benefits in comparison with SHS systems. The most important is the much higher energy density per unit mass and per unit volume (Lane, 1983, 1986).

The storage capacity of the LHS with a PCM medium is given by

$$Q_s = mc_s(t_f - t_0) + fmh_f + fmc_l(t_{end} - t_f) \quad \text{kJ} \tag{10.43}$$

where Q_s is amount of heat stored in kJ, m is mass of PCM in kg, c_s is average specific heat of solid phase between t_0 and t_f in kJ/(kg K), t_f is fusion (melting) temperature in °C, t_0 is initial temperature of PCM in the solid state in °C, f is fraction of PCM in the liquid state, h_f is heat of fusion in kJ/kg, t_{end} is final temperature of PCM in the liquid state in °C, and c_l is average specific heat of the liquid phase between t_f and t_{end} in kJ/(kg K).

The first term represents the sensible heat below the melting point, the second term the latent heat, and the third term the sensible heat above the melting point. In practical LHS systems, the second term predominates.

The energy density per unit mass of a PCM storage medium is given by

$$q = Q_s / m = c_s(t_f - t_0) + fh_f + fc_l(t_{end} - t_f) \quad \text{kJ/kg} \tag{10.44}$$

Then the energy density per unit volume of a PCM storage medium is

$$q_v = q/\rho \quad \text{kJ/m}^3 \tag{10.45}$$

where ρ is the PCM density in the solid state in kg/m^3. The important criteria for selecting a suitable PCM are as follows (Khartchenko, 1995; Lane, 1983):

1. Thermal properties
   - suitable phase-change temperature
   - high heat of fusion
   - favorable thermophysical properties (high specific heat and thermal conductivity in both solid and liquid states, and low viscosity in liquid state) to enable good heat transfer

2. Physical properties
   - favorable phase equilibria
   - low vapor pressure
   - small volume change
   - high density
   - compatibility with materials used to construct the containment vessel and heat exchangers

3. Chemical and kinetic properties
   - long-term chemical stability
   - sufficient crystallization rate
   - freezing (crystallizing) without subcooling

4. Economic properties
   - availability
   - cost-effectiveness

However, there is no PCM that can simultaneously fulfill all these requirements. PCMs store both sensible and latent heat. For an intermediate-temperature range, sodium and potassium nitrates and nitrite mixtures with fusion temperatures in the range 280–500°C are used. Properties of some PCM materials are given in Tables 10.5 and 10.6.

Table 10.5. Properties of some low-temperature PCM media

Material	t_f, °C	ρ, kg/m^3 Solid	Liquid	k, W/(m K)	c, J/(kg K) Solid	Liquid	h_f, kJ/kg
Sodium sulfate decahydrate	32.4	1460	1410	0.5/0.3	1760	3310	251
Paraffin–Wachs	46.7	786	NA	0.5	2890	NA	209
Lauric acid	49	1007	862	0.4/0.2	1.6	NA	177
Stearic acid	70.7	965	848	NA/0.2	1670	2300	200.3

Here, t_f, fusion (melting) temperature; h_f, heat of fusion; NA, not applicable.
Properties in solid/liquid state are ρ, density; k, thermal conductivity; c, specific heat.

Table 10.6. Storage medium mass and volume, and size of sensible heat and latent heat storages

	Rock	Sodium chloride
Mass, 10^3 kg	5357	604
Volume, 10^3 m^3	5.15	0.43
Storage size, m		
Length × width × height	50 × 20 × 5.15	15 × 8 × 3.58

Storage capacity is 100 MWh, and the temperature difference is 80 K. For sodium chloride, the fusion temperature is 800°C, the fusion enthalpy is 520 kJ/kg, the density is 2160 kg/m^3, and the specific heat is 0.95 kJ/(kg K). For both rock and salt storage, a void fraction of 35% is assumed.

Example 10.3

Compare the mass and volume of sensible heat storage media, water and rock bed, with those of a latent heat storage medium, lauric acid, that are required for a storage unit with a capacity of 5 MWh over a temperature range 40–60°C.

The properties of the storage media are as follows:

	Water	Rock	Lauric acid
Density ρ, kg/m^3	1000	1600	1007
Specific heat c, kJ/(kg K)	4.19	0.8	1.6
Heat of fusion h_f, kJ/kg	—	—	177
Bed void fraction ε	—	0.4	—
Fusion temperature, °C	—	—	49

Solution

Storage energy density per unit mass is

1. For lauric acid

$$q_{la} = 1.6(49 - 40) + 177 + 1.6(60 - 49) = 209 \text{ kJ/kg}$$

2. For water

$$q_w = 4.19(60 - 40) = 83.8 \text{ kJ/kg}$$

3. For rock bed

$$q_{rb} = 0.8(60 - 40) = 16 \text{ kJ/kg}$$

Storage medium mass and volume are

1. For lauric acid

$$m_{la} = Q/q_{la} = 5 \times 3.6 \times 10^6 \text{ kJ}/209 \text{ kJ/kg} = 86,124 \text{ kg}$$
$$V_{la} = m_{la}/\rho_{la} = 86,124/1007 = 85.5 \text{ m}^3$$

2. For water

$$m_w = Q/q_w = 5 \times 3.6 \times 10^6 \text{ kJ}/83.8 \text{ kJ/kg} = 214,797 \text{ kg}$$
$$V_w = 214,797/1000 = 214.8 \text{ m}^3$$

3. For rock bed

$$m_{rb} = Q/q_{rb} = 5 \times 3.6 \times 10^6 \text{ kJ}/16 \text{ kJ/kg} = 1,125,000 \text{ kg}$$
$$V_{rb} = m_{rb}/[\rho_{rb}(1 - \varepsilon)] = 1,125,000/[1600(1 - 0.4)] = 1171.9 \text{ m}^3$$

PCM may have some undesirable properties such as low physical and chemical stability, incongruent melting behavior, undercooling, and especially, unfavorable thermophysical properties. Special additives are required to improve behavior of the PCM. Low thermal conductivity of the PCM results in low rates of heat transfer to and from the LHS during charging and discharging. In order to partially overcome this difficulty, special techniques and designs are used such as PCM encapsulation and extended heat transfer surfaces (Khartchenko, 1995; Garg et al., 1985). Figure 10.7 shows an LHS design with a finned heat pipe as a heat transfer surface.

THERMOCHEMICAL ENERGY STORAGE

Reversible Chemical Reactions

Reversible chemical reactions that can proceed in both directions may be used for energy storage. When thermal energy is added to reactants, they combine to form products that can, in turn, react

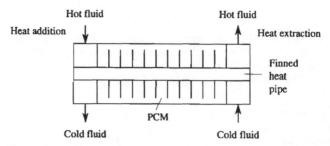

Figure 10.7. Latent heat storage with a PCM embedded heat pipe heat exchanger.

to form the initial species while releasing heat at a lower temperature. Thus the products of the forward endothermic reaction store heat as chemical energy. This energy can be recovered as heat when the reverse exothermic reaction proceeds under certain conditions.

The possible reactions suitable for energy storage are either catalytic, dissociation, or adsorption reactions. The thermochemical energy storage system should use inexpensive materials that have a high reaction heat and are capable of proceeding at reasonable speed in both forward and reverse directions at temperatures that match both the energy source and the energy utilization (Khartchenko, 1995; Garg et al., 1985).

In general, the cycle of a thermochemical energy storage can be presented as a sequence of the following reactions.

1. Catalytic reaction during energy storage charging

$$A + B + \text{Catalyst} + \text{Heat (higher temperature)} \rightarrow C + D \tag{10.46}$$

2. Storing of the product of the above reaction practically without energy dissipation over a long period of time.
3. Catalytic reaction during energy storage charging

$$C + D + \text{Catalyst} \rightarrow A + B + \text{Heat (lower temperature)} \tag{10.47}$$

The forward reaction, according to Eq. (10.46), predominates at the higher temperature and thus it is accompanied by the absorption of heat. Hence this reaction suits for storage charging. The reverse reaction, according to Eq. (10.47), proceeds preferentially at a lower temperature with the evolution of heat and therefore it is used for recovery of heat and reactants.

The following reversible reaction types can be used for energy storage (Khartchenko, 1995; Garg et al., 1985):

- dehydration of salt hydrates and acids
- deoxygenation of metal oxides
- thermal dissociation of gases such as SO_3
- decomposition of metal carbonates and other salts such as $CaCO_3$, $FeCl_2 \cdot 6NH_3$ or $CaCl_2 \cdot 8NH_3$

Some reactions with a significant energy storage potential are listed below:

$$Na_2S \cdot 5H_2O \leftrightarrow Na_2S + 5H_2O \text{ (energy density 500 Wh/m}^3) \tag{10.48}$$

$$H_2SO_4 \cdot H_2O \leftrightarrow H_2SO_4 + H_2O \text{ (energy density 300 Wh/m}^3) \tag{10.49}$$

$$Ca(OH)_2 \leftrightarrow CaO + H_2O \text{ (energy density 250 Wh/m}^3) \tag{10.50}$$

$$CaCl_2 \cdot 8NH_3 \leftrightarrow CaCl_2 \cdot 4NH_3 + 4NH_3 \text{ (energy density 100 Wh/m}^3) \tag{10.51}$$

Adsorption reactions in a system such as water-zeolith can also be used.

The main advantage of thermochemical energy storage is high energy storage density. Another important advantage is the possibility of storing the products of forward reaction at room temperature for long-term periods without energy losses. The following reaction may be used for storage and long-distance transport of solar energy:

$$CH_4 + H_2O \text{ (steam)} + \text{Concentrated solar flux} + \text{Catalyst} \leftrightarrow CO + 3H_2 \qquad (10.52)$$

Heat is stored by absorbing it in the endothermic reaction of steam reforming of methane. The enthalpy of formation of $CO + 3H_2$ in the forward reaction at 25°C is $(-110.6) + 0 = -110.6$ MJ/kmol, that of $CH_4 + H_2O$ (liquid) in the reverse reaction at 25°C is $(-14.9) + (-286) = -360.9$ MJ/kmol. Thus the forward endothermic reaction with both reactants and products maintained at 25°C requires a net addition of energy to the system of $(-110.6) - (-360.9) = +250.3$ MJ/kmol. In the reverse exothermic reaction, an equal amount of energy, i.e., 250.3 MJ/kmol, is released, referred to 25°C (Khartchenko, 1995).

Thus the theoretical energy storage density of this reversible reaction can be estimated as follows: 250.3 MJ/kmol/16 kg CH_4/kmol = 15.64 MJ/kg CH_4. Although the actual heat effect is slightly less, it is much higher than the energy storage density of storage media used for sensible or latent heat storage. However, thermochemical energy storage is more complex than sensible heat or latent heat storage, and materials often have high costs. Therefore there is no commercial storage so far. Thermochemical energy storage can potentially be used in low-, middle-, and high-temperature applications.

The turnaround efficiency of an energy storage is the ratio of the energy output to the energy input during a charge-discharge cycle. It is less than 100% because of energy loss and is about 85–90% for thermochemical energy storage compared to 90% for LHS and 75–85% for hot water TES.

Hydrogen Storage

Energy can be both stored and transported as hydrogen or hydrogen hydrides. Hydrogen is commercially produced by steam reforming of hydrocarbons (e.g. CH_4) or from the products of coal gasification with a subsequent conversion to hydrogen. An alternative method is water electrolysis, whereby the input energy is used to decompose water into hydrogen and oxygen. Hydrogen can be transported either as compressed hydrogen gas, as liquid hydrogen, or in the form of a solid compound with certain metals or alloys.

The chemical energy in hydrogen can be converted into thermal, mechanical, or electrical energy. There are several possibilities: burning hydrogen in air to produce heat and power in a conventional steam turbine or a gas (combustion) turbine. The highest efficiency is expected from integration of the hydrogen-producing facility into a combined cycle based on high-temperature solid-oxide fuel cells. Overall efficiencies of more than 60% are predicted for such a combined cycle. In addition, waste heat can be used, and thus even higher fuel energy utilization factors will be attained.

Thermochemical Energy Storage and Transport

The chemical heat pipe is a system that can be used for energy storage and long-distance transport, especially of solar energy. Thermochemical energy storage with a chemical heat pipe is depicted in Figure 10.8.

Solar energy absorbed at high temperatures in a solar furnace or in a central receiver is used for catalytic reforming of methane in a high-temperature reactor at 960°C according to the reaction presented by Eq. (10.52). The product of the reforming process is a gas mixture of CO and H_2, which is used as a synthesis gas or an energy carrier gas. It is transported in a pipeline to the location of energy use. At a temperature of about 500–700°C, a methanation reaction can proceed with heat release. Rhodium or nickel is used as the catalyst in both reactors. The actual energy effect of this reaction is 6.02 MJ/kg CH_4 (Khartchenko, 1995). Thus thermochemical energy transport occurs with a high energy density without energy losses.

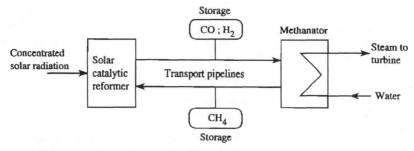

Figure 10.8. Thermochemical energy storage with chemical heat pipe.

ELECTROCHEMICAL AND ELECTROMAGNETIC ENERGY STORAGE

Storage Batteries

Storage batteries are one of the three energy storage technologies that have ever greater importance for utilities. The other two are PHPS and CAES discussed above. When a storage battery is charged, by connecting it to a source of direct electric current, electrical energy is converted into stored chemical energy. When the battery is discharged, the stored chemical energy is converted into electrical energy.

Potential applications of batteries are utility peak shaving, load leveling, and storage of electrical energy generated by wind turbine or photovoltaic plants. The capability of rapid change in operation from charge to discharge makes batteries suitable for electric utility applications (Elliot, 1995). The most common type of storage battery is the lead-acid battery.

The voltage U at external resistance (load R_e) of a lead-acid battery is given by the emf E less the voltage drop in battery internal resistance R_i:

$$U = E - IR_i \quad V \tag{10.53}$$

The current delivered is given by

$$I = E/(R_i + R_e) = nF \quad A \tag{10.54}$$

where n is the number of moles of electrons per mole of sulfuric acid ($=2$) and $F = 96,487$ C/mole (Faraday's constant). Thus the power output is

$$P = UI \quad W \tag{10.55}$$

Charging of a lead-acid battery occurs according to the following reaction:

$$2PbSO_4 + H_2O \rightarrow PbO_2 + 2H_2SO_4 + Pb \tag{10.56}$$

The acid-unsoluble lead sulfate of a positive electrode is converted to porous lead oxide, and the lead sulfate of a negative electrode is converted to pure lead. During discharging of a lead-acid battery, the above reaction proceeds in the reverse direction. Thus

$$PbO_2 + 2H_2SO_4 + Pb \rightarrow 2PbSO_4 + H_2O \tag{10.57}$$

Similarly, charging of a Ni-Cd battery occurs according to

$$2Ni(OH)_2 + Cd(OH)_2 \rightarrow 2NiO(OH) + Cd + H_2O \tag{10.58}$$

Discharging of a Ni-Cd battery proceeds as

$$2NiO(OH) + Cd + H_2O \rightarrow 2Ni(OH)_2 + Cd(OH)_2 \tag{10.59}$$

Major performance parameters of a storage battery are

- capacity C in ampere-hours (Ah)
- nominal voltage U_n in V

- nominal current I_n in A
- energy recovery efficiency
- charge factor

Typical values of U_n are 2.0 V for a lead-acid battery and 1.2 V for a Ni-Cd battery. The energy recovery efficiency of a storage battery is the ratio of the useful recovered energy to the energy input:

$$\eta = E_{out}/E_{in} \tag{10.60}$$

The maximum value of η reaches 75–90%. The energy recovery efficiency of a storage battery varies with the type of battery and the rate of discharge, but 75% should be attainable. However, the efficiencies are often lower.

The charge factor is the ratio of the electric charge input to the useful electric charge:

$$f_{ch} = E_{in}/E_{out} \tag{10.61}$$

It lies between 1.05 and 1.2 and is a reciprocal of the battery efficiency η.

Example 10.4

Calculate power output and sulfuric acid consumption at the anode of a 12-V lead-acid battery that delivers a current of 60 A.

Solution

Power output

$$P = UI = 12\ V \times 60\ A = 720\ W$$

At the cathode of the battery, 2 mol of electrons are released per mole of sulfuric acid. The number of moles of sulfuric acid required for anode and cathode reactions is

$$N_{acid} = 2I/(nF) = 2 \times 60\ A/(2 \times 94687\ coulombs/mol) = 6.34 \times 10^{-4}\ mol/s$$

where $F = 96{,}487$ coulombs/mol (Faraday's constant) and $n = 2$ mol electrons/mol sulfuric acid. Mass flow rate of sulfuric acid (molar mass $M_{acid} = 98$ g/mol) is

$$m = N_{acid}M_{acid} = 6.34 \times 10^{-4}\ mol/s \times 98\ g/mol = 0.06\ g/s = 0.22\ kg/h$$

Lead-acid batteries have some basic limitations, e.g., relatively low energy density (79 kWh/m³) and specific power (33 W/kg) and short life (only 300 cycles), especially when subjected to repeated cycling. Nickel-cadmium batteries have lower specific power (26 W/kg) and energy density (55 kWh/m³) but longer service life (2000 cycles) than lead-acid batteries. Silver oxide–cadmium batteries achieve an energy density of 146 kWh/m³ (Elliot, 1995).

The ac-to-dc conversion efficiency of battery systems is more than 75%. Batteries for storing large quantities of energy must meet the following requirements: specific power of about 50 W/kg, energy density of about 200 Wh/kg, 1000 full cycles, service life 4–6 years. For stationary applications, a battery energy storage system (BESS) should be capable of at least 3000 deep discharges over a lifetime of 10–15 years. The discharge time for peak power supply would be 8–10 hours, and the charge time roughly 10 hours. Utility-battery energy storage should have costs of less than $500/kW (Elliot, 1995). No existing storage battery can meet these requirements as yet. Existing utility storage systems rely on improved lead-acid batteries. A utility-operated reliable BESS in Berlin, Germany, has a capacity of 17 MW and a reserve-power discharge rate of 9.6 MWh. The expected BESS service life is 6–8 years.

Batteries for photovoltaic (PV) plants have the following life at an 80% depth of discharge: 3–7 years or 500 cycles for batteries for small-scale PV plants and 7–15 years or 1500 cycles for batteries for large-scale PV plants. The life of Ni-Cd batteries is much longer—15–25 years, or 2800 cycles. Changing-out individual batteries in a BESS, the storage system has an expected service life of about 30 years. Several innovative types of storage batteries, such as sodium-sulfur and zinc-bromine, some operating at high temperatures, are now under development.

Table 10.7. Comparison of different electrical energy storage technologies

Storage technology	Capacity, MW	Efficiency, %
CAES	50–350	81
PHPS	1000–2000	72
Battery	up to 1000	75
SMES	1–2000	91

Table 10.8. Comparison of various energy storage technologies

Storage	Energy density, Wh/kg	Efficiency, %	Cycle duration	Service life or number of cycles
Mechanical energy storage				
Pumped hydro, 300 m	0.81	65–75	Several days or months	50–80 years
Flywheel	20–30	75–85	Several minutes or hours	20 years
CAES (100 bars)	4	70–80	Several hours	20 years
Thermal energy storage $(\Delta t = 50 \text{ K})$				
Water	58	80–90	Several hours, days, or months	20 years
Concrete, rock	13	80–90	Several hours, days, or months	20 years
Electrical energy storage				
Lead-acid battery	40	80–90	Several hours or days	up to 1500 cycles
Nickel-cadmium battery	100	50–60	Several hours or days	up to 3000 cycles

Electromagnetic Energy Storage

Electromagnetic energy storage requires the use of a superconducting material that can carry strong electric currents with little or no loss when cooled below −255°C for a compound of niobium and tin (Nb_3Sn) or even below −263°C for a niobium-titanium (Nb-Ti) alloy. Electrical energy supplied as direct current to a superconducting wire coil would be stored in the electromagnetic field. Subsequently, the stored energy could be recovered as dc electrical energy by attaching the coil to a load. Provided that a number of serious problems are resolved, such a superconducting magnetic energy storage (SMES) system would be a long-term storage technology for eventual use in load leveling and/or peaking applications in large electric utility installations.

The energy capacity of a superconducting coil is given by

$$P = 1/2LI^2 \quad \text{J} \tag{10.62}$$

where L is coil inductance in H and I is current in A.

Four energy storage technologies for large utility utilization, namely, CAES, PHPS, electrochemical batteries, and SMES, are compared in Table 10.7. General comparison of various energy storage technologies is given in Table 10.8.

ENERGY STORAGE TECHNOLOGIES FOR SOLAR POWER PLANTS

Solar power plants need TES systems for the intermediate-temperature range of 300–500°C and the high-temperature range up to 1200°C. Because of large heat losses at such temperatures, only short-term TES (from 0.5 to 3 hours) can be economically feasible.

Storage Concepts for Central Receiver Power Plants

Suitable storage concepts depend on the type of heat transfer fluid (HTF) used and its pressure and temperature levels. Air, HTF salts such as HITEC, synthetic oils, and water/steam are employed as HTF in solar power plants.

Energy storage systems increase the power generation of solar power plants by extending daily operation times. They allow power generation to be adapted to power demand and simplify the

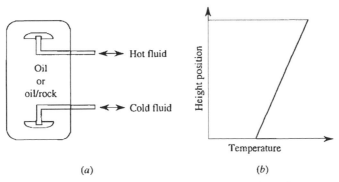

Figure 10.9. Advanced single-tank thermocline storage system of a central receiver solar power plant: (*a*) schematic and (*b*) temperature versus height position.

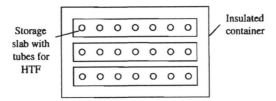

Figure 10.10. Tubes-in-concrete sensible heat storage for solar power plants.

operation of the plant. However, storage systems are expensive. Advanced solar power plants must be equipped with a storage system to optimize the solar power generation, to reduce or to avoid supplementary fossil fuel firing. Solar/fossil hybrid power plants without storage systems may be attractive for near-term use.

Three different storage technologies are considered for utilization in central receiver systems (Elliot, 1995):

- sensible heat storage system using either receiver HTF (molten salt, liquid sodium) or solids (ceramic bricks or spheres, oil/rock, steel plates, concrete slabs, etc.) as the storage material
- phase-change storage system
- thermochemical storage system

Figure 10.9a depicts a schematic of the advanced single-tank thermocline storage system of the central receiver solar power plant SSPS-DCS, with a temperature profile in the vertical direction shown in Figure 10.9b. An SHS concept using multiple tubes embedded in concrete slabs is shown in Figure 10.10. The tubes carry hot HTF for storage charging and cold fluid for storage discharging.

Experimental heat storage systems have been built and operated or are currently under development (particularly advanced storage systems for use in volumetric air receiver plants). The phase-change and the thermochemical storage systems, which offer technical and operating advantages, are in the early stages of development for future applications. Molten-salt storage will be applied for commercial molten-salt receiver systems (e.g., Solar Two project in the United States). Ceramic pebble-bed storage will be applied in near-term projects using advanced volumetric air receivers.

The optimum storage capacity depends on the plant configuration, the application, and the specific storage capital costs. The storage capacity may be chosen so that a large power output, e.g., 1200 MW, will be provided for a very short period of time, e.g., 5–10 s, to guarantee transmission stability. A intermediate storage duration may correspond to 1–3 hours of plant power output. In some cases it may be extended up to 10–15 hours in order to increase the plant capacity factor. As

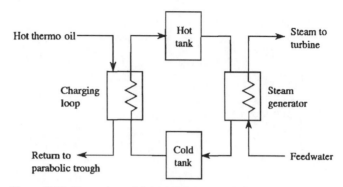

Figure 10.11. Two-tank sensible heat TES for a parabolic trough solar power plant.

expected, solar power plants with an optimum storage capacity have a better cost-effectiveness ratio than solar power plants without energy storage. The response time of a storage system depends on the storage type. It is very short (0.02 s) for electric batteries and superconducting coils and is about 10 min for potential energy storage like PHPS and CAES.

Several TES concepts have been tested in the experimental solar central receiver plants built so far. They include a single-tank storage using a receiver HTF such as liquid sodium or molten HITEC salt as the storage medium. An example is the 18-MWht molten HITEC salt storage of the CESA-1 plant (Meinecke and Bohn, 1995). A thermocline storage concept used a storage vessel filled with a dual-medium storage system containing oil and rock (see Figure 10.9). Yet another storage type used in the earlier central receiver plants is a water/steam storage in pressurized vessels.

Figure 10.11 shows the two-tank SHS concept applicable for parabolic trough solar power plants. This type of storage has been implemented in the 7-MWh advanced molten-nitrate-salt storage system successfully operated at Sandia National Laboratory (Meinecke and Bohn, 1995). In the European 30-MWe PHOEBUS central receiver project, a 250-MWht storage system will be used. An advanced experimental storage with a 0.5–1 MWht useful capacity contains a package of ceramic spheres, ~10 mm in diameter, as the storage medium (Meinecke and Bohn, 1995).

Storage Concepts for Parabolic Trough Power Plants

Storage systems are less economically feasible for parabolic trough plants than for central receiver plants. Nevertheless, some storage systems have been used in experimental parabolic trough plants. They are based on either the single-tank thermocline principle (see Figure 10.9) or the two-tank principle (see Figure 10.10). This is because of the relatively low temperature of the synthetic oil HTF, which reaches only 390°C. Unlike molten salt, the thermo oil has poor properties if it is used as a storage medium. The oil temperature drop by passing the heat transfer circuitry in the storage tank is relatively small, and this results in much larger storage vessels.

There are some examples of these storage systems used in experimental solar power plants. In the 13.8-MWe SEGS 1 power plant, in the United States, a 117-MWht two-tank storage system used thermo oil as the storage medium. In the 500-kWe IEA-SSPS-DCS plant in Almeria, Spain, an advanced single-tank thermocline 4-MWht dual-medium storage system is employed. The storage tank contains cast iron plates and is filled with the heat transfer oil. Thus the storage capacity is enhanced, and the expensive oil is saved.

Thus there are basically three SHS concepts for central receiver and parabolic trough power plants:

1. single-tank dual-medium thermocline storage (oil + rock)
2. two-tank single-medium storage (molten salt, thermo oil)
3. packed-bed storage (rock, pebbles, ceramic spheres)

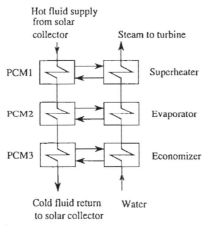

Figure 10.12. Cascaded TES with three PCMs
for a solar power plant.

Volumetric energy density in SHS is typically about 540 MJ/m^3 for packed-bed storage with
25% void fraction and about 790 MJ/m^3 for molten-salt (nitrates) storage.

More Advanced Concepts

In mid- and long-term, advanced energy storage concepts based on a combination of ceramic
and phase-change (e.g. nitrate salts) materials are anticipated to be integrated into solar power
generation systems. Such storage systems may be applied to future commercialized parabolic
trough plants based on direct steam generation, which allows the generation of slightly superheated
steam during discharging for improved system efficiency.

Metal hydrides, e.g., MgH$_2$, can be used as storage media for 300–400°C. When heat is added,
they decompose in metal and hydrogen, so that a thermochemical energy storage is possible. For
the temperature range 500–1300°C, the sensible storage concept with ceramic material such as
Al$_2$O$_3$, SiO$_2$, or MgO is applicable.

In advanced solar power plant projects, PCM-based TES will also be used. Some prospective
PCMs for potential use in storage systems are eutectic salt (70% NaF + 30% FeF$_2$) with a fusion
temperature of about 680°C and the highest (of all PCMs) volumetric energy storage density of
about 1500 MJ/m^3, and zink chloride with a fusion temperature of about 370°C and a volumetric
energy storage density of about 400 MJ/m^3. Figure 10.12 shows the cascaded TES concept with
three PCMs, each having a different fusion temperature. In this concept the desired temperature
stratification in the thermal storage is achieved. The heat is extracted from corresponding sections
of the storage for feedwater heating in the economizer, evaporating and superheating the generated
steam to be further led to the steam turbine of a hybrid solar–fossil power plant.

High energy density is achieved in a dual-medium LHS. Thereby, the pores of ceramic spheres
are filled with a PCM salt, and thus a hybrid sensible–latent heat storage medium will be made.
In addition, effective heat transfer occurs when the HTF gas passes through the channels in the
storage material (Meinecke and Bohn, 1995).

CLOSURE

The emphasis in this chapter is on TES technology. Sensible and latent heat storage are successfully used in low-temperature solar plants. Major types of TES for short-term and long-term
applications are thoroughly discussed. The problem of efficient high-temperature energy storage,
which is required for solar thermal power generation technology, has not yet been solved in an
economically viable way.

PROBLEMS

10.1. Calculate the pumping and generation powers and the energy storage efficiency for a pumped hydropower storage (PHPS) system if the elevation difference between the two reservoirs is 250 m, the volumetric flow rates in pumping and generating modes are 100 and 180 m^3/s, the pumping and generation period duration are 12.5 and 9.5 h/d, and the pump, turbine, and generator-motor efficiencies are 0.77, 0.9, and 0.98, respectively. Assume a head loss in the water channel of 4% and a water density of 1000 kg/m^3.

10.2. Consider compressed-air energy storage (CAES) for a 150-MW peaking compressor/gas turbine facility. The compressor and gas turbine isentropic efficiencies are 0.86 and 0.82, respectively. The generator-motor efficiency is 0.98. The other energy losses over the storage cycle are 6%. The two-stage intercooled compressor intake is at 20°C and 0.1 MPa and discharge is at 4.9 MPa. Intercooling occurs at an optimal pressure back to the intake temperature. The reheat expansion turbine operates with a supplementary firing, which raises the compressed-air temperature in the first stage to a TIT of 1100°C by burning natural gas. After expansion in the HP portion of the gas turbine to a pressure of 0.7 MPa, the gas is reheated to the same temperature. Assuming a lower heating value of 49.7 MJ/kg for the natural gas and a constant specific heat for the air of 1.05 kJ/(kg K) in the entire temperature range, calculate (a) the total fuel rate of supplementary firing and reheating in kg/s, (b) the CAES storage capacity in MWh, and (c) the average air flow rate during a 5-hour cavern charging duty in m^3/h.

10.3. Compare the mass and volume of a storage wall of a passively heated office building made of (a) concrete and (b) water reservoirs if the required storage capacity is 320 kWh and the maximum storage temperature change is 13 K. The specific heat of water is 1.16 Wh/(kg K) or 1160 Wh/(m^3 K) and that of concrete is 0.28 Wh/(kg K) or 672 Wh/(m^3 K).

10.4. Calculate the mass of a nitrate salt latent heat storage (LHS) required for a 3-hour full-capacity operation of a parabolic trough solar power plant with an electric power output of 80 MW and an overall plant efficiency of 14%. The storage medium is a phase-change material (PCM) that consists of 60% $NaNO_3$ and 40% KNO_3 (% by volume) and operating within a temperature range of 280–350°C. The storage efficiency is 0.95. Assume for the PCM the melting temperature of 323°C, density of 1898 kg/m^3, specific heat of 1.83 kJ/(kg K), and heat of fusion of 169.2 kJ/kg.

REFERENCES

Beckman, G., and Gilli, P. V. 1984. *Thermal energy storage*. Wien: Springer-Verlag.

Bitterlich, W. 1987. *Speicher für thermische Energie*. Düsseldorf: VDI-Verlag.

Burnham, L., ed. 1993. *Renewable energy sources for fuels and electricity*. Washington, D.C.: Island Press.

Dalenbäck, J.-O. 1990. Central solar heating plants with seasonal storage. Status Report Doc. D14, Swedish Council for Building Research, Stockholm.

Dinter, F. 1992. *Thermische Energiespeicher in Solarfarmkraftwerken und ihre Bewertung*. Aachen: Shaker.

Duffie, J. A., and Beckman, W. A. 1991. *Solar engineering of thermal processes*, 2nd ed. New York: Wiley.

Elliot, T. 1995. Electrical energy storage hinges on three leading technologies. *Power* 139(8):42–45.

Garg, H. P., Mullick, S. K., and Bhargava, A. K. 1985. *Solar thermal energy storage*. Dordrecht: Reidel.

Hoogendoorn, C. J., and Bart, G. C. J. 1992. Performance and modelling of latent heat stores. *Solar Energy* 48(1):53.

Jensen, J., and Sorensen, B. 1981. *Fundamentals of energy storage*. New York: Wiley.

Khartchenko, N. V. 1995. *Thermische Solaranlagen*. Berlin: Springer.

Lane, G. A. 1983. *Solar heat storage: Latent heat materials*, vol. I. Boca Raton, Fla.: CRC Press.

Lane, G. A. 1986. *Solar heat storage: Latent heat materials*, vol. II. Boca Raton, Fla.: CRC Press.

Meinecke, W., and Bohn, M. 1995. *Solar energy concentrating systems*, eds., M. Becker and P.C. Klimas. Heidelberg: Müller.

LIST OF ABBREVIATIONS

AC	alternating current
AF	air-fuel ratio
AF	as fired (fuel ratio)
AFBC	atmospheric fluidized bed combustion
AFC	alkaline fuel cell
ASU	air separation unit
BGL	British Gas/Lurgi
CAES	compressed-air energy storage
CASH	compressed-air energy storage with air humidification
CC	combined-cycle power plant
CCCS	clean coal combustion system
CCOFA	close-coupled overfire air
CF	capacity factor
CFBC	circulating fluidized bed combustion
CFS	concentric firing system
CG	cogeneration
CGCU	cold gas cleanup
CHAT	chemical humid air turbine cycle
COS	carbon oxysulfide
DAF	dry ash free (solid fuel)
DC	direct current
DCCS	direct coal combustion system
DCFD	direct coal-fired diesel engine
DCFGT	direct coal-fired gas turbine
DLN	dry, low-NO_x
DOE	U.S. Department of Energy
EFCC	externally fired combined-cycle power plant
EFCC/HIPPS	externally fired combined-cycle/high-performance power system
EMF	electro-motive force
EPA	U.S. Environmental Protection Agency
ESP	electrostatic precipitator
EUF	energy utilization factor
FBC	fluidized bed combustion
FESR	fuel energy savings ratio
FGD	flue gas desulfurization
FGR	flue gas recirculation
FWH	feedwater heater
GT	gas turbine
HAT	humid air turbine cycle
HGCU	hot gas cleanup
HHV	higher heating value
HIPPS	high-performance power system

HP	high pressure
HRSG	heat recovery steam generator
HRU	heat recovery unit
HTAF	high-temperature air heater
HTAH	high-temperature air heater
HTF	heat transfer fluid
HTW	high temperature Winkler
ICC	intercooled combined cycle
ICR	intercooled recuperated cycle
IEA	International Energy Agency
IGAC	integrated gasification advanced cycle
IGCC	integrated gasification combined cycle
IGFC	integrated gasification fuel-cell power plant
IHR	incremental heat rate
IP	intermediate pressure
ISTIG	intercooled steam injected cycle
LEA	low excess air
LEAB	low excess air boiler
LHS	latent heat storage
LHV	lower heating value
LP	low pressure
LPC	low-pressure compressor
MCFC	molten-carbonate fuel cell
MHD	magnetohydrodynamic generator
OFA	overfire air
PAFC	phosphoric acid fuel cell
PCCG	pulverized coal fired cogeneration
PCFBC	pressurized circulating fluidized bed combustion
PCLEB	pulverized coal low-emission boiler
PCM	phase-change material
PFBC	pressurized fluidized bed combustion
PHPS	pumped hydro power storage
PPM	parts per million
PPMV	parts per million by volume
RWI	recuperated water injected cycle
SCR	selective catalytic reduction
SEGS	solar electric generating system
SHS	sensible heat storage
SMES	superconducting magnetic energy storage
SNCR	selective noncatalytic reduction
SNG	substitute natural gas
SNRB	SO_x-NO_x Rox-Box
SOFA	separated overfire air
SOFC	solid electrolyte fuel cell
STAG	gas and steam turbine combined-cycle
STIG	steam-injected gas turbine cycle
TES	thermal energy storage
TIT	gas turbine inlet temperature
TOT	gas turbine outlet temperature
UHC	unburned hydrocarbons
WHB	waste-heat boiler
WHR	waste-heat recuperator

APPENDICES

APPENDIX A: UNIT CONVERSION FACTORS

Length
1 m = 3.281 ft
1 ft = 0.3048 m
1 in = 25.4 mm
1 mile = 1.609 km

Area
1 ft^2 = 0.09290304 m^2
1 in^2 = 0.00064516 m^2

Volume
1 m^3 = 10^3 liter = 35.31 ft^3 = 264.2 US gal
1 ft^3 = 0.0283 m^3
1 U.K. gal = 4.546 liters
1 U.S. gal = 3.785 liters

Mass
1 lb = 0.45359231 kg
1 kg = 2.205 lbm
1 metric ton = 1000 kg
1 short ton = 2000 lbm

Density
1 kg/m^3 = 0.06243 lbm/ft^3
1 lbm/ft^3 = 16.02 kg/m^3

Force
1 N = 1 kg m/s^2 = 0.22481 lb$_f$
1 lb$_f$ = 4.4482 N

Pressure
1 Pa = 1 N/m^2
1 bar = 10^5 Pa = 0.9869 atm
1 psi = 1 lbf/in^2 = 6.894 Pa
1 atm = 1.01325 bar = 14.696 lbf/in

Energy, heat
1 J = 1 N m = 1 W s
1 kJ = 0.94783 Btu
1 MJ = 0.27767 kWh
1 kWh = 3.6 MJ = 3412.11 BTU

Mass flow rate
1 kg/s = 1632.93 lb$_m$/h
1 lb$_m$/h = 0.0006124 kg/s = 2.20464 kg/h
1 BTU = 1.055 kJ = 0.292963 × 10^{-3} kWh
1 MBTU = 1.055 GJ = 292.963 kWh
1 J = 6.242 × 10^{18} eV = 2.778 × 10^{-7} kWh

Power
1 W = 1 J/s
1 kW = 3412.2 BTU/h
1 BTU/h = 0.2929 W

Specific heat
1 kJ/(kg K) = 0.2388 BTU/lbm °F
1 BTU/(lbm °F) = 4.1868 kJ/kg K

Thermal conductivity
1 W/(m K) = 0.5778 BTU/h ft °F
1 BTU/(h ft °F) = 1.7307 W/m K

Heat transfer coefficient
1 BTU/h ft^2 °F = 5.6783 W/m^2 K
1 W/(m^2 K) = 0.17611 BTU/h ft^2 °F

Viscosity
1 cP = 10^3 Pa s
1 lbf h/ft^2 = 0.1124 MPa s

Temperature
1 K = 1.8 °R
T [K] = t[°C] + 273.15
t°C = (t[°F] − 32)/1.8
T[°R] = t[°F] + 459.67

Electrical and magnetic units
1 A = 1 W/V = 1 C/s
1 V = 1 J/C = 1 A • ohm
1 ohm = 1 V/A
1 F = 1 A • s/V
1 Wb = 1 V • s
1 tesla = 1 Wb/m^2

APPENDIX B: PHYSICAL CONSTANTS

Boltzmann constant: k = 1.381 × 10^{-23} J/K = 8.618 × 10^{-11} MeV/K
Electron charge: e = 1.602 × 10^{-19} C (1 C = J/V)
Faraday constant: F = 9.649 × 10^7 coulomb /(kg mol of electron)
Planck constant: h = 6.626 × 10^{-34} J s = 4.136 × 10^{-21} MeV s
Stefan-Boltzmann constant: σ = 5.67 × 10^{-8} W/(m^2 K^4)
Universal gas constant: R = 8.314 kJ/(kg mol K)
Velocity of light: c = 2.998 × 10^8 m/s
Avogadro's number: Av = 6.022 × 10^{26} molecules (or atoms)/(kg mol)
Solar constant: Isc = 1.367 kW/m^2 at an average distance between the sun and the Earth of
    149.5 × 10^6 km
Mass of the Earth: M_E = 5.979 × 10^{24} kg
Radius (average) of the Earth: r_E = 6.371 × 10^6 m
Standard atmospheric pressure: 1 atm = 1.013 × 10^5 Pa = 1.013 bars = 14.696 lbf/in
Standard gravitational constant: g = 9.807 m/s^2

APPENDIX C: TEMPERATURE CONVERSION FROM CELSIUS SCALE TO FAHRENHEIT SCALE

$$t[°C] = (t[°F] - 32)/1.8; \ t[°F] = 1.8t[°C] + 32$$

$$\Delta t[°F] = 1.8\Delta t[°C]; \ \Delta t[°C] = \Delta t[°F]/1.8$$

t, °C	t, °F	t, °C	t, °F	t, °C	t, °F	t, °C	t, °F
0	32	100	212	450	842	1400	2552
10	50	110	230	500	932	1500	2732
20	68	120	248	600	1112	1600	2912
30	86	140	284	700	1292	1700	3092
40	104	160	320	800	1472	1800	3272
50	122	180	356	900	1652	1900	3452
60	140	200	392	1000	1832	2000	3622
70	158	250	482	1100	2012	2200	3992
80	176	300	572	1200	2192	2600	4712
90	194	400	752	1300	2372	3000	5432

APPENDIX D: AVERAGE CONSTANT PRESSURE MOLAR SPECIFIC HEAT OF IDEAL GASES, IN kJ/(kMOL K)

t, °C	N_2	O_2	H_2	Air	H_2O	CO_2	CO	NH_3	CH_4	SO_2
0	29.09	29.26	28.62	29.08	33.47	35.92	29.11	34.99	34.59	38.91
100	29.12	29.53	28.94	29.15	33.71	38.17	29.16	36.37	37.02	40.71
200	29.20	29.92	29.07	29.30	34.08	40.13	29.29	38.13	39.54	42.43
300	29.35	30.39	29.14	29.52	34.54	41.83	29.50	40.04	42.34	43.99
400	29.56	30.87	29.19	29.79	35.05	43.33	29.77	41.98	45.23	45.35
500	29.82	31.32	29.25	30.09	35.59	44.66	30.08	44.04	48.20	46.53
600	30.11	31.75	29.32	30.41	36.15	45.85	30.41	46.09	50.70	47.55
700	30.40	32.14	29.41	30.72	36.74	46.91	30.74	48.01	53.34	48.43
800	30.69	32.49	29.52	31.03	37.34	47.86	31.05	49.85	55.77	49.20
900	30.98	32.82	29.65	31.32	37.95	48.72	31.36	51.53	58.03	49.88
1000	31.25	33.11	29.79	31.60	38.56	49.50	31.65	53.08	60.25	50.47
1100	31.52	33.38	29.95	31.86	39.16	50.21	31.92	54.50	62.29	51.01
1200	31.77	33.62	30.12	32.11	39.76	50.85	32.17	55.84	64.13	51.49
1300	32.00	33.85	30.29	32.35	40.34	51.44	32.41	57.06		51.92
1400	32.22	34.07	30.47	32.57	40.91	51.98	32.63	58.14		52.31
1500	32.43	34.28	30.65	32.77	41.47	52.47	32.84	59.19		52.67
1600	32.62	34.47	30.84	32.97	42.00	52.93	33.03	60.20		53.00
1700	32.80	34.65	31.02	33.15	42.52	53.35	33.21	61.12		53.31
1800	32.97	34.83	31.21	33.32	43.03	53.74	33.38	61.95		53.59
1900	33.12	35.00	31.39	33.48	43.51	54.10	33.54	62.75		53.85
2000	33.28	35.17	31.58	33.64	43.97	54.44	33.69	63.46		54.09
2100	33.42	35.33	31.75	33.79	44.42	54.76	33.83	64.13		54.32
2200	33.55	35.48	31.93	33.93	44.86	55.06	33.96	64.76		54.54
2300	33.68	35.64	32.10	34.06	45.27	55.34	34.08	65.35		54.75
2400	33.80	35.78	32.27	34.19	45.68	55.60	34.20	65.93		54.94
2500	33.91	35.93	32.44	34.31	46.07	55.85	34.31	66.48		55.13
2600	34.02	36.07	32.60	34.42	46.44	56.09	34.42	66.98		55.31
2700	34.12	36.21	32.76	34.54	46.80	56.31	34.52	67.44		55.47
2800	34.22	36.35	32.91		47.15	56.52	34.62	67.86		55.64
2900	34.31	36.48	33.07		47.49	56.72	34.71	68.28		55.79
3000	34.40	36.62	33.22		47.82	56.91	34.79	68.70		55.95
3100	34.48	36.75	33.36		48.13	57.10	34.88			56.09
3200	34.56	36.87	33.51		48.44	57.27	34.96			56.24
3300	34.64	37.00	33.65		48.73	57.44	35.03			56.37
M	28.01	32.00	2.016	28.95	18.02	44.01	28.01	17.03	16.04	64.06

M is molar mass in kg/kmol.

APPENDIX E: AVERAGE CONSTANT PRESSURE MOLAR SPECIFIC HEAT OF IDEAL GASES AND WATER VAPOR, IN kJ/(kg K)

t, °C	CO_2	O_2	N_2	Air	CO	H_2	H_2O	SO_2
0	0.8165	0.9148	1.0387	1.0033	1.0397	14.07	1.8584	0.6083
25	0.8299	0.9164	1.0387	1.0036	1.0399	14.17	1.8608	0.6153
50	0.8429	0.9182	1.0389	1.0042	1.0403	14.28	1.8640	0.6224
100	0.8677	0.9230	1.0396	1.0059	1.0416	14.40	1.8718	0.6365
150	0.8907	0.9288	1.0408	1.0081	1.0435	14.41	1.8814	0.6503
200	0.9122	0.9354	1.0426	1.0111	1.0462	14.43	1.8924	0.6634
250	0.9321	0.9425	1.0450	1.0145	1.0496	14.44	1.9046	0.6700
300	0.9509	0.9499	1.0480	1.0185	1.0537	14.45	1.9177	0.6878
350	0.9685	0.9574	1.0516	1.0229	1.0583	14.46	1.9316	0.6988
400	0.9850	0.9649	1.0556	1.0278	1.0634	14.48	1.9460	0.7090
450	1.0005	0.9721	1.0601	1.0328	1.0688	14.50	1.9608	0.7186
500	1.0152	0.9792	1.0648	1.0380	1.0745	14.51	1.9760	0.7274
550	1.0291	0.9860	1.0698	1.0434	1.0803	14.52	1.9915	0.7357
600	1.0422	0.9925	1.0750	1.0488	1.0862	14.55	2.0074	0.7434
650	1.0546	0.9987	1.0802	1.0542	1.0921	14.57	2.0236	0.7505
700	1.0663	1.0047	1.0855	1.0595	1.0979	14.59	2.0400	0.7572
750	1.0775	1.0103	1.0907	1.0648	1.1036	14.62	2.0566	0.7634
800	1.0880	1.0157	1.0960	1.0700	1.1092	14.64	2.0733	0.7692
850	1.0981	1.0209	1.1011	1.0751	1.1148	14.67	2.0901	0.7747
900	1.1076	1.0258	1.1062	1.0800	1.1201	14.71	2.1070	0.7798
950	1.1167	1.0305	1.1112	1.0848	1.1254	14.74	2.1239	0.7846
1000	1.1253	1.0350	1.1160	1.0895	1.1304	14.78	2.1408	0.7891
1100	1.1413	1.0434	1.1254	1.0985	1.1402	14.85	2.1744	0.7975
1200	1.1560	1.0511	1.1343	1.1069	1.1492	14.94	2.2075	0.8050
1300	1.1693	1.0583	1.1426	1.1173	1.1577	15.03	2.2399	0.8117
1400	1.1816	1.0651	1.1504	1.1242	1.1656	15.12	2.2716	0.8179
1500	1.1928	1.0715	1.1578	1.1309	1.1730	15.21	2.3024	0.8235
1600	1.2032	1.0775	1.1647	1.1378	1.1799	15.30	2.3322	0.8286
1700	1.2128	1.0832	1.1711	1.1442	1.1863	15.39	2.3610	0.8334
1800	1.2217	1.0888	1.1772	1.1504	1.1924	15.48	2.3889	0.8378
1900	1.2299	1.0941	1.1829	1.1562	1.1980	15.56	2.4157	0.8419
2000	1.2376	1.0993	1.1883	1.1608	1.2034	15.65	2.4416	0.8457
2100	1.2449	1.1043	1.1934	1.1659	1.2084		2.4666	0.8493
2200	1.2516	1.1092	1.1981	1.1712	1.2131		2.4906	0.8527
2300	1.2580	1.1140	1.2026	1.1758	1.2175		2.5138	0.8559
2400	1.2640	1.1186	1.2069	1.1799	1.2217		2.5362	0.8590
2500	1.2696	1.1232	1.2109	1.1844	1.2257		2.5577	0.8619
2600	1.2750	1.1276	1.2147		1.2295		2.5785	0.8647
2700	1.2800	1.1320	1.2184		1.2331		2.5986	0.8673
2800	1.2848	1.1363	1.2218		1.2365		2.6180	0.8699
2900	1.2894	1.1405	1.2251		1.2398		2.6368	0.8723
3000	1.2938	1.1446	1.2282		1.2429		2.6549	0.8747

Source: Baehr, H. D., Hartmann, H., Pohl, H.-Chr. and Schomächer, H. 1968. Thermodynamische Funktionen idealer Gase für Temperaturen bis 6000 K. Berlin.

APPENDIX F: THERMODYNAMIC PROPERTIES OF SATURATED WATER AND SATURATED WATER VAPOR: TEMPERATURE TABLE

t, °C	p, bar	v', dm³/kg	v'', m³/kg	h', kJ/kg	h'', kJ/kg	r, kJ/kg	s', kJ/(kg K)	s'', kJ/(kg K)
0.00	0.006108	1.0002	206.3	−0.04	2502	2502	−0.0002	9.158
5	0.008718	1.0000	147.2	21.01	2511	2490	0.0762	9.027
10	0.012270	1.0003	106.4	41.99	2520	2478	0.1510	8.902
15	0.01704	1.0008	77.98	62.94	2529	2466	0.2243	8.783
20	0.02337	1.0017	57.84	83.86	2538	2454	0.2963	8.668
25	0.03166	1.0029	43.40	104.77	2547	2443	0.3670	8.559
30	0.04241	1.0043	32.93	125.7	2556	2431	0.4365	8.455
35	0.05622	1.0060	25.24	146.6	2565	2419	0.5049	8.354
40	0.07375	1.0078	19.55	167.5	2574	2407	0.5721	8.258
45	0.09582	1.0099	15.28	188.4	2583	2395	0.6383	8.166
50	0.1234	1.0121	12.05	209.3	2592	2383	0.7035	8.078
55	0.1574	1.0145	9.579	230.2	2601	2371	0.7677	7.993
60	0.1992	1.0171	7.679	251.1	2610	2359	0.8310	7.911
65	0.2501	1.0199	6.202	272.0	2618	2346	0.8933	7.832
70	0.3116	1.0228	5.046	293.0	2627	2334	0.9548	7.757
75	0.3855	1.0259	4.134	313.9	2635	2322	1.0154	7.684
80	0.4736	1.0292	3.409	334.9	2644	2309	1.0753	7.613
85	0.5780	1.0326	2.829	355.9	2652	2297	1.134	7.545
90	0.7011	1.0361	2.361	376.9	2660	2283	1.193	7.480
95	0.8453	1.0399	1.982	398.0	2668	2270	1.250	7.417
100	1.0133	1.0437	1.673	419.1	2676	2257	1.307	7.355
105	1.2080	1.0477	1.419	440.2	2684	2244	1.363	7.296
110	1.433	1.0519	1.210	461.3	2691	2230	1.419	7.239
115	1.691	1.0562	1.036	482.5	2699	2216	1.473	7.183
120	1.985	1.0606	0.8915	503.7	2706	2202	1.528	7.129
125	2.321	1.0652	0.7702	525.0	2713	2188	1.581	7.077
130	2.701	1.0700	0.6681	546.3	2720	2174	1.634	7.026
135	3.131	1.0750	0.5818	567.7	2727	2159	1.687	6.977
140	3.614	1.0801	0.5085	589.1	2733	2144	1.739	6.928
145	4.155	1.0853	0.4460	610.6	2739	2129	1.791	6.882
150	4.760	1.0908	0.3924	632.2	2745	2113	1.842	6.836
155	5.433	1.0964	0.3464	653.8	2751	2097	1.892	6.791
160	6.181	1.1022	0.3068	675.5	2757	2081	1.943	6.748
165	7.008	1.1082	0.2724	697.3	2762	2065	1.992	6.705
170	7.920	1.1145	0.2426	719.1	2767	2048	2.042	6.663
175	8.924	1.1209	0.2165	741.1	2772	2031	2.091	6.622
180	10.027	1.1275	0.1938	763.1	2776	2013	2.139	6.582
185	11.23	1.1344	0.1739	785.3	2780	1995	2.188	6.542
190	12.55	1.1415	0.1563	807.5	2784	1977	2.236	6.504
200	15.55	1.1565	0.1272	852.4	2791	1939	2.331	6.428
210	19.08	1.1726	0.1042	897.7	2796	1899	2.425	6.354
215	21.06	1.1811	0.09463	920.6	2798	1878	2.471	6.318
220	23.20	1.1900	0.08604	943.7	2800	1856	2.518	6.282
225	25.50	1.1992	0.07835	966.9	2801	1834	2.564	6.246
230	27.98	1.2087	0.07145	990.3	2802	1812	2.610	6.211
235	30.63	1.2187	0.06525	1013.8	2802	1789	2.656	6.176
240	33.48	1.2291	0.05965	1037.6	2802	1765	2.702	6.141
245	36.52	1.2399	0.05461	1061.6	2802	1740	2.748	6.106
250	39.78	1.2513	0.05004	1085	2800	1715	2.794	6.071
255	43.25	1.2632	0.04590	1110	2799	1689	2.839	6.036
260	46.94	1.2756	0.04213	1135	2796	1662	2.885	6.001
265	50.88	1.2887	0.03871	1160	2794	1634	2.931	5.966
270	55.06	1.3025	0.03559	1185	2790	1605	2.976	5.930
275	59.50	1.3170	0.03274	1211	2786	1575	3.022	5.895

(Continued)

APPENDIX F *(Continued)*

t, °C	p, bar	v', dm³/kg	v'', m³/kg	h', kJ/kg	h'', kJ/kg	r, kJ/kg	s', kJ/(kg K)	s'', kJ/(kg K)
280	64.20	1.3324	0.03013	1237	2780	1544	3.068	5.859
285	69.19	1.3487	0.02773	1263	2775	1511	3.115	5.822
290	74.46	1.3659	0.02554	1290	2768	1478	3.161	5.785
295	80.04	1.3844	0.02351	1317	2760	1443	3.208	5.747
300	85.93	1.4041	0.02165	1345	2751	1406	3.255	5.708
305	92.14	1.4252	0.01993	1373	2741	1368	3.303	5.669
310	98.70	1.4480	0.01833	1402	2730	1328	3.351	5.628
315	105.61	1.4726	0.01686	1432	2718	1286	3.400	5.586
320	112.9	1.4995	0.01548	1463	2704	1241	3.450	5.542
325	120.6	1.5289	0.01419	1494	2688	1194	3.501	5.497
330	128.6	1.5615	0.01299	1527	2670	1144	3.553	5.449
335	137.1	1.5978	0.01185	1560	2650	1090.5	3.606	5.398
340	146.1	1.6387	0.01078	1596	2626	1030.7	3.662	5.343
350	165.4	1.7411	0.008799	1672	2568	895.7	3.780	5.218
360	186.8	1.8959	0.006940	1764	2485	721.3	3.921	5.060
370	210.5	2.2136	0.004973	1890	2343	452.6	4.111	4.814
371	213.1	2.2778	0.004723	1911	2318	407.4	4.141	4.774
372	215.6	2.3636	0.004439	1936	2287	351.4	4.179	4.724
373	218.2	2.4963	0.004084	1971	2244	273.5	4.233	4.656
374	220.8	2.8407	0.003458	2046	2155	108.6	4.249	4.517
374.15	221.2	3.17	0.00317	2107	2107	0	4.443	4.443

APPENDIX G: THERMODYNAMIC PROPERTIES OF SATURATED WATER AND SATURATED WATER VAPOR: PRESSURE TABLE

p, bars	t, °C	u', dm³/kg	u'', m³/kg	ϱ'', kg/m³	h', kJ/kg	h'', kJ/kg	r, kJ/kg	s', kJ/(kg K)	s'', kJ/(kg K)
0.010	6.9808	1.0001	129.20	0.007739	29.34	2514.4	2485.0	0.1060	8.9767
0.020	17.513	1.0012	67.01	0.01492	73.46	2533.6	2460.2	0.2607	8.7246
0.030	24.100	1.0027	45.67	0.02190	101.00	2545.6	2444.6	0.3544	8.5785
0.040	28.983	1.0040	34.80	0.02873	121.41	2554.5	2433.1	0.4225	8.4755
0.050	32.898	1.0052	28.19	0.03547	137.77	2561.6	2423.8	0.4763	8.3960
0.060	36.183	1.0064	23.74	0.04212	151.50	2567.5	2416.0	0.5209	8.3312
0.070	39.025	1.0074	20.53	0.04871	163.38	2572.6	2409.2	0.5591	8.2767
0.080	41.534	1.0084	18.10	0.05523	173.86	2577.1	2403.2	0.5925	8.2296
0.090	43.787	1.0094	16.20	0.06171	183.28	2581.1	2397.9	0.6224	8.1881
0.10	45.833	1.0102	14.67	0.06814	191.83	2584.8	2392.9	0.6493	8.1511
0.20	60.086	1.0172	7.650	0.1307	251.45	2609.9	2358.4	0.8321	7.9094
0.30	69.124	1.0223	5.229	0.1912	289.30	2625.4	2336.1	0.9441	7.7695
0.40	75.886	1.0265	3.993	0.2504	317.65	2636.9	2319.2	1.0261	7.6709
0.50	81.345	1.0301	3.240	0.3086	340.56	2646.0	2305.4	1.0912	7.5947
0.60	85.954	1.0333	2.732	0.3661	359.93	2653.6	2293.6	1.1454	7.5327
0.70	89.959	1.0361	2.365	0.4229	376.77	2660.1	2283.3	1.1921	7.4804
0.80	93.512	1.0387	2.087	0.4792	391.72	2665.8	2274.0	1.2330	7.4352
0.90	96.713	1.0412	1.869	0.5350	405.21	2670.9	2265.6	1.2696	7.3954
1.0	99.632	1.0434	1.694	0.5904	417.51	2675.4	2257.9	1.3027	7.3598
1.5	111.37	1.0530	1.159	0.8628	467.13	2693.4	2226.2	1.4336	7.2234
2.0	120.23	1.0608	0.8854	1.129	504.70	2706.3	2201.6	1.5301	7.1268
2.5	127.43	1.0675	0.7184	1.392	535.34	2716.4	2181.0	1.6071	7.0520
3.0	133.54	1.0735	0.6056	1.651	561.43	2724.7	2163.2	1.6716	6.9909
3.5	138.87	1.0789	0.5240	1.908	584.27	2731.6	2147.4	1.7273	6.9392
4.0	143.62	1.0839	0.4622	2.163	604.67	2737.6	2133.0	1.7764	6.8943
4.5	147.92	1.0885	0.4138	2.417	623.16	2742.9	2119.7	1.8204	6.8547

(Continued)

APPENDIX G *(Continued)*

p, bars	t, °C	u', dm³/kg	u'', m³/kg	ϱ'', kg/m³	h', kJ/kg	h'', kJ/kg	r, kJ/kg	s', kJ/(kg K)	s'', kJ/(kg K)
5.0	151.84	1.0928	0.3747	2.669	640.12	2747.5	2107.4	1.8604	6.8192
6.0	158.84	1.1009	0.3155	3.170	670.42	2755.5	2085.0	1.9308	6.7575
7.0	164.96	1.1082	0.2727	3.667	697.06	2762.0	2064.9	1.9918	6.7052
8.0	170.41	1.1150	0.2403	4.162	720.94	2767.5	2046.5	2.0457	6.6596
9.0	175.36	1.1213	0.2148	4.655	742.64	2772.1	2029.5	2.0941	6.6192
10.0	179.88	1.1274	0.1943	5.147	762.61	2776.2	2013.6	2.1382	6.5828
11	184.07	1.1331	0.1774	5.637	781.13	2779.7	1998.5	2.1786	6.5497
12	187.96	1.1386	0.1632	6.127	798.43	2782.7	1984.3	2.2161	6.5194
13	191.61	1.1438	0.1511	6.617	814.70	2785.4	1970.7	2.2510	6.4913
14	195.04	1.1489	0.1407	7.106	830.08	2787.8	1957.7	2.2837	6.4651
15	198.29	1.1539	0.1317	7.596	844.67	2789.9	1945.2	2.3145	6.4406
16	201.37	1.1586	0.1237	8.085	858.56	2791.7	1933.2	2.3436	6.4175
17	204.31	1.1633	0.1166	8.575	871.84	2793.4	1921.5	2.3713	6.3957
18	207.11	1.1678	0.1103	9.065	884.58	2794.8	1910.3	2.3976	6.3751
19	209.80	1.1723	0.1047	9.555	896.81	2796.1	1899.3	2.4228	6.3554
20	212.37	1.1766	0.09954	10.05	908.59	2792.2	1888.6	2.4469	6.3367
25	223.94	1.1972	0.07991	12.51	961.96	2800.9	1839.0	2.5543	6.2536
30	233.84	1.2163	0.06663	15.01	1008.4	2802.3	1793.9	2.6455	6.1837
40	250.33	1.2521	0.04975	20.10	1087.4	2800.3	1712.9	2.7965	6.0685
50	263.91	1.2858	0.03943	25.36	1154.5	2794.2	1639.7	2.9206	5.9735
60	275.55	1.3187	0.03244	30.83	1213.7	2785.0	1571.3	3.0273	5.8908
70	285.79	1.3513	0.02737	36.53	1267.4	2773.5	1506.0	3.1219	5.8162
80	294.97	1.3842	0.02353	42.51	1317.1	2759.9	1442.8	3.2076	5.7471
90	303.31	1.4179	0.02050	48.79	1363.7	2744.6	1380.9	3.2867	5.6820
100	310.96	1.4526	0.01804	55.43	1408.0	2727.7	1319.7	3.3605	5.6198
110	318.05	1.4887	0.01601	62.48	1450.6	2709.3	1258.7	3.4304	5.5595
120	324.65	1.5268	0.01428	70.01	1491.8	2689.2	1197.4	3.4972	5.5002
130	330.83	1.5672	0.01280	78.14	1532.0	2667.0	1135.0	3.5616	5.4408
140	336.64	1.6106	0.01150	86.99	1571.6	2642.4	1070.7	3.6242	5.3803
150	342.13	1.6579	0.01034	96.71	1611.0	2615.0	1004.0	3.6859	5.3178
200	365.70	2.0370	0.005877	170.2	1826.5	2418.4	591.9	4.0149	4.9412
220	373.69	2.6714	0.003728	268.3	2011.1	2195.6	184.5	4.2947	4.5799
221.2	374.15	3.17	0.00317	315.5	2107.4	2107.4	0	4.4429	4.4429

APPENDIX H: THERMOPHYSICAL PROPERTIES OF WATER

t, °C	p, bars	ϱ, kg/m³	c_p, kJ/(kg K)	k, W/(m K)	$10^3\beta$, 1/K	$10^3\eta$, kg/(m s)	$10^6\nu$, m²/s	$10^6\alpha$, m²/s	Pr
0	0.9807	999.8	4.218	0.552	−0.07	1.792	1.792	0.131	13.67
10		999.7	4.192	0.578	+0.088	1.307	1.307	0.138	9.47
20		998.2	4.182	0.598	0.206	1.002	1.004	0.143	7.01
30		995.7	4.178	0.614	0.303	0.797	0.801	0.148	5.43
40		992.2	4.178	0.628	0.385	0.653	0.658	0.151	4.35
50		988.0	4.181	0.641	0.457	0.548	0.554	0.155	3.57
60		983.2	4.184	0.652	0.523	0.467	0.475	0.158	3.00
70		977.8	4.190	0.661	0.585	0.404	0.413	0.161	2.56
80		971.8	4.196	0.669	0.643	0.355	0.365	0.164	2.23
90		965.3	4.205	0.676	0.698	0.315	0.326	0.166	1.96
100	1.0132	958.4	4.216	0.682	0.752	0.282	0.295	0.169	1.75
120	1.9854	943.1	4.245	0.686	0.860	0.235	0.2485	0.171	1.45

(Continued)

APPENDIX H *(Continued)*

t, °C	p, bars	ϱ, kg/m³	c_p, kJ/(kg K)	k, W/(m K)	$10^3\beta$, 1/K	$10^3\eta$, kg/(m s)	$10^6\nu$, m²/s	$10^6\alpha$, m²/s	Pr
140	3.6136	926.1	4.287	0.684	0.957	0.199	0.215	0.172	1.25
160	6.1804	907.4	4.324	0.682	1.098	0.172	0.1890	0.173	1.09
180	10.027	886.9	4.409	0.676	1.233	0.151	0.1697	0.172	0.98
200	15.550	864.7	4.497	0.666	1.392	0.136	0.1579	0.171	0.92
220	23.202	840.3	4.610	0.653	1.597	0.125	0.1488	0.168	0.88
240	33.480	813.6	4.760	0.636	1.862	0.116	0.1420	0.164	0.87
260	46.491	784.0	4.978	0.612	2.21	0.107	0.1365	0.157	0.87
280	64.191	750.7	5.309	0.581	2.70	0.0994	0.1325	0.145	0.91
300	85.917	712.5	5.86	0.541	3.46	0.0935	0.1298	0.129	1.00
320	112.89	667.0	6.62	0.491	4.60	0.0856	0.1282	0.111	1.15
340	146.08	609.5	8.37	0.430	8.25	0.0775	0.1272	0.0844	1.5
360	186.74	524.5	13.4	0.349		0.0683	0.1306	0.0500	2.6
374.2	221.24	326	∞	0.209	∞	0.0506	0.155	0	∞

APPENDIX I: THERMOPHYSICAL PROPERTIES OF AIR AT 101.3 kPa

t, °C	c_p, kJ/(kg K)	k, W/(m K)	$10^5\eta$, kg/(m s)	$10^6\nu$, m²/s	$10^6\alpha$, m²/s	Pr
−150	1.026	0.0120	0.870	3.11	4.19	0.74
−100	1.009	0.0165	1.18	5.96	8.28	0.72
−50	1.005	0.0206	1.47	9.55	13.4	0.715
0	1.005	0.0243	1.72	13.30	18.7	0.711
20	1.005	0.0257	1.82	15.11	21.4	0.713
40	1.009	0.0271	1.91	16.97	23.9	0.711
60	1.009	0.0285	2.00	18.90	26.7	0.709
80	1.009	0.0299	2.10	20.94	29.6	0.708
100	1.013	0.0314	2.18	23.06	32.8	0.704
120	1.013	0.0328	2.27	25.23	36.1	0.70
140	1.013	0.0343	2.35	27.55	39.7	0.694
160	1.017	0.0358	2.43	29.85	43.0	0.693
180	1.022	0.0372	2.51	32.29	46.7	0.69
200	1.026	0.0386	2.58	34.63	50.5	0.685
250	1.034	0.0421	2.78	41.17	60.3	0.68
300	1.047	0.0454	2.95	47.85	70.3	0.68
350	1.055	0.0485	3.12	55.05	81.1	0.68
400	1.068	0.0516	3.28	62.53	91.9	0.68
450	1.080	0.0543	3.44	70.54	103.1	0.685
500	1.093	0.0570	3.58	78.48	114.2	0.69
600	1.114	0.0621	3.86	95.57	138.2	0.69
700	1.135	0.0667	4.12	113.7	162.2	0.70
800	1.156	0.0706	4.37	132.8	185.8	0.715
900	1.172	0.0741	4.59	152.5	210	0.725
1000	1.185	0.0770	4.80	173	235	0.735

APPENDIX J: THERMOPHYSICAL PROPERTIES OF SATURATED WATER VAPOR

t, °C	p, bars	ϱ, kg/m^3	c_p, kJ/(kg K)	$10^3\beta$, 1/K	10^3k, W/(m K)	$10^3\eta$, kg/(m s)	$10^6\nu$, m^2/s	$10^6\alpha$, m^2/s	Pr,	$10^3\sigma$, N/m
0.01	0.006112	0.004850	1.864	3.669	18.2	8.02	1650	2029	0.815	75.60
10	0.012271	0.009397	1.868	3.544	18.8	8.42	896	1080	0.831	74.24
20	0.023368	0.01729	1.874	3.431	19.4	8.82	510	602	0.847	72.78
30	0.042417	0.03037	1.883	3.327	20.0	9.22	304	352	0.863	71.23
40	0.073749	0.05116	1.894	3.233	20.6	9.62	188	213	0.883	69.61
50	0.12334	0.08300	1.907	3.150	21.2	10.02	121	135	0.896	67.93
60	0.19919	0.1302	1.924	3.076	21.9	10.42	80.0	87.6	0.913	66.19
70	0.31161	0.1981	1.944	3.012	22.5	10.82	54.6	58.7	0.930	64.40
80	0.47359	0.2932	1.969	2.958	23.2	11.22	38.3	40.4	0.947	62.57
90	0.70108	0.4233	1.999	2.915	24.0	11.62	27.5	28.5	0.966	60.69
100	1.0132	0.5974	2.034	2.882	24.8	12.02	20.1	20.4	0.984	58.78
110	1.4326	0.8260	2.075	2.861	25.6	12.42	15.0	15.0	1.00	56.83
120	1.9854	1.121	2.124	2.851	26.5	12.80	11.4	11.2	1.02	54.85
130	2.7012	1.496	2.180	2.853	27.5	13.17	8.80	8.46	1.04	52.83
140	3.6136	1.966	2.245	2.868	28.5	13.54	6.89	6.50	1.06	50.79
150	4.7597	2.547	2.320	2.897	29.6	13.90	5.46	5.06	1.08	48.70
160	6.1804	3.259	2.406	2.941	30.8	14.25	4.37	3.94	1.11	46.59
170	7.9202	4.122	2.504	3.001	32.1	14.61	3.54	3.13	1.13	44.44
180	10.003	5.160	2.615	3.078	33.6	14.96	2.90	2.52	1.15	42.26
190	12.552	6.398	2.741	3.174	35.1	15.30	2.39	2.03	1.18	40.05
200	15.551	7.865	2.883	3.291	36.8	15.65	1.99	1.64	1.21	37.81
210	19.080	9.596	3.043	3.432	38.7	15.99	1.67	1.35	1.24	35.53
220	23.201	11.63	3.222	3.599	40.7	16.34	1.40	1.11	1.26	33.23
230	27.979	14.00	3.426	3.798	43.0	16.70	1.19	0.922	1.29	30.90
240	33.480	16.77	3.656	4.036	45.5	17.07	1.02	0.767	1.33	28.56
250	39.776	19.99	3.918	4.321	48.4	17.45	0.873	0.642	1.36	26.19
260	46.940	23.74	4.221	4.665	51.7	17.85	0.752	0.537	1.40	23.82
270	55.051	28.11	4.574	5.086	55.5	18.28	0.650	0.451	1.44	21.44
280	64.191	33.21	4.996	5.608	60.0	18.75	0.565	0.379	1.49	19.07
290	74.448	39.20	5.507	6.267	65.5	19.27	0.492	0.319	1.54	16.71
300	85.917	46.25	6.144	7.117	72.2	19.84	0.429	0.266	1.61	14.39
310	98.697	54.64	6.962	8.242	80.6	20.7	0.379	0.222	1.71	12.11
320	112.90	64.75	8.053	9.785	86.5	21.7	0.335	0.173	1.94	9.89
330	128.65	77.15	9.589	12.02	96.0	23.1	0.299	0.133	2.24	7.75
340	146.08	92.76	11.92	15.50	107	24.7	0.266	0.0943	2.82	5.71
350	165.37	113.4	15.95	21.73	119	26.6	0.235	0.0613	3.83	3.79
360	186.74	143.5	26.79	38.99	137	29.2	0.203	0.0380	5.34	2.03
370	210.53	201.7	112.9	170.9	166	34.0	0.169	0.0107	15.7	0.47
374.15	221.20	315.5	∞	∞	238	45.0	0.143	0	∞	0

APPENDIX K: MOLLIER ENTHALPY-ENTROPY DIAGRAM FOR STEAM: CRITICAL POINT 374.15°C AND 221.2 bars

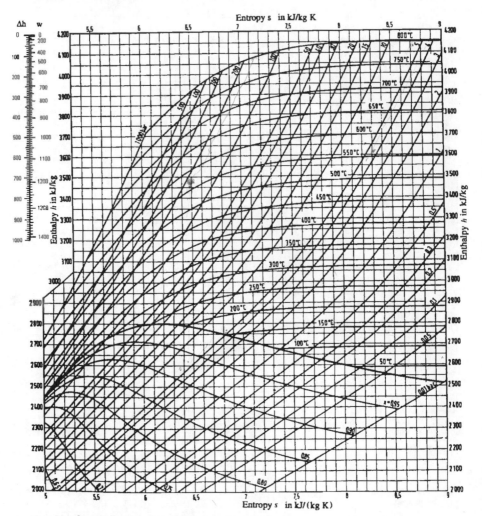

Note. Nozzle: enthalpy drop Δh in kJ/kg, steam velocity w in m/s.

INDEX

(Continued on next page)

(*Continued on next page*)